# AQA BIOLOGY
## Specification A

*Further Studies in*

# BIOLOGY

**Margaret Baker**
**Bill Indge**
**Martin Rowland**

ghton
E GROUP

The Publishers would like to thank the authors of this book for providing the following photographs:

Introduction (whale), 3.17 & 3.18 p69, 4.1 p73, 4.2 p23, 4.3 p74, 4.4 p75, 4.6 p76, 4.9 p79, 4.10 p81, 4.11 p81, 4.14 p83, 4.15 p84, 9.22 p189, 11.9 p217, 11.10 p217, 11.14 p219, 12.6 p233, 12.7 p234, 12.11 p237, 12.12 p237, 12.17 p240, 12.19 and 12.20 p243.

The Publishers would like to thank the following individuals and institutions for permission to reproduce photographs in this book:

**Ardea:** 1.12a–b p15 (Steve Hopkin), 2.18 p47 (P Morris), 3.16a p67 (Steve Hopkin), 3.16b p67 (Alan Weaving), 3.16c p67 (Ake Lindau), 3.16f p67 (Ken Lucas), 3.166h p67 (Jim Zipp), 3.20 p103 (Ian Beanes), 6.1 p113 (Anthony & Elizabeth Bomford); 7.21b p146 (Don Hadden), 9.2 p174 (Graham Robertson), 10.2b p195 (Pat Morris), 10.2c p195 (Kurt Amsler), 10.2d p195 (John Clegg), 10.11 p300 (Pat Morris), 10.13 p202 (Pascal Goetgheluck); **Anthony Blake Photo Library:** 3.2 p54 (Gerrit Buntrock); **John Cleare/Mountain Camera Picture Library:** 11.1 p212; **Bruce Coleman Collection:** 2.8 p37; **Collections:** 6.6 p117 (Ray Farrar); **FLPA:** 2.16 p46 (David Hosking); 10.1b p194 (H M Wellman); 13.18 p268 (Dick Jones), 13.19 p268 (E & D Hosking); **Garden & Wildlife Matters:** 7.1 p132 & 7.21a p146 (John Feltwell); **Holt Studios International:** 3.3a p55 (Sarah Rowland), 3.3b p55 (Nigel Cattlin), 6.11 p120, 6.12 p121; **International Artists Ltd:** 1.1 p7; **NHPA:** 1.21 p27, 2.11 p43 (Daniel Heuclin), 2.15 p46 (Stephen Dalton), 3.1 p54 (Kevin Schafer), 3.7 p62 (J & M Bain), 3.9 p63 (Stephen Dalton), 3.13a–b p65 (Stephen Dalton), 3.13c p65 (G J Cambridge), 3.16g & i p67 (John Shaw), 3.16h p67 (Stephen Dalton), 8.17 p168 (Robert Erwin); **Oxford Scientific Films:** 2.1 p31 (Doug Allan), 3.16a p67 (Mark Deeble & Victoria Stone), 3.16e p67 (Zig Leszczynski), 5.24 p105 (Harold Taylor), 7.21c p146 (Chris Hawes), 9.1 p174 (David Macdonald), 10.1a p94 (Doug Allan), 13.3e p65 (Pammy Gardiner); **Science Photo Library:** 1.2 p7, 1.3 p8, 1.7 p10, 2.19 p47, 3.3 p55, 3.6 p61, 3.7a–c p62, 5.6 p96, 5.19 p102, 7.2 p133, 7.4 a–b p134, 7.10a–b p137, 7.12a–b p137, 7.16a–b p141, 8.1 p152, 9.12 p181, 9.16 p184, 10.2a p195, 10.13 p204, 11.3 p213, 11.6 p216, 11.8 p216, 12.1 p228, 12.13b–d p218, 13.9 p259.

Orders: please contact Bookpoint Ltd, 130 Milton Park, Abingdon, Oxon OX14 4SB.
Telephone: (44) 01235 827720, Fax: (44) 01235 400454. Lines are open from 9.00–6.00, Monday to Saturday, with a 24 hour message answering service.
Email address: orders@bookpoint.co.uk

*British Library Cataloguing in Publication Data*
A catalogue record for this title is available from The British Library

ISBN 0 340 802448

First published 2001
Impression number 10 9 8 7 6 5 4 3 2
Year 2007 2006 2005 2004 2003 2002 2001



Typeset by Cambridge Publishing Management Ltd

Printed in Italy for Hodder & Stoughton Educational, a division of Hodder Headline Ltd, 338 Euston Road, London NW1 3BH by Printer Trento.

# Contents

# Introduction

The photograph shows a southern right whale. It is a marine mammal and can be seen off the coasts of Australia, South Africa and South America. There is a northern right whale as well. It is found in the northern Atlantic and Pacific oceans. These whales are different species. They are similar in appearance because it is thought that they share a common ancestor, but there are also slight differences between them which have arisen as the result of the two species being isolated from each other for thousands of years. The first three chapters in this book look at genetics and evolution and should help you to understand how species such as the southern right whale evolved.

A southern right whale surfacing after being under water for 30 minutes

If you look at the photograph carefully you will see a number of lumps on the whale's head. These are structures called callosities and they have a life of their own! They are covered with small animals called barnacles. Barnacles feed on tiny photosynthetic organisms which they filter out from the surrounding water. Whale lice and parasitic worms live among the barnacles. The relationships between the whale and the animals that live on its callosities form part of its ecology. Ecology is the study of the relationships between organisms and their environment and we shall look at this aspect of biology in chapters 4–6.

A whale can remain underwater for a long time. In southern right whales, dives may last for over 30 minutes. These animals have many adaptations which allow them to do this. Their gas exchange and blood systems, for example, keep vital organs such as the brain supplied with oxygen all the time the whale is underwater. The way in which different systems of the body function in different animals is the subject of the remaining chapters in this book.

So, if you want a broad picture of the biology of the southern right whale, or any other organism, you will need to study its genetics, its ecology and its physiology, and understand how all these different aspects are linked together. Although this book is divided into chapters and each chapter concentrates on a particular topic, it is important to appreciate that all these areas of biology are related to each other. We have tried to help you to understand the relationships between topics by using text questions, extension boxes and the assignments which are found at the end of individual chapters.

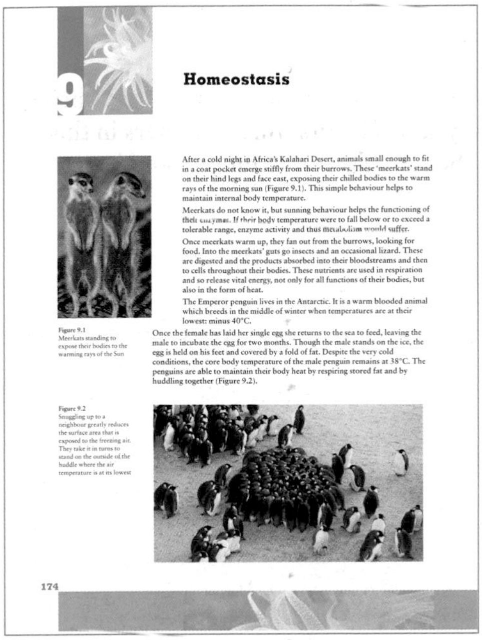

This book has been written in a similar way to the AS textbook, *A New Introduction to Biology*. Each chapter shares a number of features.

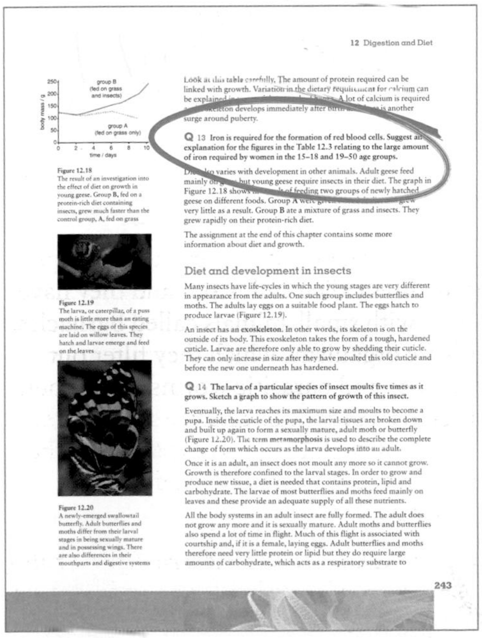

## The chapter opening

Chapters start with a topical introduction. In most cases, this is an interesting application or an unusual aspect of biology related to the content of the chapter.

## The text

Your AS course should have provided you with a sound understanding of many fundamental biological ideas. The material in this book builds on these concepts and develops many of them further. In writing each chapter, we have again tried to help you to gain a good understanding of basic principles rather than providing you with a lot of unnecessary detail and too much technical language. The book has been illustrated in colour throughout. The drawings, photographs and information contained in the captions should provide you with additional help in gaining an understanding of the subject.

## Text questions

The text questions should help you to understand what you have just been reading. They are meant to be answered as you go along. Many of them are straightforward and can be attempted from the information in the paragraph or two which come immediately before. Some of them, however, require you to apply this knowledge to a new situation or link it with material covered earlier in your course. We haven't included any answers to these questions. If you get stuck, try reading the previous paragraph again. If you still have difficulties, make a note of the question and get some help.

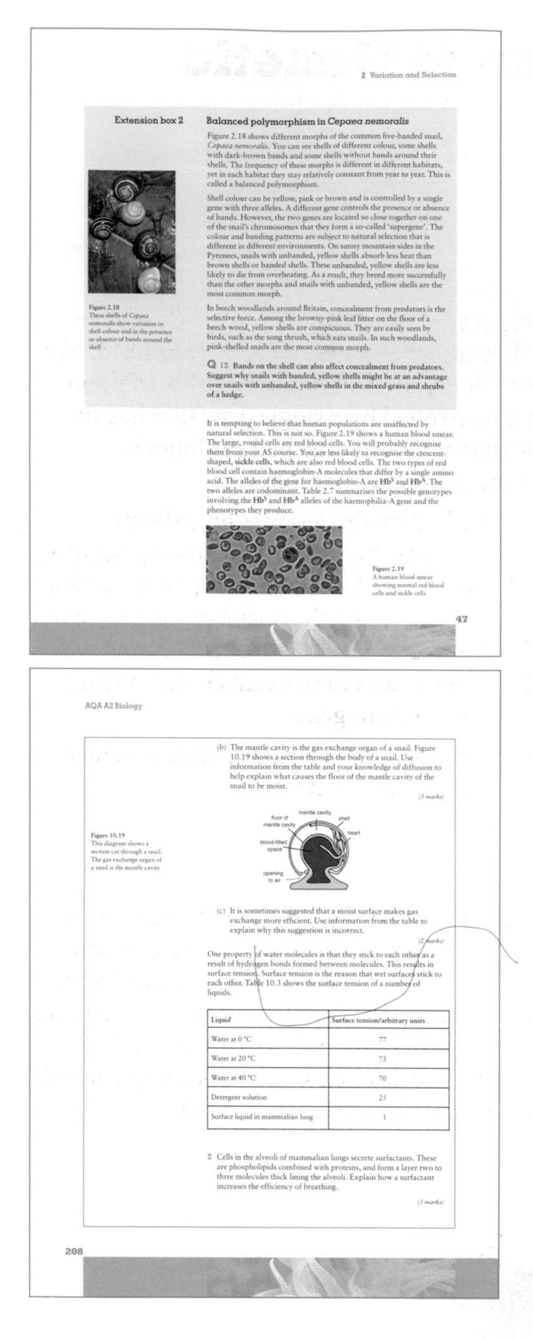

2 Variation and Selection

Extension box 2

**Balanced polymorphism in *Cepaea nemoralis***

Figure 2.18 shows different morphs of the common five-banded snail, *Cepaea nemoralis*. You can see shells of different colour, some shells with dark-brown bands and some shells without bands around their shells. The frequency of these morphs is different in different habitats, yet in each habitat they stay relatively constant from year to year. This is called a balanced polymorphism.

Shell colour can be yellow, pink or brown and is controlled by a single gene with three alleles. A different gene controls the presence or absence of bands. However, the two genes are located so close together on one of the snail's chromosomes that they form a so-called 'supergene'. The colour and banding patterns are subject to natural selection that is different in different environments. On sunny mountain sides in the Pyrenees, snails with unbanded, yellow shells absorb less heat than brown shells or banded shells. These unbanded, yellow shells are less likely to die from overheating. As a result, they breed more successfully than the other morphs and snails with unbanded, yellow shells are the most common morph.

In beech woodlands around Britain, concealment from predators is the selective force. Among the browny-pink leaf litter on the floor of a beech wood, yellow shells are conspicuous. They are easily seen by birds, such as the song thrush, which eats snails. In such woodlands, pink-shelled snails are the most common morph.

Figure 2.18
These shells of *Cepaea nemoralis* show variation in shell colour and in the presence or absence of bands around the shell

**Q 12 Bands on the shell can also affect concealment from predators. Suggest why snails with banded, yellow shells might be at an advantage over snails with unbanded, yellow shells in the mixed grass and shrubs of a hedge.**

It is tempting to believe that human populations are unaffected by natural selection. This is not so. Figure 2.19 shows a human blood smear. The large, round cells are red blood cells. You will probably recognise them from your AS course. You are less likely to recognise the crescent-shaped, **sickle cells**, which are also red blood cells. The two types of red blood cell contain haemoglobin-A molecules that differ by a single amino acid. The alleles of the gene for haemoglobin-A are $\mathbf{Hb^S}$ and $\mathbf{Hb^A}$. The two alleles are codominant. Table 2.7 summarises the possible genotypes involving the $\mathbf{Hb^S}$ and $\mathbf{Hb^A}$ alleles of the haemophilia-A gene and the phenotypes they produce.

Figure 2.19
A human blood smear showing normal red blood cells and sickle cells

47

AQA A2 Biology

(b) The mantle cavity is the gas exchange organ of a snail. Figure 10.19 shows a section through the body of a snail. Use information from the table and your knowledge of diffusion to help explain what causes the floor of the mantle cavity of the snail to be moist.

*(3 marks)*



Figure 10.19
This diagram shows a section cut through a snail. The gas exchange organ of a snail is the mantle cavity

(c) It is sometimes suggested that a moist surface makes gas exchange more efficient. Use information from the table to explain why this suggestion is incorrect.

*(2 marks)*

One property of water molecules is that they stick to each other as a result of hydrogen bonds formed between molecules. This results in surface tension. Surface tension is the reason that wet surfaces stick to each other. Table 10.3 shows the surface tension of a number of liquids.

| Liquid | Surface tension/arbitrary units |
|---|---|
| Water at 0 °C | 77 |
| Water at 20 °C | 73 |
| Water at 40 °C | 70 |
| Detergent solution | 25 |
| Surface liquid in mammalian lung | 1 |

2 Cells in the alveoli of mammalian lungs secrete surfactants. These are phospholipids combined with proteins, and form a layer two to three molecules thick lining the alveoli. Explain how a surfactant increases the efficiency of breathing.

*(3 marks)*

208

## Extension boxes

The extension boxes have several functions. Where we have a worked example of a calculation or a problem, we have used a box to separate it from the main text. This should make the chapter easier to read. Extension boxes will also tell you more about some topics and help you to appreciate some of the links between different aspects of the subject.

## Summary

At the end of each chapter is a summary. You can use it as a checklist to make sure that you have done everything your specification requires. It summarises what you need to know when you have finished a particular topic.

## Examination questions

The examination questions are reproduced or adapted by permission of the Assessment and Qualifications Alliance (AQA). It should be noted that these questions are based on the previous syllabus and are only intended to give you an idea of what might be asked.

## Assignment

Studying biology entails a lot more than learning facts. It involves acquiring a range of skills. As a biologist, you should be able to apply knowledge to new situations; interpret drawings and photographs, graphs and tables; and show an understanding of how the different aspects of the subject link together. In addition, you should be able to use mathematical skills to carry out a range of calculations and you must be able to communicate your knowledge and ideas effectively using appropriate biological language. If you want to get high grades in your unit tests, you will have to learn the necessary facts but you will also need a range of other important skills. The purpose of the text in this book is to provide you with the necessary understanding of the facts that you will need; the purpose of the assignments is to help you to master the skills you require.

# Transmission of Genetic Information

**Figure 1.1**
The comedian Paul Merton has attached ear lobes. Whether your ear lobes are attached to the side of your head or not is controlled by a single gene

The comedian Paul Merton (Figure 1.1) has them. Do you? Check the fleshy lobes of your ears. Are they attached to the sides of your head, like Paul's, or are they unattached so that you can flip them backwards and forwards? This characteristic is controlled by a single gene that has two forms, called alleles. The allele that causes unattached ear lobes is dominant over the allele that causes attached ear lobes.

Other facial features that are controlled by single genes include: the presence or absence of freckles; a hairline that forms a widow's peak or is continuous across the forehead; dimples in the cheeks or smooth cheeks; a cleft chin or a smooth chin; long eyelashes or short eyelashes; a Roman nose or a straight nose. In each case, the first characteristic is controlled by a dominant allele of the relevant gene and the second characteristic is controlled by a recessive allele of the relevant gene.

This chapter explains how genes are inherited and, once inherited, how they interact to produce an observable characteristic.

## Chromosomes and cell division

Figure 1.2 shows a single human chromosome. It has two 'arms', called **chromatids**, held together by a region, called the **centromere**. The two chromatids are identical copies of the chromosome that are made by the semi-conservative replication of DNA, prior to cell division. You can also see dark and light bands along the chromosome. These are stained regions of DNA that are 6 to 10 Mb long (1 Mb represents a sequence of 1 megabase, i.e. 1000 organic bases in the nucleotides of DNA).

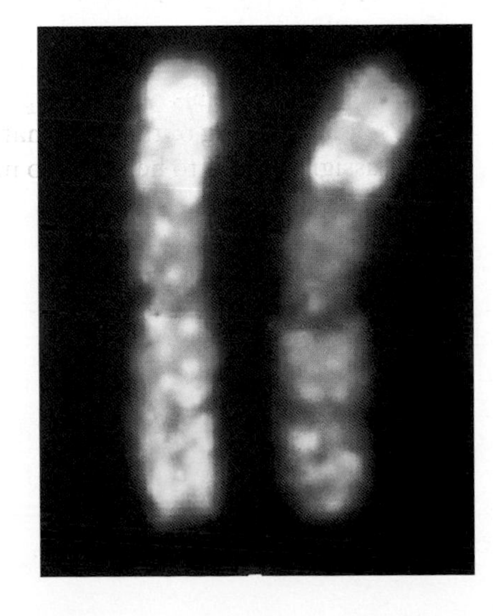

**Figure 1.2**
A pair of human chromosomes, showing the dark and light bands that result from specific staining. These bands represent similarities at the 6–10 Mb level of chromosome structure

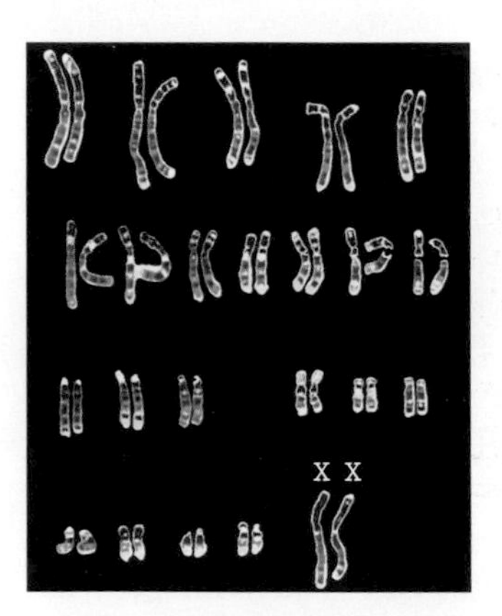


**Figure 1.3**
The chromosomes in a human female cell. The chromosomes in each pair are called homologous chromosomes. They contain genes controlling the same characteristics arranged in the same sequence

Figure 1.3 shows all the chromosomes in a female human cell. The biologist who made this illustration first took a photograph of a dividing body cell. She then cut each chromosome from the photograph and arranged the cut chromosomes in descending order of length. You can see that:

- there are 46 chromosomes. With some exceptions, which we will examine later, this is the usual number of chromosomes in human cells.
- the chromosomes have been arranged as 23 pairs. The members of each pair are called **homologous chromosomes**. They not only look the same – they contain genes controlling the same characteristics, arranged in the same sequence.

**Q** 1 **Suggest why the chromatids in Figure 1.2 have an identical pattern of dark and light bands.**

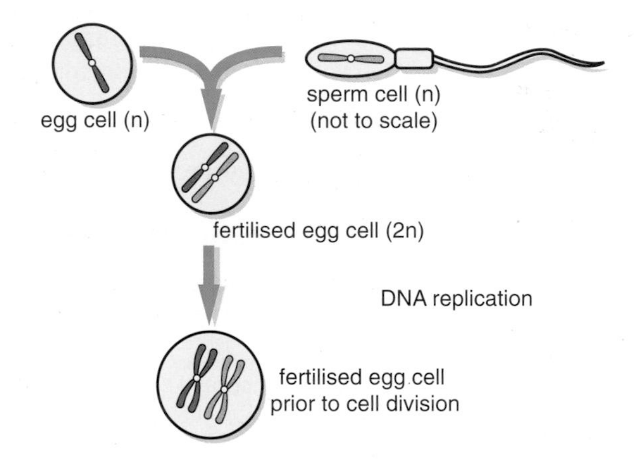


**Figure 1.4**
Human egg and sperm cells are haploid. At fertilisation a diploid cell is formed

We need to understand why cells, such as the one shown in Figure 1.3, have pairs of homologous chromosomes. Figure 1.4 shows what happens when a human egg cell is fertilised by a human sperm cell. To make the drawing easier to follow, only one pair of homologous chromosomes is represented. The egg cell and sperm cell each have only one chromosome from each homologous pair: cells like this are called **haploid** and their chromosome number is represented as (n). When the sperm cell fertilises the egg cell, its chromosome is incorporated into the nucleus of the fertilised egg. This cell now has a pair of homologous chromosomes. Cells like this are called **diploid** and their chromosome number is represented as (2n).

During your AS course, you learned how cells replicate their chromosomes and separate them during mitosis forming new cells with the same chromosome number as the parent cell (i.e. diploid parent cells → diploid daughter cells or haploid parent cells → haploid daughter cells). Eggs and sperms are haploid cells formed by the division of diploid parent cells. The cell division involved is meiosis, which we will now study in more detail.

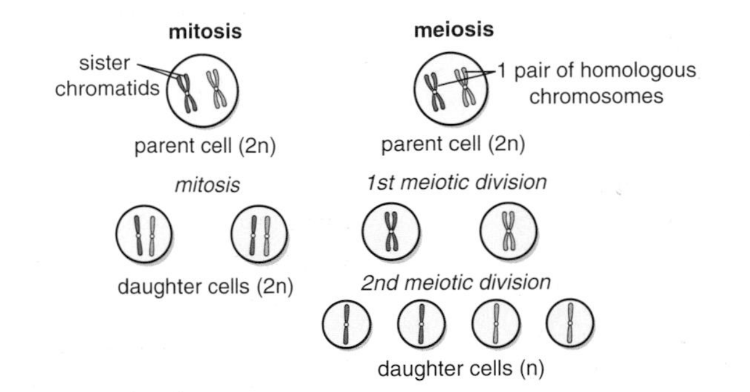


**Figure 1.5**
From a diploid parent cell, mitosis produces genetically identical diploid daughter cells whereas meiosis produces haploid cells that are genetically different from each other. To make the diagrams simple, only one pair of homologous chromosomes has been shown

## The process of meiosis

Figure 1.5 shows that meiosis occurs as two separate divisions. The first meiotic division (meiosis 1) separates the two homologous chromosomes in each pair, producing haploid cells. Like mitosis, the second meiotic division (meiosis 2) separates the sister chromatids in each chromosome. Just as in mitosis, the two divisions of meiosis involve several stages. The first meiotic division involves prophase 1, metaphase 1, anaphase 1 and telophase 1. The second meiotic division involves prophase 2, metaphase 2, anaphase 2 and telophase 2. These stages are represented diagrammatically in Figure 1.6.

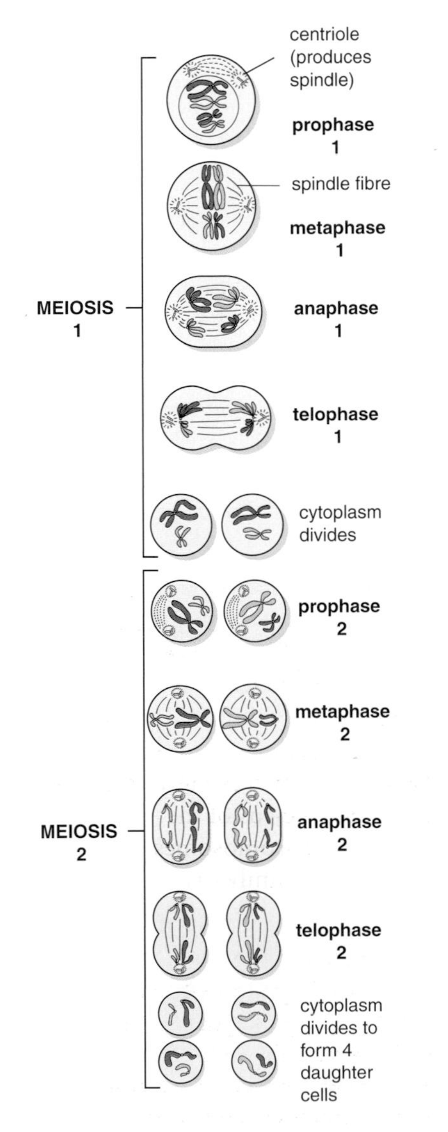


**Figure 1.6**
The behaviour of chromosomes during meiosis

### First meiotic division

During prophase 1, the chromosomes, which are normally long and thin, condense to become short and fat: they can now be seen with an optical microscope. Unique to meiosis, the homologous chromosomes pair together. This pairing is very precise, so that the bands of coding regions, shown in Figure 1.2, lie opposite each other. At the same time, the nuclear envelope disperses and a spindle of protein fibres forms in the cytoplasm of the cell. These fibres attach to the centromeres of the chromosomes in each pair.

**Q** 2 **Name one event that occurs:**

**(a) in both prophase of mitosis and prophase 1 of meiosis**
**(b) in prophase 1 of meiosis but not in prophase of mitosis.**

During metaphase 1, the pairs of homologous chromosomes, still lying next to each other, line up on the equator of the spindle. Seen from a suitable angle, this produces a line of chromosomes. By this stage, the forces holding the homologous chromosomes together have weakened, so that they have begun to drift apart. They are held together at points along their length, called **chiasmata** (singular: **chiasma**), where chromatids have become entangled. We will examine the significance of this later in the chapter.

The homologous chromosomes separate in anaphase 1, when the protein fibres of the spindle contract, pulling the chromosomes by their centromeres to opposite poles of the cell. In anaphase 1 each chromosome still has its sister chromatids together.

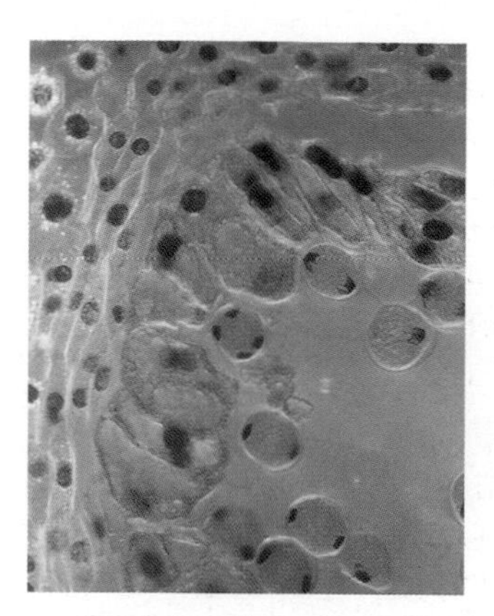

**Figure 1.7**
Most of the cells here are at Anaphase I with chromosomes at the spindle poles. The cell divides and the nuclear membrane reforms (Telophase I). These stages are repeated (Prophase II–Telophase II) to produce 4 haploid cells. At centre is a cell at Telophase II

During telophase 1, the chromosomes bunch up together, the cytoplasm of the cell begins to divide into two and the spindle disintegrates. Sometimes, after this, the chromosomes become long and thin again but usually the second meiotic division occurs straight away.

## Second meiotic division

At the start of the second division, the two daughter cells from meiosis 1 have one chromosome from each homologous pair, i.e. they are haploid. During prophase 2, a spindle forms in each cell at right angles to the now disintegrated spindle from prophase 1. If the chromosomes became long and thin at the end of meiosis 1, they now condense to become short and fat again.

During metaphase 2, the chromosomes in each cell line up on the equator of the new spindle. In metaphase 2 you can see the lines of chromosomes, each made of two sister chromatids, on the equators of the spindles. After a while, the centromeres holding the sister chromatids together divide and contraction of the spindle fibres pulls sister chromatids to opposite poles of the spindles.

At the poles of the spindles, the chromosomes bunch up into a tight ball and a nuclear envelope forms around each new nucleus. This is telophase 2 and marks the end of meiosis. The two divisions have produced four daughter cells, each of which contains only one chromosome from each homologous pair, i.e. is haploid.

**Q** 3 **During which of the two meiotic divisions are homologous chromosomes separated?**

### Extension box 1

### When homologous chromosomes fail to separate

Very occasionally, the homologous chromosomes of one pair are pulled to the same pole of the spindle during anaphase 1 of meiosis. This is known as **chromosome non-disjunction** and results in one daughter cell with two copies of the chromosome and one daughter cell with no copies of the chromosome. If the gamete that contains both homologous chromosomes of a pair is fertilised by a gamete that also has one copy of that chromosome, we end up with a zygote that has three copies of the same chromosome, i.e. is trisomic for that chromosome.

Non-disjunction of the X chromosomes in a human female can lead to eggs that have two copies of the X chromosome (XX) or eggs that lack an X chromosome (designated O). If the XX egg cell is fertilised by an X-carrying sperm, an XXX zygote is formed. This develops into a phenotypically normal female. The effects are more serious if the XX egg cell is fertilised by a Y-carrying sperm. The resulting XXY zygote develops into a sterile male, a condition known as **Klinefelter's syndrome** that occurs in about 1 in every 500 live male births in the UK.

If the egg cell lacking an **X** chromosome is fertilised by a **Y**-carrying sperm, the resulting **OY** zygote fails to develop. Apparently, humans cannot survive without the genes on the **X** chromosome. If the **O** egg cell is fertilised by an **X**-carrying sperm, an **XO** zygote is produced that develops into a female who shows stunted physical growth and whose sex organs do not mature. This is called Turner's syndrome and occurs in about 1 in every 2000 live female births in the UK.

Chromosome non-disjunction can also occur with **autosomes** (chromosomes other than the sex chromosomes). A case with which you are likely to be familiar involves non-disjunction of chromosome 21 during the production of human eggs. After fertilisation, a zygote with three copies of chromosome 21 is produced. This zygote develops into a person who has Down's syndrome.

## Meiosis and genetic variation

During your AS course, you saw that mitosis produces cells that are genetically identical to each other and to the parent cell. This is important during growth and cell repair. Any organism that reproduces by mitosis will produce offspring that are genetically identical to itself and to each other. This is not the case when meiosis is involved. Two events that occur during meiosis result in genetic differences between the resulting daughter cells: independent assortment of homologous chromosomes and crossing over.

### Independent assortment of homologous chromosomes

If you look back to Figure 1.4, you can see that, in a human cell, the homologous chromosomes in each pair were derived from either the egg cell or the sperm cell. Although the homologous chromosomes in each pair have genes controlling the same characteristics in the same sequence, those genes do not necessarily have the same sequence of bases. Different versions of the same gene are called **alleles**, which we will consider later in this chapter. The point to realise here is that the two homologous chromosomes of a pair are not genetically identical to each other.

So far, we have considered only one of the pairs of homologous chromosomes in a cell. We have seen that the homologous chromosomes are separated during anaphase 1. The pole to which each chromosome from the pair is pulled is completely random. For example, the chromosome being pulled to the left pole of the cell during anaphase 1 in Figure 1.6 would be just as likely to have been pulled to the right pole: whichever pole it had been pulled to, its homologous partner would have been pulled to the opposite pole.

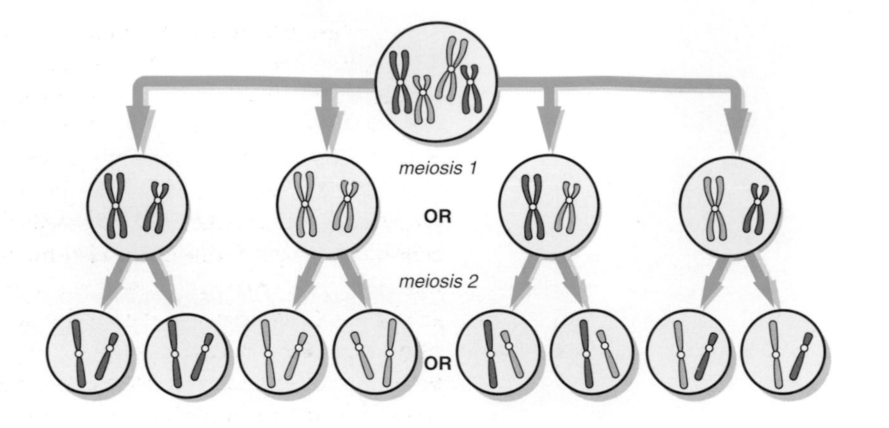


**Figure 1.8**
Independent assortment of chromosomes during meiosis results in genetically different daughter cells

Figure 1.8 shows two pairs of homologous chromosomes in a cell: those in red were derived from the egg cell and those in blue were derived from the sperm cell. The chromosomes in each pair are pulled at random to one of the poles of the spindle during anaphase 1. The movement of the chromosomes in one pair has no effect on the direction in which chromosomes of the second pair are pulled, i.e. they are pulled independently of each other. This shows independent assortment of homologous chromosomes. As a result of **independent assortment**, parent cells like those in Figure 1.8 could produce four kinds of genetically different daughter cells. As a general rule, an independent assortment of chromosomes in a diploid cell (2n chromosomes) can result in any of $2^n$ different possible combinations of chromosomes in the four haploid cells produced by meiosis.

**Q 4 A human male has 23 pairs of chromosomes. How many different combinations of these chromosomes are possible in the sperm he produces?**

## Chiasmata and crossing over

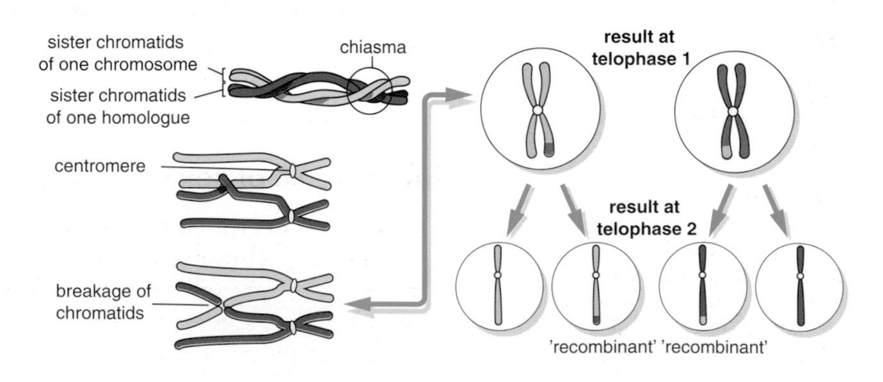


**Figure 1.9**
A chiasma shows where chromatids become entangled during prophase 1 of meiosis. Often the chromatids break at these points. When the broken fragments join the 'wrong' chromosome, genetic crossing over takes place

So far, we have assumed that chromatids never break. As they condense in prophase 1 of meiosis, the paired arms of the sister chromatids in each pair of homologous chromosomes often become intertwined to form one

Figure 1.12
Fruit flies of the species *Drosophila melanogaster*. Some of these flies have long wings and some have vestigial wings. This phenotype is controlled by a single gene

## Monohybrid inheritance with dominance

In **monohybrid inheritance**, we consider the effects of a single gene. Figure 1.12 shows flies belonging to the species *Drosophila melanogaster*. Because they are easy to keep in the laboratory and have a relatively short life cycle, *D. melanogaster* have been used extensively to study inheritance. In the UK, *D. melanogaster* can be found in summer around ripe, or rotting fruit, hence their common name of fruit flies.

The fruit flies in Figure 1.12 show variation in one phenotypic characteristic: some have long wings and some have short (vestigial) wings. This characteristic is controlled by a single gene that is located at a locus on one of the chromosomes in the cells of *D. melanogaster*. There are two varieties of this gene, called **alleles.** One allele of the gene for wing length (represented as L) carries a base sequence that results in long wings, the other allele of the gene for wing length (represented as l) carries a base sequence that can result in vestigial wings. Table 1.1 shows the possible genotypes involving these alleles and the phenotypes that result. Notice that the l allele has no effect unless it is in the homozygous form. This allele is called **recessive** and the L allele, that always shows its effect if present in the genotype, is called the **dominant** allele.

| Genotype | Description of genotype | Phenotype |
|---|---|---|
| LL | Homozygous L<br>(Homozygous dominant) | Long wings |
| Ll | Heterozygous | Long wings |
| ll | Homozygous l<br>(Homozygous recessive) | Vestigial wings |

Table 1.1
The genotypes and phenotypes of wing length in *Drosophila melanogaster*

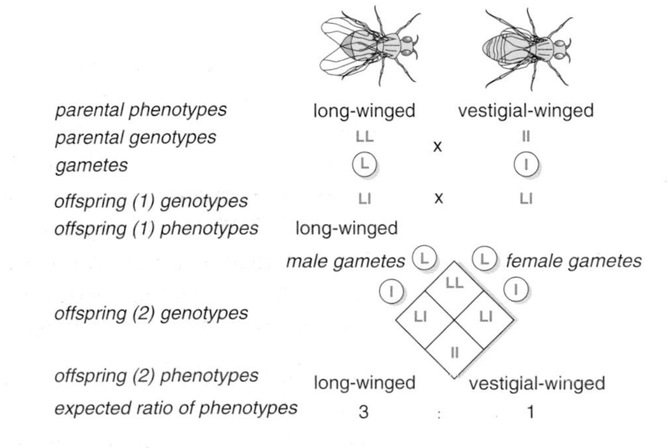


Figure 1.13
A cross between a homozygous long-winged fruit fly and a homozygous vestigial-winged fruit fly to show the expected phenotypes in the offspring (1) and offspring (2) generations. In a Unit test, you would not be expected to draw the fruit flies

Figure 1.13 represents a cross between a homozygous long-winged fly and a homozygous vestigial-winged fly. Flies in the offspring (1) generation are all heterozygous and, since the L allele is dominant, are long-winged. Figure 1.13 also shows that if two of the heterozygous

Figure 1.14
Snapdragon plants (*Antirrhinum majus*) showing variation in flower colour. The gene for flower colour has two codominant alleles, one for red flowers and one for white flowers

offspring (1) generation are mated together, they produce an offspring (2) generation in which the expected ratio of long-winged to vestigial-winged flies is 3:1. This ratio is typical of a cross between heterozygotes in a monohybrid cross where one of the alleles is dominant. (Note that in crosses such as this, where the parents are homozygous, the offspring (1) and offspring (2) generations might also be referred to as the F1 and F2 generations in a Unit test.)

A checkerboard (called a **Punnett diamond**) has been used to represent all the possible genotypes that could result from the random fusion of gametes to form the offspring (2) generation in Figure 1.13. This notation helps to simplify the diagram: you should find it particularly helpful later in this chapter when looking at the possible genotypes that could result in the offspring (2) generation of a dihybrid cross.

**Q 7 What name refers to an allele that always shows its effect in the phenotype?**

## Monohybrid inheritance with codominance

In the example of monohybrid inheritance used above, one allele of the gene for wing length was dominant over the other allele of the gene for wing length. The alleles of some genes always show their effect in the phenotype: neither is dominant or recessive. When this occurs, the alleles are said to be **codominant**. Figure 1.14 shows flowering plants called snapdragons (*Antirrhinum majus*). Although the flowers on each plant are red, pink or white, flower colour in *Antirrhinum* is controlled by a single gene with two alleles. Figure 1.15 represents a cross between a red-flowered plant and a white-flowered plant. Just as was the case with *D. melanogaster* in Figure 1.13, the homozygous parents produce an offspring (1) generation that are all heterozygous. Because the red-flower and white-flower alleles are codominant, these heterozygotes have pink flowers. When two of these pink-flowered plants are crossed, an offspring (2) generation is produced that contains red, pink and white-flowered plants in the expected ratio of 1:2:1. This ratio is typical of a cross between two heterozygotes in a monohybrid cross with codominance.

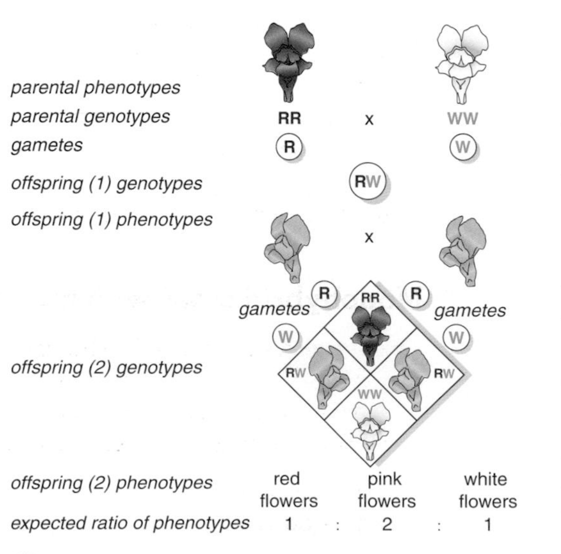


Figure 1.15
A cross between a pure-breeding (homozygous) red-flowered snapdragon and a pure-breeding (homozygous) white-flowered snapdragon to show the expected phenotypes in the offspring (1) and offspring (2) generations

**Q 8 Explain why snapdragons with the genotype RW have pink flowers and not red flowers.**

## Extension box 2 Test crosses

Sometimes, we need to know whether an organism showing the effects of the dominant allele of a particular gene is homozygous dominant or heterozygous. To find out, we perform a **test cross**. This involves crossing the organism of unknown genotype with a number of homozygous recessive individuals. The homozygous recessive individuals produce gametes that contain only recessive alleles. If the organism of unknown genotype is homozygous, all its gametes will contain the dominant allele and all the offspring will be heterozygous. We will recognise this because all the offspring of the test cross will show the effect of the dominant allele (Figure 1.16a). If the unknown organism is heterozygous, half of its gametes will contain the recessive allele and half of its offspring would be expected to be homozygous recessive and show the effects of the recessive allele (Figure 1.16b).

**(a) if long-winged parent is homozygous**

| | | | |
|---|---|---|---|
| *parental phenotypes* | long-winged | x | vestigial-winged |
| *parental genotypes* | LL | | ll |
| *gametes* | (L) | | (l) |
| *offspring (1) genotypes* | | Ll | |
| *offspring (1) phenotypes* | | long-winged | |

**(b) if long-winged parent is heterozygous**

| | | | |
|---|---|---|---|
| *parental phenotypes* | long-winged | x | vestigial-winged |
| *parental genotypes* | Ll | | ll |
| *gametes* | (L) (l) | | (l) |
| *offspring (1) genotypes* | Ll | | ll |
| *offspring (1) phenotypes* | long-winged | | vestigial-winged |
| *expected ratio of phenotypes* | 1 | : | 1 |

Figure 1.16
A test cross is used to find the genotype of an organism that shows the effects of a dominant allele. In this case, a fruit fly with long wings is crossed with a number of homozygous recessive vestigial-winged mates. The ratio of long-winged and vestigial-winged flies in the offspring show whether the long-winged fly was homozygous (a) or heterozygous (b)

A test cross is carried out to determine the genotype of organisms that are commercially important. We are unlikely to want to know whether a long-winged fruit fly is homozygous or heterozygous but we would use a test cross to determine whether a bull showing a desirable phenotype was homozygous, in which case its sperm could be used for artificial insemination, or heterozygous, which would make it less valuable for artificial insemination.

## Monohybrid cross with multiple alleles

The two genes considered so far each had two alleles. Many genes have more than two alleles. You learned about the human ABO blood groups in your AS course. These blood groups are controlled by an immunoglobulin gene (represented **I**) that has three alleles:

- $\mathbf{I^A}$ results in the formation of antigen A on the plasma membrane of red blood cells
- $\mathbf{I^B}$ results in the formation of antigen B on the plasma membrane of red blood cells
- $\mathbf{I^O}$ results in the formation of neither antigen A nor antigen B on the plasma membrane of red blood cells.

Although there are three alleles of the **I** gene, no more than two can be present in a diploid cell. The possible diploid genotypes involving these three alleles and the phenotypes that result from each are shown in Table 1.2. From this, you can see that alleles $I^A$ and $I^B$ are codominant but that both are dominant over the recessive $I^O$ allele.

| Genotype | Antigens present on plasma membrane of red blood cells | Phenotype |
|---|---|---|
| $I^A I^A$ | Antigen A | Blood group A |
| $I^A I^B$ | Antigen A and antigen B | Blood group AB |
| $I^A I^O$ | Antigen A | Blood group A |
| $I^B I^B$ | Antigen B | Blood group B |
| $I^B I^O$ | Antigen B | Blood group B |
| $I^O I^O$ | Neither antigen A nor antigen B | Blood group O |

**Table 1.2**
The genotypes that result in the human ABO blood groups

**Q 9 What is the evidence in Table 1.2 that:**
**(a) the $I^O$ allele is recessive**
**(b) the $I^A$ and $I^B$ alleles are codominant?**

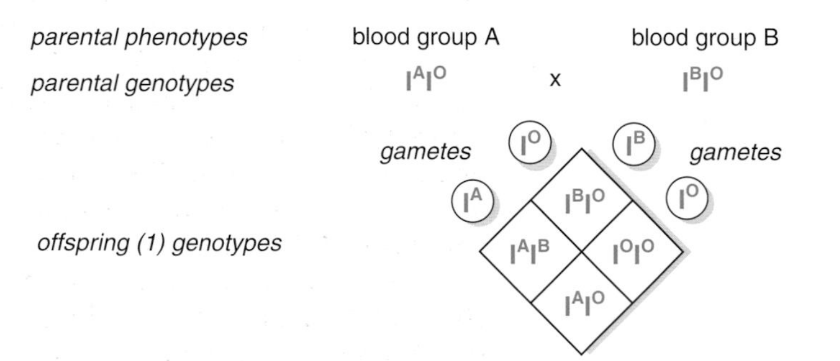


**Figure 1.17**
The human ABO blood group shows monohybrid inheritance with multiple alleles, i.e. more than two alleles of the controlling gene

Figure 1.17 represents a cross between a heterozygous person of blood group A and a heterozygous person of blood group B. The interrelationships between the $I^A$, $I^B$ and $I^O$ alleles produce an interesting range of possible phenotypes amongst the offspring. Notice that two of them are unlike either parent.

## Dihybrid inheritance

So far, we have looked at the inheritance of one gene at a time. In **dihybrid inheritance**, we consider the inheritance of two genes that occur at two different loci. This might sound difficult, but we follow exactly the same format as we have used in representing monohybrid crosses.

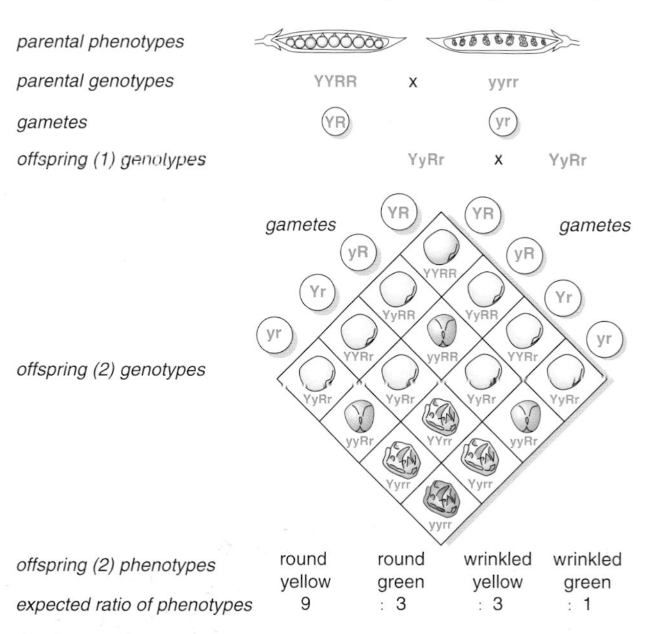


**Figure 1.18**
The inheritance of colour and shape of peas is an example of dihybrid inheritance. The colour and shape of seeds were just two of the characteristics of pea plants investigated by Gregor Mendel, the Austrian monk who discovered the principles of inheritance

Let's consider one of Mendel's crosses with pea plants. Peas are the seeds of pea plants. Mendel looked at the inheritance of two phenotypic features of peas: seed colour and seed shape. The gene for seed colour has two alleles: the allele for yellow colour (**Y**) is dominant over that for green colour (**y**). The gene for seed shape also has two alleles: the allele for round seeds (**R**) is dominant over the allele for wrinkled seeds (**r**). Figure 1.18 represents a cross between a pea plant that is homozygous dominant for both genes (i.e. has the genotype **YYRR**) and one that is homozygous recessive for both genes (i.e. has the genotype **yyrr**) and in which two of the offspring (1) plants were bred together. All the offspring (1) generation are heterozygous for the two genes (**YyRr**) and show the effects of the dominant alleles, i.e. seeds that are round and yellow. It is the offspring (2) generation that is interesting. As you might have predicted from Figure 1.8, meiosis in each offspring (1) plant will result in gametes with four different possible genotypes – **YR**, **yR**, **Yr** and **yr** – in equal numbers. The random fertilisation of these gametes results in the sixteen different combinations in the offspring (2) generation shown in Figure 1.18. Of these sixteen possible combinations, 9 are for round yellow peas, 3 are for round green peas, 3 are for wrinkled yellow peas and 1 is for wrinkled green peas. This ratio of 9:3:3:1 is characteristic of the offspring (2) or F2 generation in a dihybrid cross and will help you to recognise a dihybrid cross in a Unit test.

**Q 10 Give the ratio of F2 phenotypes you would expect from a cross involving:**
**(a) monohybrid inheritance**
**(b) dihybrid inheritance.**

**Synoptic extension box**

## A molecular model of dominance

You learned in your AS course that:

- many proteins are single polypeptides
- a gene is a length of DNA that controls the production of a particular polypeptide by a cell
- enzymes are proteins
- metabolic pathways are controlled by enzymes.

We can put this information together to produce a molecular model of dominance and codominance.

Domestic rabbits eat plants that contain a yellow, fat-soluble pigment, called xanthophyll. This yellow pigment would accumulate in the body fat of rabbits, colouring their fat yellow. This does not normally happen because rabbits produce an enzyme that breaks down xanthophyll to a colourless compound. As a result, their body fat is white. The enzyme that breaks down xanthophyll is coded by the 'normal' allele of a gene that is present in the genotype of most rabbits.

'normal' allele ➪ active enzyme

xanthophyll → colourless compound

There is also a 'defective' allele of this gene that codes for an inactive enzyme.

'defective' allele ➪ inactive enzyme

xanthophyll → no reaction

You should remember from your AS course that enzymes are used time and again in cells. Provided a rabbit has the 'normal' allele for the xanthophyll enzyme, it produces enough enzyme to break down all the xanthophyll from the rabbit's diet. Rabbits that are heterozygous for this gene, as well as rabbits that are homozygous for the 'normal' allele, will have white fat. However, rabbits that are homozygous for the 'defective' allele will produce no active enzyme. As a result, xanthophyll is not broken down and dissolves in their body fat, colouring it yellow.

This leads to a general conclusion. A dominant allele on one homologous chromosome of a pair is capable of synthesising enough of the enzyme for which it codes to result in a phenotype that is indistinguishable from the phenotype that results if the dominant allele is present on both homologous chromosomes in a pair. A recessive allele codes for the production of an inactive enzyme or no enzyme at all, i.e. some recessive alleles are deletions.

The $I^A$ and $I^B$ alleles of the human ABO blood groups code for slightly different active enzymes. The A enzyme attaches galactosamine to glycolipid molecules in the plasma membranes of red blood cells: this forms the A antigens. The B enzyme attaches galactose (a monosaccharide) to glycolipid molecules in the plasma membranes of red blood cells: this forms the B antigens. If a person has the genotype $I^AI^B$, he will produce both active enzymes so that the plasma membranes of his red blood cells will have both A and B antigens. The $I^O$ allele codes for no active enzyme. As a result, neither antigen is formed.

## Sex linkage

We saw earlier in the chapter that human sex is inherited as a result of the two sex chromosomes, **X** and **Y**. We also saw that the **X** chromosome is longer than the **Y** chromosome and carries a large number of genes that are unrelated to sex and are absent from the **Y** chromosome. **Sex linkage** is the name given to the situation in which a phenotypic characteristic is inherited as a result of a gene on the **X** chromosome that is absent from the **Y** chromosome. Sex linkage is like dihybrid inheritance, since we are looking at the inheritance of sex at the same time as that of another phenotypic feature.

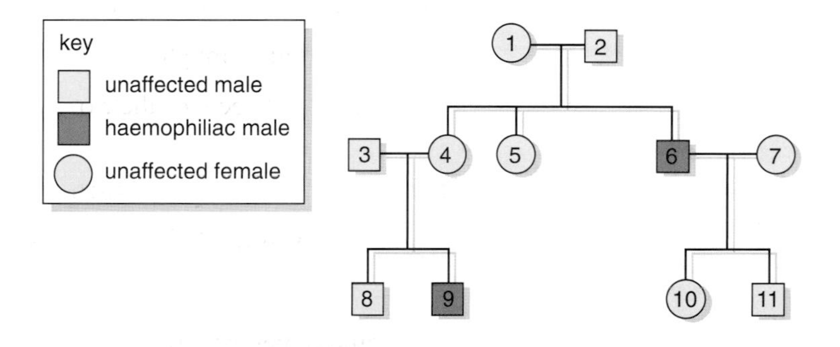


**Figure 1.19**
This pedigree shows the inheritance of haemophilia in humans. Note that haemophilia is more common in males but is never passed from father to son: these are characteristics of sex-linkage

Figure 1.19 represents part of a pedigree of a family that is affected by **haemophilia**, a disease in which the clotting time of blood is much longer than usual. You need to become familiar with interpreting diagrams like this. The circles represent females and the squares represent males. Individuals 1 and 2, both unaffected by haemophilia, had three children, two unaffected daughters (individuals 4 and 5) and a son who suffered from haemophilia (individual 6). As an adult, individual 4 had two sons by an unaffected man (individual 3): one of these was unaffected but one (individual 9) suffered from haemophilia. As an adult, individual 6 had a daughter and a son by an unaffected woman (individual 7): neither child was affected by haemophilia.

The pattern of inheritance of haemophilia shown in Figure 1.19 is typical of sex-linked inheritance. A sex-linked characteristic is:

- more common in males than in females
- never passed from father to son.

**Q 11 Why is haemophilia never passed from a father to his son?**

Table 1.3 represents the sex chromosomes involved in the inheritance of haemophilia. On each **X** chromosome is a gene for blood clotting that is totally absent from the **Y** chromosome. The gene has two alleles, the dominant **H** that results in a rapid blood clotting time and the recessive **h** that results in a slowed clotting time. The symbols from Table 1.3 have been used in Table 1.4 to represent the possible genotypes and phenotypes involved in the inheritance of haemophilia.

| Symbol | Description |
|---|---|
| $X^H$ | An **X** chromosome that carries the dominant **H** allele of the blood clotting gene |
| $X^h$ | An **X** chromosome that carries the recessive **h** allele of the blood clotting gene |
| Y | A **Y** chromosome. This small chromosome does not carry a blood clotting gene |

Table 1.3
The symbols used to represent the inheritance of haemophilia in humans

| Genotype | Description of phenotype |
|---|---|
| $X^HY$ | A male with rapid clotting time |
| $X^hY$ | A haemophiliac male |
| $X^HX^H$ | A homozygous female with rapid clotting time |
| $X^HX^h$ | A heterozygous female 'carrier' who has rapid clotting time but carries the recessive allele |
| $X^hX^h$ | A homozygous haemophiliac female. (These are rare because they arise from a haemophiliac father and a mother who is a 'carrier'.) |

Table 1.4
The genotypes and phenotypes in human haemophilia. Note that the **Y** chromosome lacks the blood clotting gene. As a result, all males are described as hemizygous for this gene - whichever they inherit on the **X** chromosome from their mother is expressed in their phenotype

We can use this information to work out the genotypes of individuals in the family pedigree in Figure 1.19. Individual 6 is shown as a haemophiliac male: we know from Table 1.4 that he must have the genotype $X^hY$. Males always inherit their **X** chromosomes from their mothers and their **Y** chromosomes from their fathers. Consequently, we know that individual 1 must have the $X^h$ chromosome in her genotype. Since she is not affected by haemophilia, she must have the genotype $X^HX^h$, in other words she is a carrier. Individual 2, an unaffected male, must have the genotype $X^HY$.

**Q 12** **Explain why each of the following must have the genotype $X^HX^h$:**
**(a) individual 4**
**(b) individual 10.**

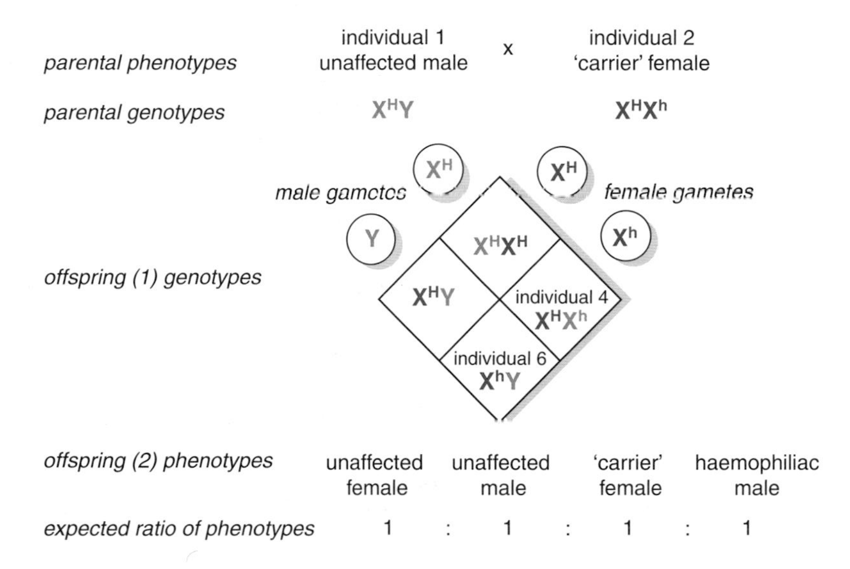


**Figure 1.20**
A cross showing the inheritance of haemophilia, using individuals 1, 2, 4 and 6 from Figure 1.19

Figure 1.20 shows how we can use our standard method for representing the cross between individuals 1 and 2. The genotype we have deduced for individual 4 ($\mathbf{X^HX^h}$) and for individual 6 ($\mathbf{X^hY}$) are shown in Figure 1.20. From the information given, we cannot deduce the genotype of individual 5.

## When offspring from experimental crosses do not exactly match expected ratios: the chi-squared ($\chi^2$) test

If you were to toss a perfectly balanced coin, there is a 50:50 chance that it will fall as 'heads'. If it fell 'tails', you would not be surprised - this could happen as a result of chance. If you tossed the coin in the air twice and on both occasions it landed 'tails', you would not be surprised - again, this could happen as a result of chance. If you tossed the coin in the air 200 times and it landed 'tails' every time, you would not accept that this was due to chance and would suspect that the coin was unevenly weighted or double-sided 'tails'. This illustrates two ideas. Firstly, we do not let chance variations from what we expect prevent us from accepting a hypothesis (in this case, that the coin is perfectly balanced and has a 'heads' side and a 'tails' side). Secondly, there is a point at which variations from what we expect are so great that we begin to doubt our hypothesis and look for another (that the coin is unevenly weighted or that both sides are the same).

Let's relate this idea to some of the crosses we have examined in this chapter. In Figure 1.11, we saw that there is a theoretical 1 in 2 chance of a human child being a boy or a girl. The chance factor here is that an egg cell with an **X** chromosome could be fertilised at random by either an **X**-carrying sperm or a **Y**-carrying sperm. You probably know families with two, three of four children who are all the same sex. This does not make us doubt our hypothesis that there is a 1 in 2 chance of each child being a boy. In fact, for every family with n children there will be $2^n$ possible permutations of births. For example, for every family of two children there will be four (= $2^2$) permutations of births: boy-boy; boy-girl; girl-boy; girl-girl. On this basis, we would expect that one quarter of couples with two children would have two daughters, one quarter would have two sons and half would have a daughter and a son.

Humans have so few children that we never begin to doubt that chance is playing a role in their inheritance. This is not so with fruit flies, where a single mating can produce over two hundred offspring. Look back at the inheritance of wing length in *Drosophila melanogaster* in Figure 1.13. If 200 offspring were produced, we would expect 150 to have long wings and 50 to have vestigial wings. If there were actually 145 long-winged and 55 vestigial-winged, we would probably accept this was a chance variation from what we expected. If there were 50 long-winged and 150 vestigial-winged offspring, we would not accept this as a chance deviation from what we expected and would begin to doubt that our explanation of inheritance was true. But where do we draw the line? Two conventions help us here.

- We need a way of measuring the probability that what we observe is a chance deviation from what we expect. This is what the chi-squared test ($\chi^2$ test) does.
- We need a commonly agreed limit on what we accept is due to chance. For our purposes, we reject a hypothesis if the probability of our observations being a chance variation of what we expected to happen is less than 1 in 20 (5% or, as a decimal fraction, once its probability is less than 0.05). We say that the probability value of 0.05 is our **level of significance.**

To calculate $\chi^2$, we use the following formula, where **O** represents the observed value and **E** represents the expected value:

$$\chi^2 = \sum \frac{(O - E)^2}{E}$$

The Greek letter Σ (pronounced 'sigma') means 'sum of' and shows that we must add up all the different calculated values of:

$$\frac{(O - E)^2}{E}$$

Suppose 200 offspring were produced in the cross shown in Figure 1.13. Anticipating a 3:1 ratio, we would expect 150 of them to be long-winged and 50 to be vestigial winged. In fact, suppose we found that only 145 were long-winged and 55 were vestigial-winged. Is this a chance variation of what we expected? Table 1.5 shows how we would perform a $\chi^2$ test of our null hypothesis, i.e. that there is no difference between the ratio we observed and the 3:1 ratio that we expected.

| Component | Long-winged offspring | Vestigial-winged offspring |
|---|---|---|
| Expected (E) | 150 | 50 |
| Observed (O) | 145 | 55 |
| O – E | 5 | 5 |
| $(O-E)^2$ | 25 | 25 |
| $\frac{(O-E)^2}{E}$ | 0.17 | 0.5 |
| $\chi^2 = \Sigma \frac{(O-E)^2}{E}$ | 0.67 | |

Table 1.5
Calculation of $\chi^2$

From Table 1.5, the calculated value of $\chi^2$ is 0.67. We now need to use Table 1.6 to see whether this value of $\chi^2$ represents a chance variation of what we expected or not. As there are 2 different categories of flies in our example ($n = 2$), there is $n - 1 = 1$ degree of freedom. The table shows that, with one degree of freedom, the probability of our observed ratio of long-winged:vestigial-winged flies occurring by chance is between 0.50 and 0.30. This is above our level of significance of 0.05, so we can accept the ratio as a chance variation of what we expected.

| Categories ($n$) | Degrees of freedom ($n - 1$) | $\chi^2$ | | | | | | | |
|---|---|---|---|---|---|---|---|---|---|
| 2 | 1 | 0.016 | 0.15 | 0.46 | 1.07 | 2.71 | 3.84 | 5.41 | 6.64 |
| 3 | 2 | 0.21 | 0.71 | 1.39 | 2.41 | 4.61 | 5.99 | 7.82 | 9.21 |
| 4 | 3 | 0.58 | 1.42 | 2.37 | 3.67 | 6.25 | 7.82 | 9.84 | 11.34 |
| Probability that chance alone could produce the deviation from what was expected | | 0.90 | 0.70 | 0.50 | 0.30 | 0.10 | 0.05 | 0.02 | 0.01 |

Table 1.6
Table of $\chi^2$ values. To accept our null hypothesis, that there is no significance between our expected and observed values, we need to find a value of $\chi^2$ that is associated with a probability of 0.05 or more. If the probability is less than 0.05, we reject our null hypothesis

## Summary

- During meiosis, cells with two copies of each chromosome (diploid cells) divide to form cells with only one copy of each chromosome (haploid cells).
- The separation of homologous chromosomes takes place during the first meiotic division. During the second meiotic division, sister chromatids of each chromosome are separated into daughter cells.
- Because of the way that the members of different pairs of homologous chromosomes separate independently of each other during the first meiotic division (independent assortment of homologous chromosomes), meiosis produces daughter cells that are genetically different from each other. In general, if there are *n* pairs of homologous chromosomes, independent assortment will produce $2^n$ different combinations of chromosomes in the daughter cells. This genetic variation is increased because chromatids break at points where they intertwine (chiasma) and join to the other homologue.
- Monohybrid crosses involve the inheritance of a character controlled by a single gene. Dihybrid crosses involve the inheritance of characters controlled by two genes that occur at two different loci.
- Any diploid cell has two copies of the gene controlling a single character. If both versions of the gene (called alleles) are the same, this individual has a homozygous genotype. If the two alleles of the gene are different, the individual has a heterozygous genotype.
- In representing genetic crosses, we use a standard format that shows: phenotype of parents, genotype of parents, gametes, genotype of offspring (1) generation, expected frequency of phenotype of offspring (1) generation. If the parents are pure-breeding (homozygous), the offspring (1) generation is sometimes called the F1 generation and the offspring (2) generation is sometimes called the F2 generation.
- An allele of a gene is said to be recessive if it fails to exert its effect in a heterozygote. The allele of the same gene that exerts its effect in the heterozygote is said to be dominant. The alleles are codominant if they both show their effect in a heterozygote.
- Sex is determined by sex chromosomes. A character unrelated to sex is sex-linked if it is controlled by a gene whose locus is on the female sex chromosome.
- The chi-squared test ($\chi^2$ test) helps us to determine whether observed results fit the expectations we had based on a null hypothesis. As a general rule, we accept any variation from what we expected so long as its probability is $\geq 0.05$.

# Assignment

Figure 1.21
Ladybirds are common insects but they are not often seen in large numbers. However, swarms can contain as many as 400 insects in a square metre. Those pictured here are 7-spot ladybirds.

## Ladybird, ladybird

Ladybirds, such as those shown in Figure 1.21, are beetles. They are common insects and their distinctive colour patterns make them quite easy to identify. Some species are also very variable in appearance. Many of these variations are genetically controlled. In this assignment we will look at the inheritance of some of the colour patterns found on the wing-cases of ladybirds. Before you start work, look at the examples of crosses shown in this chapter and make sure you know how to use a genetic diagram to explain the results of a particular cross. The crosses at the start of this assignment are very straightforward but you may find some of those towards the end much more challenging!

The 2-spot ladybird is a much more variable species than the 7-spot ladybird shown in Figure 1.21. Figure 1.22 shows two varieties of this ladybird. These varieties are controlled by a single gene with two alleles. The allele for black wing-cases, **B**, is dominant to that for red wing-cases, **b**.

Figure 1.22

1 Use a genetic diagram to show the offspring you would expect from a cross between a red ladybird and a ladybird heterozygous for these alleles.

*(3 marks)*

Three more varieties of this ladybird are shown in Figure 1.23. These varieties are controlled by another gene. It has two alleles. The typica allele ($\mathbf{W^T}$) and the annulata allele ($\mathbf{W^R}$) are codominant. The heterozygote is intermediate in appearance.

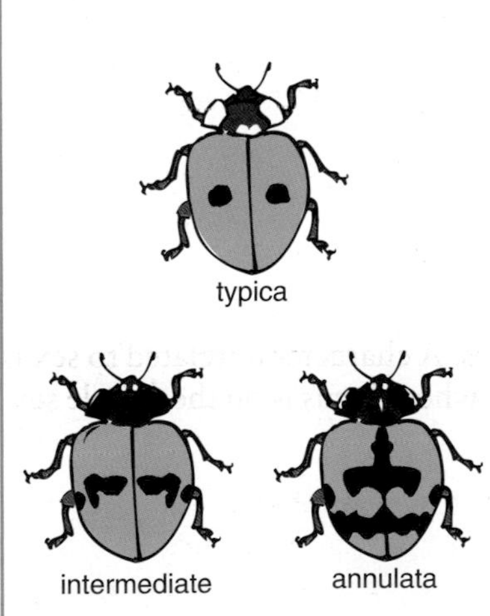



Figure 1.23

2 (a) Construct a table to show the possible genotypes of each of the three ladybirds shown in Figure 1.23.

*(2 marks)*

(b) Predict the ratio of phenotypes of the offspring from a cross between two intermediate ladybirds. Use a genetic diagram to explain your answer.

*(2 marks)*

The 2-spot ladybird also occurs in the United States. We will look now at some work carried out on American 2-spot ladybirds nearly a hundred years ago. Figure 1.24 shows five varieties of this ladybird. These varieties were found to be controlled by a single gene with five alleles: $\mathbf{W^B}$ (*bipunctata*), $\mathbf{W^M}$ (*melanopleura*), $\mathbf{W^A}$ (*annectans*), $\mathbf{W^C}$ (*coloradensis*) and $\mathbf{W^H}$ (*humeralis*).

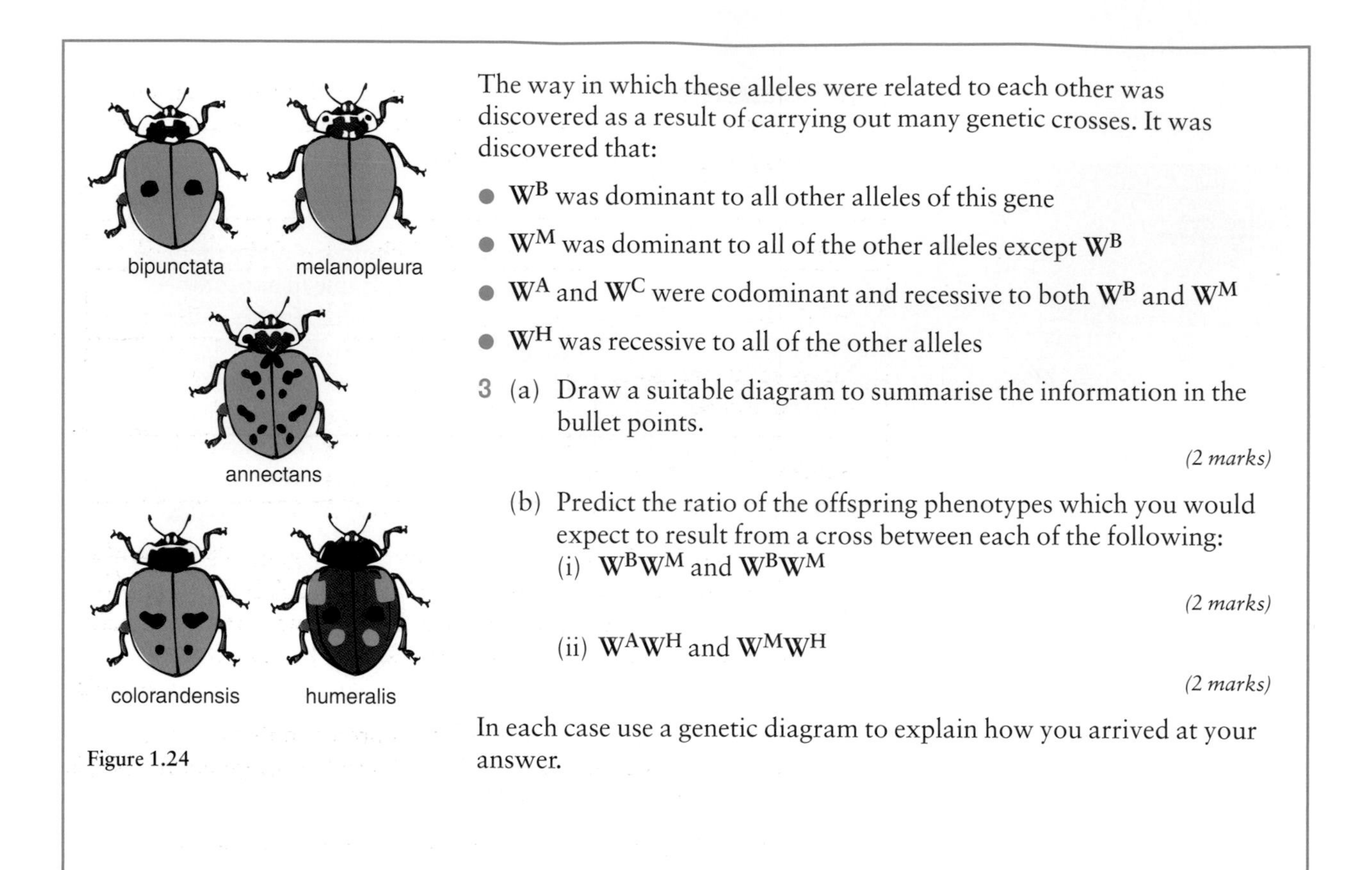


Figure 1.24

The way in which these alleles were related to each other was discovered as a result of carrying out many genetic crosses. It was discovered that:

- $W^B$ was dominant to all other alleles of this gene
- $W^M$ was dominant to all of the other alleles except $W^B$
- $W^A$ and $W^C$ were codominant and recessive to both $W^B$ and $W^M$
- $W^H$ was recessive to all of the other alleles

3 (a) Draw a suitable diagram to summarise the information in the bullet points.

*(2 marks)*

(b) Predict the ratio of the offspring phenotypes which you would expect to result from a cross between each of the following:
(i) $W^BW^M$ and $W^BW^M$

*(2 marks)*

(ii) $W^AW^H$ and $W^MW^H$

*(2 marks)*

In each case use a genetic diagram to explain how you arrived at your answer.

## Examination questions

1 A queen honey bee can lay both fertilised and unfertilised eggs. Fertilised eggs develop into diploid females and unfertilised eggs develop into haploid males. The diagram shows the formation of gametes in female bees and in male bees.

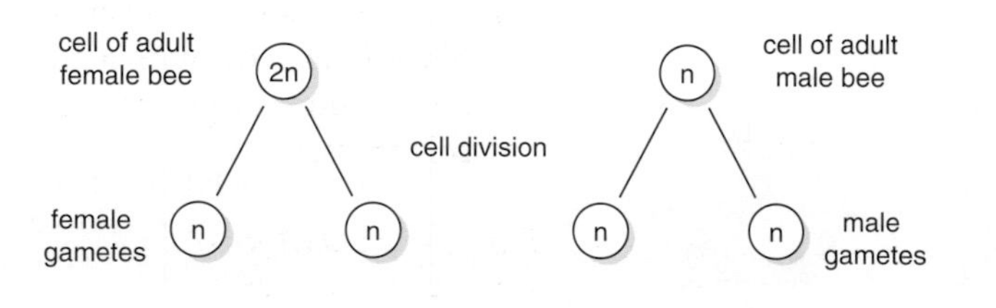


(a) Giving a reason for your answer in each case, name the type of cell division in the bee that produces:
(i) female gametes
(ii) male gametes

*(1 mark)*

(b) The table shows some features which contribute to variation in the offspring of bees. Complete the table with a tick if the feature may contribute or a cross if it does not.

*(2 marks)*

| Feature | Female offspring | Male offspring |
|---|---|---|
| Crossing over | | |
| Independent segregation of chromosomes | | |
| Random fusion of gametes | | |

(c) Body colour in bees is determined by a single gene. The allele **B** for yellow body is dominant to the allele **b** for black body. Explain why, in the offspring of a mating between a pure-breeding black female and a yellow male, all the males will be black.

*(1 mark)*

2 In cats, one of the genes for coat colour is present only on the X chromosome. This gene has two alleles. The allele for ginger fur, $\mathbf{X^B}$, is dominant to that for black fur, $\mathbf{X^b}$.

(a) All the cells in the body of a female mammal carry two X chromosomes. During an early stage of development, one of these becomes inactive and is not expressed. Therefore female mammals have patches of cells with one X chromosome expressed and patches of cells with the other chromosome expressed. Tortoiseshell cats have coats with patches of ginger and patches of black fur.

(i) What is the genotype of a tortoiseshell cat?

*(1 mark)*

(ii) Explain why there are no male tortoiseshell cats.

*(1 mark)*

(b) A cat breeder who wished to produce tortoiseshell cats crossed a black female cat with a ginger male. Complete the genetic diagram and predict the percentage of tortoiseshell kittens expected from this cross.

*Parental phenotypes:* black female ginger male
*Parental genotypes:*
*Gametes:*
*Offspring genotypes:*
Percentage of tortoiseshell kittens:

*(3 marks)*

**3** Night blindness is a condition in which affected people have difficulty seeing in dim light. The allele for night blindness, **N**, is dominant to the allele for normal vision, **n**. (These alleles are *not* on the sex chromosomes.)

The diagram shows part of a family tree showing the inheritance of night blindness.

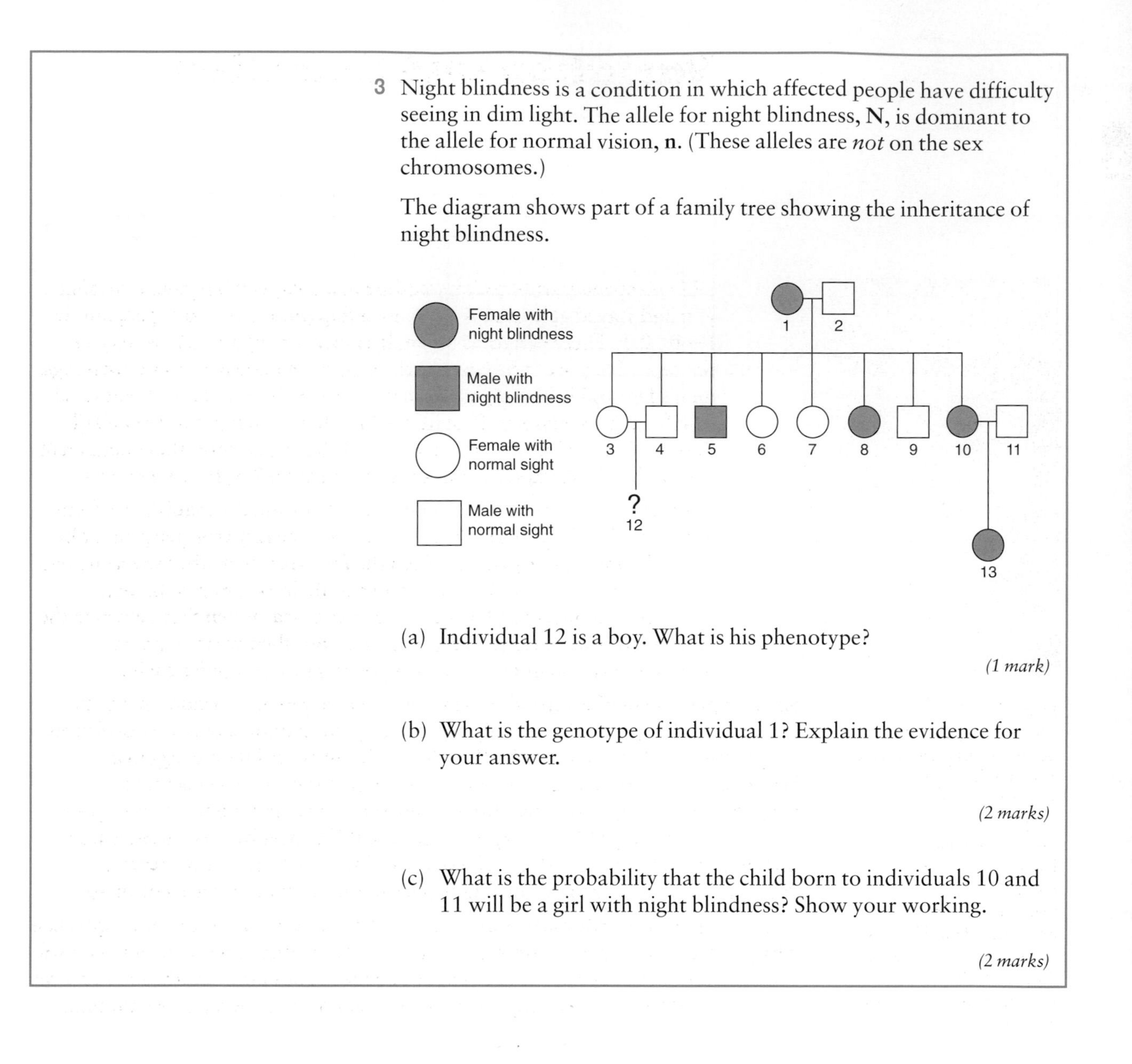


(a) Individual 12 is a boy. What is his phenotype?

*(1 mark)*

(b) What is the genotype of individual 1? Explain the evidence for your answer.

*(2 marks)*

(c) What is the probability that the child born to individuals 10 and 11 will be a girl with night blindness? Show your working.

*(2 marks)*

# 2 Variation and Selection

Figure 2.1
These male emperor penguins spend the coldest two months of the Antarctic winter incubating their eggs. Each male has a single egg balanced on his feet and covered by the folds of feathered skin from his abdomen. He will stay like this for about two months, during which time he will lose up to one-third of his body mass

Did you spend part of today standing in a cold, wet bus queue wishing you had stayed at home? If so, spare a thought for the poor penguins in Figure 2.1. These penguins are male emperor penguins (*Aptenodytes forsteri*). They are incubating their eggs. Each male balances a single egg on his feet and is keeping it warm by covering it with folds of feathered skin from his abdomen. Despite the fact that the temperature can fall to −40 °C and wind speeds can exceed 160 km per hour, these males will stay huddled together, as you see them in Figure 2.1, for two months.

Look carefully at the birds in Figure 2.1. Can you distinguish one from another? The chances are that you cannot; one emperor penguin looks much the same as any other. When the females return after spending two months feeding at sea, they will return to their own mates. In fact, emperor penguins spend most of the year at sea. When they return to the ice around Antarctica to breed, they seek out their mates from the previous year. Obviously, emperor penguins do recognise each other.

Since emperor penguins reproduce sexually, there is genetic variation between members of the species. In sexually reproducing populations, only identical twins are genetically identical to each other. A female only ever lays one egg in a breeding season, so there are no pairs of identical twins amongst emperor penguins. It is this genetic variation that enables emperor penguins to recognise each other. We are good at recognising genetic differences between humans, so that no two humans look identical to us. Our inability to recognise genetic differences between emperor penguins does not mean that they are not there.

In this chapter, you will learn about some of the causes of variation and appreciate why genetic variation is so common in sexually reproducing populations. You will also see how variation can lead to genetic changes in populations, through natural selection, and begin to use simple statistical techniques to investigate variation.

## Investigating variation

In this part of the chapter, you will learn about ways in which we can collect, represent and analyse quantitative data. You might use some of these techniques during your coursework investigations.

### Frequency distributions

In Chapter 1, we looked at the way in which some phenotypic characteristics are inherited. Look back to Figure 1.13, which shows the inheritance of wing length in *Drosophila melanogaster*. If a pair of *Drosophila* reproduced exactly as you would expect from Figure 1.13, there would be three times as many long-winged flies in the offspring (2)

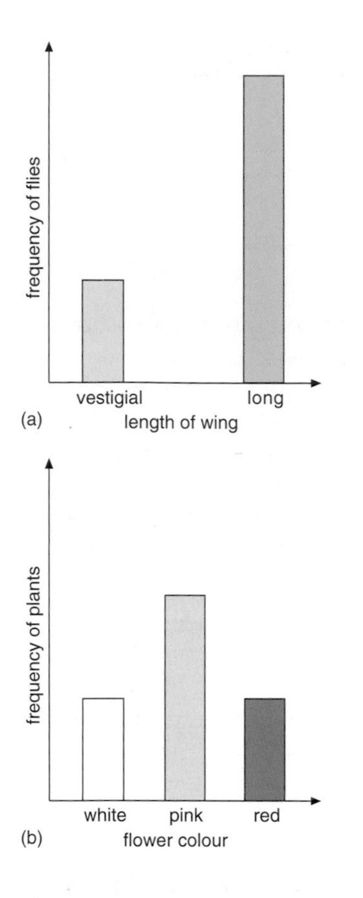


**Figure 2.2**
(a) A frequency distribution of the expected wing lengths of the offspring (2) generation of flies produced by the cross in Figure 1.13
(b) A frequency distribution of the expected flower colours in the offspring (2) generation of snapdragons produced by the cross in Figure 1.15

generation as there were vestigial-winged flies. We can plot these expected results on a bar graph. This has been done in Figure 2.2(a). On the horizontal axis, we have the characteristic being measured, wing length, and on the vertical axis we have the number of organisms in each category of wing length. Because this type of graph shows the number of organisms within each of several measured categories, it is called a **frequency distribution.** Because we have only two wing-length categories – vestigial and long – there are only two bars on our frequency distribution. Now look back to Figure 1.15, which shows the inheritance of flower colour in snapdragons (*Antirrhinum majus*). If a pair of snapdragon plants reproduced exactly as you would expect from Figure 1.15, there would be equal numbers of plants with red flowers and white flowers but twice as many plants with pink flowers. Figure 2.2(b) shows a frequency distribution of these expected results. Instead of the two categories of wing length that we had in Figure 2.2(a) we now have three categories of flower colour and the frequency distribution has a different shape.

It is easy to judge whether the wing length of a fly is vestigial or long or to judge whether a flower is red, pink or white. This is because these characteristics are quite distinct from each other. Data like these are described as **discrete data.** A frequency distribution of discrete data always shows **discontinuous variation.** However, data are not always discrete. Imagine a student wished to carry out an investigation of the height of students in two of her study groups. Although some people are tall and others are short, there is not a single 'tall' category or a single 'short' category. The person conducting the investigation would have to decide the size categories (called class intervals) she would use. In doing so, she might choose the same class intervals as you or different class intervals, it would not much matter.

Table 2.1 shows how this student might record the heights of her fellow students. The left-hand column shows the class intervals she has chosen, in this case she has chosen 2 cm intervals. In the middle column, she recorded her raw data. She has used a common convention for doing this. Each time she measured a height within a particular class interval, she wrote 'I' in the appropriate row in the middle column. However, every fifth time she drew a diagonal line through the previous four 'Is'. This divides her raw data into groups of five, making counting easier. Finally, in the right-hand column, she totalled the number of students in each class interval. Figure 2.3 shows the frequency distribution plotted from the data in Table 2.1. Once again, the bars represent the number of counts in each class interval. However, a curve has been added to the frequency distribution to show the general trend. This trend shows that measurements of human height are not discrete data; instead they are continuous data. A frequency distribution of continuous data, like the curve in Figure 2.3, shows **continuous variation.**

| Class interval/cm | Tally count of number of students | Number of students |
| --- | --- | --- |
| 1480–1499 | I | 1 |
| 1500–1519 | III | 3 |
| 1520–1539 | IIII | 4 |
| 1540–1559 | ~~IIII~~ I | 6 |
| 1560–1579 | ~~IIII~~ II | 7 |
| 1580–1599 | ~~IIII~~ ~~IIII~~ | 10 |
| 1600–1619 | ~~IIII~~ III | 8 |
| 1620–1639 | IIII | 4 |
| 1640–1659 | III | 3 |
| 1660–1679 | II | 2 |
| 1680–1699 | I | 1 |

Table 2.1
A method for recording raw data about the height of students in study groups. Before the investigation, a decision was made about the class interval. A tally was recorded in the middle column, grouping counts into fives for ease of counting. The final column was completed at the end of the investigation

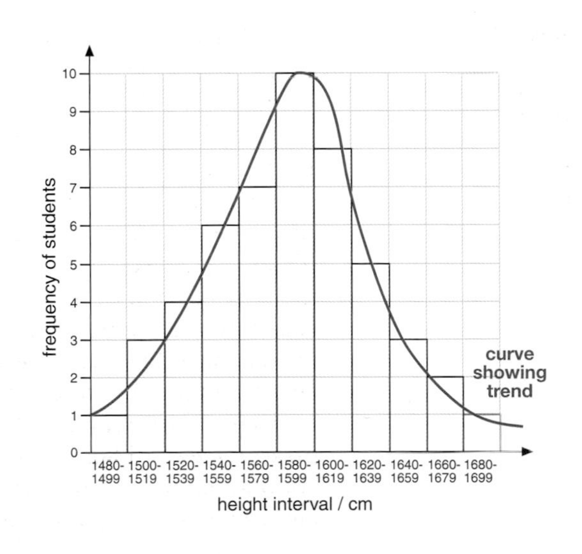


Figure 2.3
A frequency distribution plotted from the data in Table 2.1. The curve shows the general trend of the data

## The effect of chance and bias on data collection

In the examples used above, we looked at samples of a population – the offspring (2) generation from a single cross or the people who happen to be in the same study groups. These samples are sub-groups of the total population and might not be truly representative of that population. Even measuring the height of each person in her study groups was a sampling process. Had our student made the same measurement several times, she would probably have found that she recorded different heights for the same people.

Variation in sampling can come about in three ways:

- the differences between samples reflect real differences in the populations from which they came
- chance – what we would commonly call luck
- sampling bias – the investigator, knowingly or unknowingly chooses samples

When we investigate populations, we want to be sure that any differences we find reflect real differences in the populations we are sampling.

We eliminate chance by taking several samples. A single, small sample is unlikely to be representative of the population from which it came. This is why you are encouraged to use the average of several readings during practical work or to combine results from the group and take the average values. The number of samples we take is limited by the time we have available. In laboratory work, you are probably encouraged to take three samples by setting up three replicates of an experiment or by taking three readings. Biologists often use more than three samples when they want to have confidence in their data.

We eliminate bias by removing human choice from the sampling method. Methods that do this ensure **random sampling**. You are expected to carry out fieldwork, during which you collect quantitative data, at least once in your A2 course. You will use random sampling methods to collect this data. These might involve drawing a grid over a map of the sample area, numbering squares and then choosing the numbers of the squares to be sampled from a table of random numbers. It will not involve throwing a square quadrat frame over your shoulder and sampling where it falls. You can read more about sampling methods in Chapter 4.

**Q** 1 **Why would throwing a quadrat frame over your shoulder *not* be a suitable sampling method during fieldwork?**

## The normal distribution

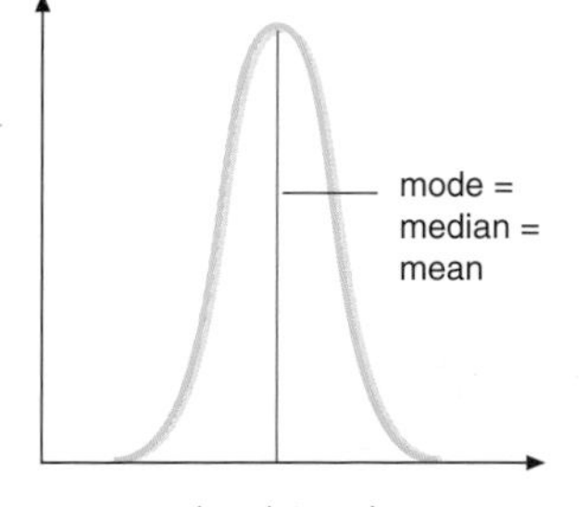


**Figure 2.4**
A normal distribution curve. This is a special type of frequency distribution in which the mode, median and mean are identical

The normal distribution is a special type of frequency distribution. Figure 2.4 shows a graph of a normal distribution, called a normal distribution curve. Like any frequency distribution, it has the class characteristic being investigated on the horizontal axis and the number of individuals on the vertical axis. The curve has a symmetrical bell shape. It also has important mathematical properties, which are:

- its most frequent value (mode); middle value (median) and average value (mean) are the same

- it is symmetrical – 50% of the values are above the mean and 50% of the values are below the mean
- 95% of its values are within two standard deviations of the mean.

Let's look at these concepts in a little more detail. Table 2.2 shows measurements that were made on the length of twenty-seven plant leaves. The lengths have been arranged in rank order. The **mode** is the measurement that occurs most often. If plotted on a frequency distribution, it will be the highest point in the bar chart or graph. In Table 2.2, the most common leaf length is 3.8 cm – this is the mode of these data. The **median** is the middle number in the ranked list. In Table 2.2, we have twenty-seven measurements. The middle measurement, with thirteen above it and thirteen below it, is 3.8 cm. In this case, as well as being the modal value, 3.8 cm is also the median.

The arithmetic mean is written as $\overline{x}$ (pronounced 'x bar'). It is calculated as

$$\text{Mean} = \frac{\text{sum of all the measurements}}{\text{number of measurements}}$$

In mathematical notation this is written as

$$\overline{x} = \frac{\sum x}{n}$$

We can use this formula to calculate the arithmetic mean of the measurements in Table 2.2.

Sum of all the measurements = 100.7

Number of measurements = 27

Mean length = 100.7 / 27 = 3.7

**Q 2 Do the data in Table 2.2 follow a normal distribution? Explain your answer.**

| Length of leaf/cm | | | | | | | | |
|---|---|---|---|---|---|---|---|---|
| 2.9 | 3.0 | 3.1 | 3.1 | 3.3 | 3.4 | 3.5 | 3.6 | 3.6 |
| 3.7 | 3.7 | 3.7 | 3.8 | 3.8 | 3.8 | 3.8 | 3.8 | 3.9 |
| 3.9 | 3.9 | 4.0 | 4.0 | 4.1 | 4.2 | 4.3 | 4.3 | 4.5 |

**Table 2.2**
The length of the leaves in a sample of plant leaves

## Extension box 1

## Standard deviation

Although you will not be expected to calculate standard deviations and standard errors in a Unit Test; you are expected to be familiar with their meaning and their use. You might also wish to calculate these measures using data that you collect in your coursework.

Figure 2.5 shows two normal distribution curves. Although they have the same mean, median and mode, they have a different spread of measurements. We represent the spread of measurements about the mean of a normal distribution curve by a value called the **standard deviation.** We calculate the standard deviation in the following way:

1. square each measurement ($x^2$)
2. add all the squared values together ($\Sigma x^2$)
3. divide this total by the number of measurements made $\frac{\Sigma x^2}{n}$
4. square the mean of the measurements ( $\bar{x}^2$)
5. subtract the result of step 4 from the result of step 3 $\left(\frac{\Sigma x^2}{n} - \bar{x}^2\right)$
6. take the square root of step 5. This is the standard deviation. Written as mathematical equation, the above steps becomes:

$$\text{standard deviation, s} = \sqrt{\left(\frac{\Sigma x^2}{n} - \bar{x}^2\right)}$$

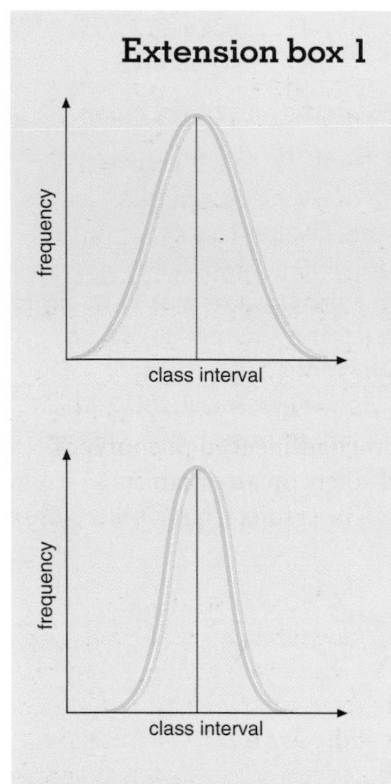


**Figure 2.5**
Although these are both normal distribution curves, one has a greater spread than the other. The curve with the greater spread has a greater standard deviation than the thinner curve

When given together, the mean and standard deviation of a series of measurements is useful to us. It tells us the central value of the distribution and how spread out the individual measurements are from the central value. Whatever the value of the mean and standard deviation, 95% of the data lie within two standard deviations above or below the mean.

Because we calculate means and standard deviations from small samples, they are only estimates of the means and standard deviations of whole populations. We take account of this using another measure, called the standard error of the mean (or **standard error**, for short). The standard error is calculated using the standard deviation (s) and the number of measurements ($n$):

$$\text{Standard error (SE)} = \frac{s}{\sqrt{n}}$$

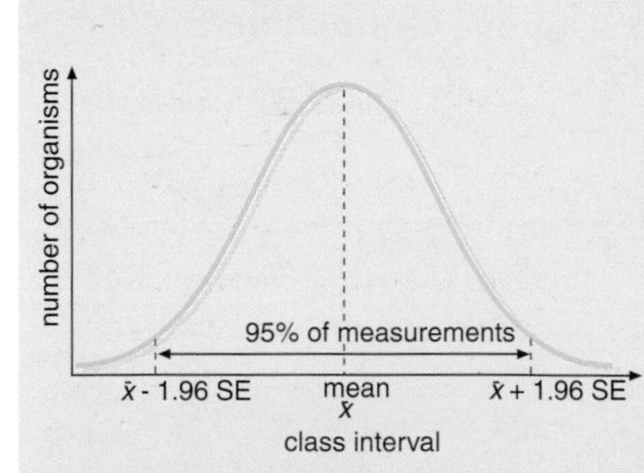


**Figure 2.6**
A normal distribution curve showing the mean ($\bar{x}$) and 1.96 standard errors (SE) either side of the mean. 95% of the measurements lie within the range mean ± 1.96 SE

The standard error is important because it is used to define confidence limits. The true mean of a whole population falls within the range ± 1.96 SE of the sample mean (Figure 2.6). You would be 95% confident that the mean of the population fell within the range ± 1.96 SE. If you look back to Table 1.6, you will see that we have already used this idea. Without knowing it then, the reason that we used the 5% (0.05) probability level in Table 1.6 was because we were rejecting any values that fell outside 1.96 SE of the mean of our sample.

**Q** 3 **Does a normal distribution curve that is thin have a greater or smaller standard deviation than one that is broad?**

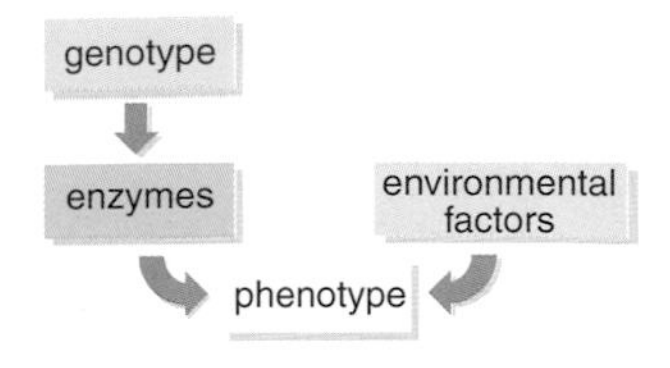


Figure 2.7
An organism's phenotype results from an interaction of genetic and environmental factors

## The causes of variation

Variation exists between the members of any population. Variation is greater in populations of species that reproduce sexually than in those that reproduce asexually, but it is always there.

The differences we see between individuals in a population are differences in phenotype. In humans, characteristics such as height, eye colour, ABO blood group and presence or absence of cystic fibrosis are aspects of our phenotype. In other words, the **phenotype** describes the characteristics of an organism that we can see or detect. We learned about a number of phenotypes in Chapter 1, where we looked at inheritance. In Chapter 1, we used diagrams to show how alleles of genes were inherited to form genotypes, which then influenced phenotypes. However, the phenotype does not depend solely on an organism's genotype. As Figure 2.7 shows, the phenotype results from an interaction between genetic and environmental factors.

Figure 2.8
The fur colour of this Himalayan rabbit shows how genetic and environmental factors interact to form the phenotype. The gene for black fur is active only in the colder parts of the rabbit's body

Look at the Himalayan rabbit in Figure 2.8, which illustrates this interaction. This rabbit has white fur over most of its body but has black fur on its nose, ears, tail and feet. Crossing two pure-breeding Himalayan rabbits will always result in offspring with these characteristics. Clearly, Himalayan rabbits must have a gene that codes for black fur, but why does it only show up in the nose, ears, tail and feet? The answer is simple. The gene for black fur codes for an enzyme that is inactivate at temperatures above 34 °C. Over most of the body, temperatures are above 34 °C so that the fur is white (albino). Only the nose, ears, tail and feet are cold enough for the enzyme to work.

**Q** 4 **Will cells that produce white hairs in the fur of Himalayan rabbits possess the gene for black colour?**

### Meiosis as a source of genetic variation

You learned about the process of meiosis in Chapter 1. On pages 11–13, you saw how meiosis generates genetic variation in several ways.

- **Independent assortment** of the alleles of genes on different pairs of homologous chromosomes results in genetic differences amongst the gametes produced by meiosis. Look back to Figure 1.6, which shows independent assortment of two pairs of homologous chromosomes. Figure 2.9 shows a similar diagram in which a single gene with two different alleles had been included on each chromosome. The parent cell, with a genotype of **AaBb** produces four genetically different daughter cells with genotypes **AB**, **Ab**, **aB** and **aB**.
- **Random fertilisation** of gametes from a single pair of parents results in genetic differences amongst the zygotes. Use Figure 1.18 and Figure 2.9 to help you work out the possible genotypes of offspring produced by two parents of genotype **AaBb**.
- **Crossing over** between members of a single pair of homologous chromosomes changes the combination of the alleles of genes found on those chromosomes. Look back to Figure 1.9 which shows chiasma

formation during meiosis. Figure 2.10 shows the effect of crossing over between the chromosomes from Figure 2.9. Genetic combinations are formed amongst the daughter cells that were not present in Figure 2.9.

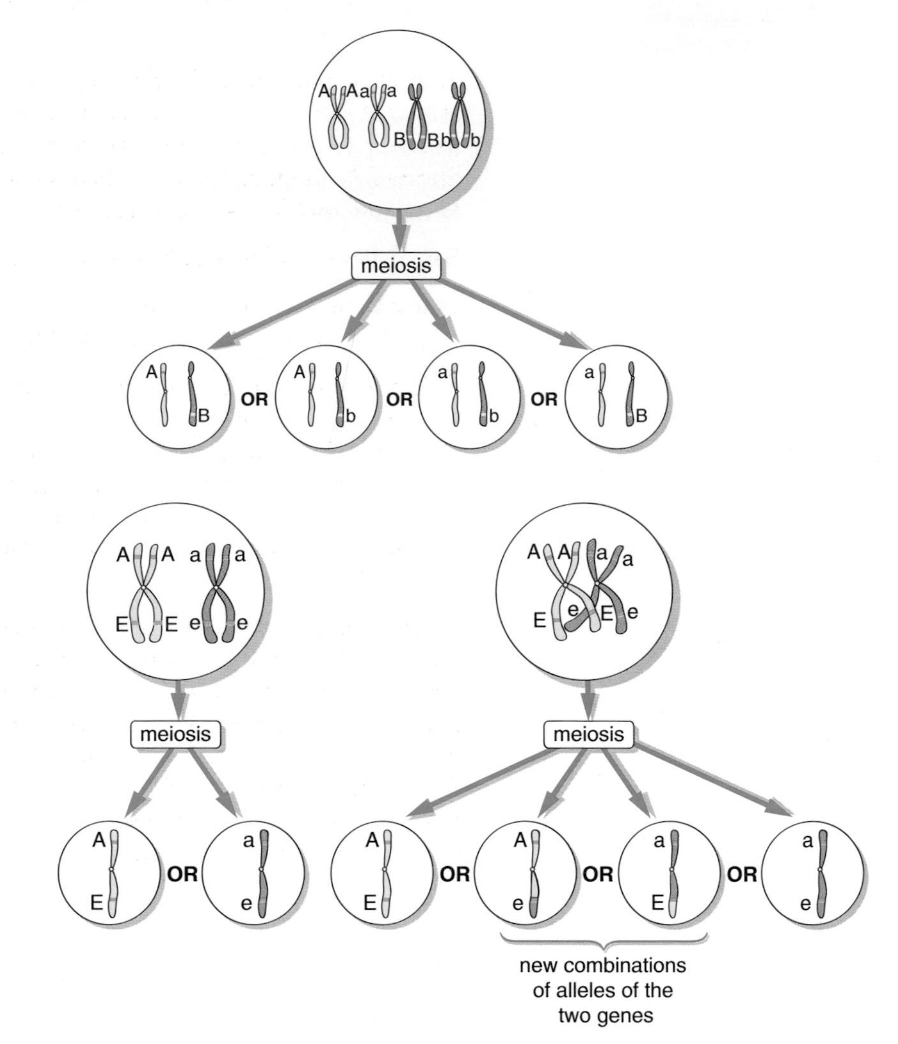


**Figure 2.9**
During meiosis, an independent assortment of pairs of homologous chromosomes produces genetic differences amongst the daughter cells. In this diagram, chromosomes are represented as single lines carrying two genes, each of which has two alleles. Four different daughter cells are produced. Crossing over between pairs of homologous chromosomes can increase genetic variation

**Figure 2.10**
This diagram shows the effect of crossing over on the genetic combination of the chromosomes from Figure 2.9. Notice that new combinations (recombinants) are formed

**Q 5 What genotypes would you expect among the offspring of a cross between two parents of genotype AaBb if the two genes are located on different chromosomes and no crossing over occurs?**

## Gene mutation

During your AS course, you learned that:

- a gene is a length of DNA that carries the genetic code for the production of a particular polypeptide
- DNA is copied during the cell cycle, so that all cells produced by mitosis contain the same DNA as their parent cells

- when transcribed to a messenger RNA molecule, the genetic code is a sequence of bases (adenine, cytosine, guanine and uracil) that determines the sequence of amino acids in a polypeptide
- a combination of three mRNA bases (a codon) codes for a specific amino acid.

From this, you will realise that any change in a DNA base sequence can affect the polypeptide for which it carries the code. Changes in the DNA base sequence of genes are called **gene mutations**. They occur naturally during DNA replication at a rate of about 1 in $10^6$ bases copied. Some environmental factors, such as radiation, increase the gene mutation rate.

Base deletion and base substitution are two types of gene mutation.

- In a **base deletion**, at least one base is not copied during DNA replication. As a result, the new DNA lacks the deleted base or bases.
- In a **base substitution**, at least one base is copied wrongly during DNA replication. Remember that the DNA bases pair adenine with thymine and cytosine with guanine. These base pairs help to ensure the correct base sequence is maintained during DNA replication. During substitution, something interferes with this normal base pairing, so that the 'wrong' base becomes incorporated in a developing DNA strand.

Before looking at mRNA codons and the amino acids for which they code, let us use three-letter words to represent codons and sentences to represent genes. In Table 2.3, we have a sentence of three-letter words that makes sense to us. This represents a normal gene. Look at the third line in Table 2.3. A deletion has occurred, so that the combinations of three letters after the deletion no longer carry the intended meaning. The fourth line in Table 2.3 represents a substitution. One of the letters has been changed, so that the sentence no longer carries the intended meaning.

| Type of error | Combination of three-letter 'words' |
|---|---|
| None (normal message) | The cat saw the dog and ran off |
| Deletion (one letter lost) | The cat sat hed oga ndr ano ff |
| Substitution (wrong letter included) | The cat saw the log and ran off |

**Table 2.3**
This table represents two types of gene mutation. The three-letter 'words' represent base triplets and the 'sentences' represent genes. In a deletion mutation, the loss of one letter changes the meaning of the triplets that follow it. In a substitution mutation, only one of the base triplets is affected. The letters in red show affected triplets

Just as the 'sentences' were changed in Table 2.3, base deletions and base substitutions change the genetic code. Table 2.4 shows some of the mRNA codons and the amino acids for which they code. The mRNA base triplet GGG codes for the amino acid glycine. A single change in the triplet can have no effect (e.g. GGA still codes for glycine) or can result in the incorporation of completely different amino acids in a polypeptide chain. Sickle-cell anaemia (which is described later in this chapter) and cystic fibrosis result from differences in protein structure of only one amino acid. Clearly, deletion and substitution can have profound effects on phenotypes.

| mRNA codon | Amino acid coded |
|---|---|
| GGG | glycine |
| GGA | glycine |
| GAG | glutamic acid |
| GCG | alanine |
| CGG | arginine |

Table 2.4
A few of the 64 possible mRNA codons and the amino acids for which they carry the code. A, C and G represent the organic bases adenine, cytosine and guanine, respectively

### Polygenic inheritance

In Chapter 1, we learned about the inheritance of discrete characteristics, for example wing length in *Drosophila melanogaster* and human ABO blood groups. As you saw, these characteristics are controlled by a single gene, with one or more alleles.

Many characteristics are not controlled by a single gene but by many genes, often located on different chromosomes. This is called **polygenic inheritance.** The different genes each contribute to the final phenotype. As a result, phenotypic characteristics controlled by polygenes have a different pattern of distribution from those controlled by a single gene. We saw these earlier in the chapter. Figure 2.2(a) showed the frequency distribution of a characteristic controlled by a single gene. An individual in the population of *Drosophila* either shows the dominant characteristic or the recessive characteristic. As a result, the frequency distribution has only two bars, representing discontinuous variation. Figure 2.4 showed a normal frequency distribution. This is typical of a characteristic controlled by many genes and represents continuous variation. Polygenic inheritance is involved in characteristics such as human height.

## Population genetics and allele frequencies

In Chapter 1, we considered the inheritance of genes from a single pair of parents to their offspring. In population genetics, we consider the genes in an entire population of organisms. Look back at Figure 1.13 which represents a monohybrid cross involving wing length in *Drosophila*. We used the symbol **L** to represent the dominant, long-wings allele of the gene for wing length and the symbol **l** to represent the recessive, vestigial-wings allele of the gene for wing length. In the offspring (2) generation the flies had one of three genotypes, **LL**, **Ll** or **ll**. Since these are the only possible genotypes for wing length all the *Drosophila* in a population must have one of these three genotype. To represent the genotypes of all the individuals in a population of, say, 1000 *Drosophila* we would need to use 2000 symbols. These symbols would represent all the genes for wing length in that population. This is called the gene pool. In general, we can define a **gene pool** as the total number of alleles of a particular gene that are present in a population at a particular time.

If all the *Drosophila* in a population of 1000 individuals had the genotype **Ll**, there would be 1000 **L** alleles and 1000 **l** alleles in the gene pool. This is a 50:50 split or, put another way, the frequency of the **L** allele would be 0.5 and the frequency of the **l** allele would be 0.5. Notice that frequencies are always expressed as decimal fractions. In the next section, we will examine allele frequencies further.

**Q** 6 What is a gene pool?

## The Hardy-Weinberg principle and the Hardy-Weinberg equation

The Hardy-Weinberg principle involves the frequencies of alleles, genotypes and phenotypes in a given population. Suppose an imaginary gene had two alleles, a dominant **A** and a recessive **a**. Three genotypes would be possible for this gene, **AA**, **Aa** and **aa**, and would be present amongst the individuals of a population. We can use symbols to represent the frequency of these alleles in the gene pool.

$p$ = the frequency of the allele **A**

$q$ = the frequency of the allele **a** $(p + q = 1)$

Since there are only two alleles of this gene, the sum of the two frequencies must have a value of 1, i.e. the total gene pool.

**Q** 7 **If the frequency of the long-wings allele (L) in a population of *Drosophila* is 0.6, what is the frequency of the vestigial-wings allele (l)?**

**The Hardy-Weinberg principle** predicts that the frequencies of the alleles of a particular gene in a particular population will stay constant from generation to generation. In other words, the values of $p$ and $q$ will be the same from one generation to the next. This principle involves a number of assumptions, which you must be able to recall in a Unit test. These assumptions are as follows.

- **The population must be large** – in small populations, chance events can cause large swings in frequencies.
- **There must be no migration into the population (immigration) or out of the population (emigration)** – migration adds new alleles to, or removes alleles from, the population.

- **Mating between individuals in the population must be completely at random** – this ensures an equal chance of the alleles being passed on to the next generation.
- **No genetic mutations must occur** – gene mutations will change allele frequencies.
- **All genotypes must be equally fertile** – this ensures an equal chance of the alleles being passed on to the next generation.

We will see later in the chapter what happens if any of these conditions are not met.

**Q 8 What does the Hardy-Weinberg principle tell us about the frequency of alleles of a particular gene?**

We cannot tell which alleles of a gene an organism possesses. What we can see, or detect, is its phenotype. **The Hardy-Weinberg equation** enables us to calculate allele frequencies by observing phenotypes. Using the symbols that we have used so far, $p$ to represent the frequency of allele **A** and $q$ to represent the frequency of allele **a**, we can represent the frequency of genotypes in our imaginary population.

The frequency of the genotype **AA** $= p^2$

The frequency of the genotype **Aa** $= 2pq$

The frequency of the genotype **aa** $= q^2$ $(p^2 + 2pq + q^2 = 1)$

You do not need to understand how these frequencies are derived or why $p^2 + 2pq + q^2 = 1$, but you must learn how to use this formula. Let's use the formula in a specific example involving wing length in *Drosophila*. In a sample of 500 fruit flies, there were found to be 480 long-winged flies and 20 vestigial-winged flies. Table 2.5 shows the steps we should always follow in calculations involving the Hardy-Weinberg equation using the values from this sample of *Drosophila*.

| Step in calculation | Calculation from example in text |
|---|---|
| 1. Work out the frequency of the homozygous recessives. (This is $q^2$.) | There are 20 vestigial-winged flies in the sample of 500. Their frequency is 20/500 = 0.04 |
| 2. Take the square root of the above value to get the frequency of the recessive allele ($q$) | $q^2 = 0.04$, so $q = \sqrt{0.04} = 0.2$ |
| 3. Find the frequency of the dominant allele (p) using the equation $p + q = 1$ | $p + 0.2 = 1$, so $p = 1 - 0.2 = 0.8$ |
| 4. Put these values of $p$ and $q$ into the Hardy-Weinberg equation to get the genotype frequencies | Frequency of **ll** $= q^2 = 0.04$<br>Frequency of **LL** $= p^2 = 0.8^2 = 0.64$<br>Frequency of **Ll** $= 2pq = 2 \times 0.2 \times 0.8 = 0.32$ |
| 5. Check that the frequencies total 1, to make sure you are right | $0.04 + 0.64 + 0.32 = 1$, so we know our calculation is correct |

**Table 2.5**
The steps you should follow in calculations involving the Hardy-Weinberg equation. Note that you always start with the homozygous recessive individuals

**Q 9 Suggest why you should always start calculations involving the Hardy-Weinberg equation with the homozygous recessive individuals rather than those showing the dominant characteristic.**

## Natural selection

We have already learned in this chapter that one of the conditions of the Hardy-Weinberg principle is not met - gene mutations occur at a predictable rate. In addition we know that some populations are, or have been, very small and that mating is not always random. All of these factors will lead to changes in frequency of the alleles of a particular gene in a population.

One of the most important factors that can lead to a change in allele frequency is natural selection. **Natural selection** results from the differential fertility of organisms that, in turn, results from their genotype. In its natural environment, the size of a population is limited by a number of environmental factors. For example, animals compete for food and plants compete for space, water and access to light. Some organisms in each population will have inherited characteristics that ensure they are more successful than others in obtaining scarce resources. These successful organisms will, consequently, have more of the resources they need for growth and reproduction: they are said to be biologically **fit**. The less successful organisms will have less of the resources they need for growth and reproduction. As a consequence, they are likely to die younger or to have fewer offspring than their more successful competitors. Whether they die or simply have fewer offspring, the less successful (less fit) organisms will not pass on their alleles to the next generation as much as the fit organisms will. This causes a change in the allele frequency of that population. Notice that natural selection is only effective in changing allele frequencies if the advantageous characteristics possessed by successful organisms are inherited.

**Figure 2.11**
Brown rats eat human food stores and spread disease. Since 1950, warfarin has been used to kill rats in areas where they are a serious pest. Through natural selection, warfarin-resistant populations of rats have evolved

Before going further, we will look at an example of natural selection to reinforce your understanding of the stages involved. The animal in Figure 2.11 is a brown rat, which is common in Britain. Rats are a pest because they eat food intended for humans and spread disease. Since the 1950s, a substance called warfarin has been used to kill rats. Food containing warfarin is placed where rats will find it and eat it. Once in their bodies, warfarin interferes with the rats' blood clotting, so that badly poisoned rats suffer fatal haemorrhages. A few years after warfarin was first used as a pesticide against rats, resistant populations were found in parts of Britain. What seems to have happened is that a chance mutation occurred in one rat, producing a new gene allele that reduced the effectiveness of warfarin. As a result of its resistance to warfarin, this rat competed more successfully than other rats for food and mates. As a result, it produced more offspring than susceptible rats. As it passed on its allele for resistance to many of its offspring, the resistance allele became more common in the new generation. In other words, a change occurred in the allele frequency of this rat population. Eventually, most of the rats in the population became resistant to warfarin.

Table 2.6 summarises the stages in the evolution of warfarin-resistant populations of rats. You can use the layout of this table as a plan when answering questions about natural selection in a Unit Test. Notice that, in this example of natural selection, there has been a shift in the frequency distribution of the rat population with respect to susceptibility to warfarin. We will now look at the way that natural selection changes frequency distributions.

| Sequence of events leading to natural selection | Application of these events to the evolution of warfarin resistance in populations of brown rats |
|---|---|
| 1. There is competition within the population for environmental resources that are limited | Members of a rat population compete for food |
| 2. There is genetic variation within the population | A random gene mutation caused one rat to become resistant to the pesticide, warfarin |
| 3. One genetic variant confers an advantage on the possessor(s) of a favourable allele of a gene | The rat with the allele for warfarin resistance would not suffer haemorrhages where warfarin was used, whereas a rat with only alleles for warfarin susceptibility would suffer haemorrhages. |
| 4. The possessor of the favourable allele will have more offspring than the possessor of only the less favourable alleles. | Being stronger and healthier than its susceptible competitors, the resistant rat will successfully raise more offspring. |
| 5. The possessor of the favourable allele will pass this allele on to some of its offspring. As a result, the frequency of this allele in the population will increase, i.e. natural selection has occurred. | The resistant rat passed on the resistance allele to some of its offspring. As a result, there were more resistant rats in the offspring generation. Repeated over several generations, the warfarin-resistance allele became very common in the rat population. |

**Table 2.6**
The evolution of warfarin resistance in populations of rats. You can apply the scheme in the left-hand column of the table to any question in a Unit Test that is about natural selection

**Stabilising selection** acts against both the extremes in a range of variation. Look at Figure 2.12. The upper graph represents the frequency distribution of phenotypes in a population before stabilising selection has occurred. The graph has a fairly wide bell-shaped curve, with its mode marked in red. The lower graph has the same mode but the curve is now a much thinner bell-shape. Selection has eliminated those organisms at the extremes of the variation. This type of selection occurs on birth mass in humans. Despite medical advances, babies with very high, or very low, birth mass have a higher infant mortality rate than those at the mode for birth mass.

**Q 10 Which curve in Figure 2.12 has the smaller standard deviation?**

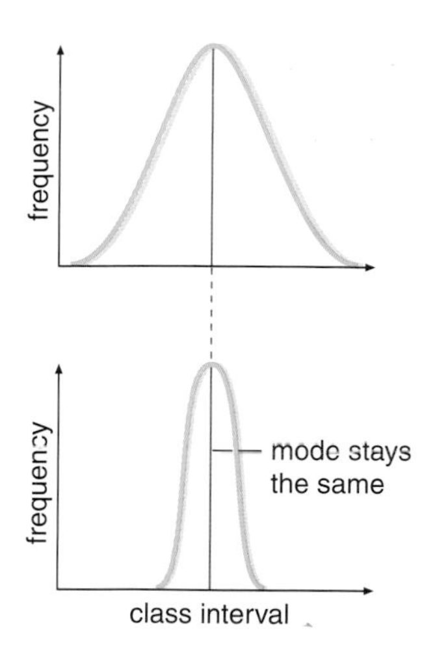


Figure 2.12
The effect of stabilising selection is to reduce variation in a population about its modal value. The upper graph represents the range of phenotypes in a population before selection and the lower graph represents the range after selection

**Directional selection** acts against one of the extremes in a range of variation. As a result, it moves phenotypic variation away from its old modal value. This is shown in Figure 2.13. The upper and lower graphs are similar to Figure 2.12, but notice how the lower graph has shifted to the right. This type of selection has occurred in populations of the myxoma virus in Australia. This virus causes myxomatosis, a disease of rabbits, and was introduced into Australia by European settlers with the intention of killing rabbits that were pests on farms. At first, all rabbits died within a short time of infection. However, the disease is now less virulent and infected rabbits survive longer. This is thought to relate to the method of transmission of the virus from one rabbit to another. In Australia, the virus is spread by mosquitoes. Since mosquitoes will only bite a live rabbit, a myxoma virus that killed its host too quickly was likely to get stranded in a dead rabbit. Viruses that were less virulent enabled their hosts to live longer, increasing their chances of being passed to another rabbit by a biting mosquito. As a result, directional selection acted against virulent myxoma viruses and selected less virulent myxoma viruses.

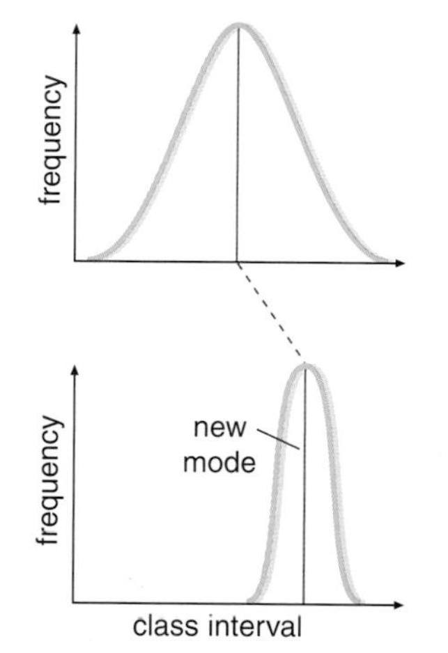


Figure 2.13
The effect of directional selection is to move the modal value. This usually occurs following a change in the environment and the new mode coincides with the new optimum value

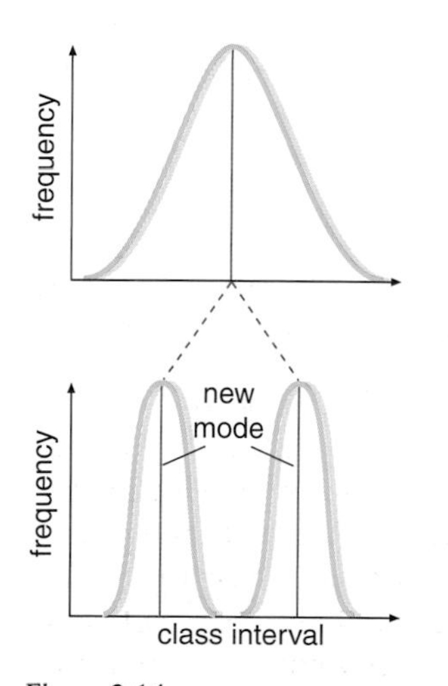


Figure 2.14
The effect of disruptive selection is to produce a bimodal distribution. Each of the new modes represents a different form (morph) of the organism that is best suited to a slightly different habitat

**Disruptive selection** acts against the mode in a range of variation. Figure 2.14 shows how this type of selection produces a bimodal distribution, which might eventually result in two distinct forms of the species, called **morphs**. This type of selection occurs on the wing colour of many insects that rest during the day, when they are susceptible to predation. Figure 2.15 shows two morphs of the peppered moth, *Biston betularia*. This moth is common in Britain. It rests on tree trunks during the day, when it is preyed upon by birds, such as the blue tit in Figure 2.16. In areas that are free of pollution, tree trunks have a blotchy pattern caused by the colour of bark and of the mosses and lichens that grow on the bark. Here, the peppered morph is less conspicuous than the black morph. As a result, birds eat more black morphs than peppered morphs and the allele for peppered wings increases in frequency. In badly polluted areas, sulphur dioxide kills lichens and mosses and soot blackens tree trunks. Here, the black morph is the less conspicuous on tree trunks. As a result, birds eat more peppered morphs than black morphs and the allele for black wings increases in frequency.

**Figure 2.15**
Two morphs of the peppered moth. Resting on tree trunks that are mottled with patches of lichen and moss, the peppered morph is less conspicuous to predatory birds than the black morph. In polluted areas, lichen and moss are killed and tree trunks become blackened by soot. Here, the black morph is less conspicuous

**Figure 2.16**
This bird has caught the conspicuous morph of the peppered moth and will eat it or feed it to its young

**Q 11 Is the evolution of warfarin resistance in rats an example of stabilising, directional or disruptive selection?**

Mutation plays an important part in natural selection, since it is a source of new genes. Where meiosis does not occur in the life cycle of a species, mutation is its only source of genetic variation. This is the case with bacteria. Many pathogenic bacteria have become resistant to the antibiotics that were used to treat infected humans. Figure 2.17 shows how this is thought to have occurred. Quite by chance, a gene mutation occurred in a single bacterial cell in a population, conferring resistance to an antibiotic. In the presence of the antibiotic, all susceptible cells in the population were killed but the resistant cell survived. Following reproduction of this cell, the entire population became resistant to the antibiotic.

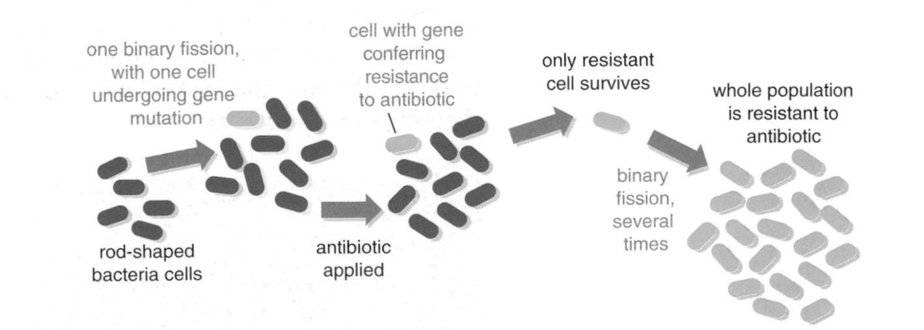


**Figure 2.17**
The evolution of antibiotic-resistant populations of bacteria. A chance gene mutation enabled one bacterial cell to survive when the rest of its population was killed by antibiotic. The new population formed from this one survivor all carried the allele for antibiotic resistance

## Extension box 2

### Balanced polymorphism in *Cepaea nemoralis*

Figure 2.18 shows different morphs of the common five-banded snail, *Cepaea nemoralis*. You can see shells of different colour, some shells with dark-brown bands and some shells without bands around their shells. The frequency of these morphs is different in different habitats, yet in each habitat they stay relatively constant from year to year. This is called a balanced polymorphism.

**Figure 2.18**
These shells of *Cepaea nemoralis* show variation in shell colour and in the presence or absence of bands around the shell

Shell colour can be yellow, pink or brown and is controlled by a single gene with three alleles. A different gene controls the presence or absence of bands. However, the two genes are located so close together on one of the snail's chromosomes that they form a so-called 'supergene'. The colour and banding patterns are subject to natural selection that is different in different environments. On sunny mountain sides in the Pyrenees, snails with unbanded, yellow shells absorb less heat than brown shells or banded shells. These unbanded, yellow shells are less likely to die from overheating. As a result, they breed more successfully than the other morphs and snails with unbanded, yellow shells are the most common morph.

In beech woodlands around Britain, concealment from predators is the selective force. Among the browny-pink leaf litter on the floor of a beech wood, yellow shells are conspicuous. They are easily seen by birds, such as the song thrush, which eats snails. In such woodlands, pink-shelled snails are the most common morph.

**Q** 12 **Bands on the shell can also affect concealment from predators. Suggest why snails with banded, yellow shells might be at an advantage over snails with unbanded, yellow shells in the mixed grass and shrubs of a hedge.**

It is tempting to believe that human populations are unaffected by natural selection. This is not so. Figure 2.19 shows a human blood smear. The large, round cells are red blood cells. You will probably recognise them from your AS course. You are less likely to recognise the crescent-shaped, **sickle cells**, which are also red blood cells. The two types of red blood cell contain haemoglobin-A molecules that differ by a single amino acid. The alleles of the gene for haemoglobin-A are $\mathbf{Hb^S}$ and $\mathbf{Hb^A}$. The two alleles are codominant. Table 2.7 summarises the possible genotypes involving the $\mathbf{Hb^S}$ and $\mathbf{Hb^A}$ alleles of the haemophilia-A gene and the phenotypes they produce.

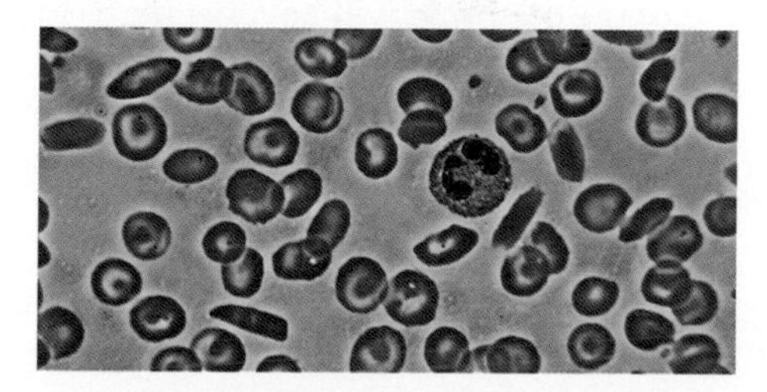

**Figure 2.19**
A human blood smear showing normal red blood cells and sickle cells

Table 2.7
The genotypes and phenotypes resulting from the alleles of the gene for haemoglobin-A

| Genotype | Phenotype |
|---|---|
| $Hb^AHb^A$ | All red blood cells are round and carry oxygen well |
| $Hb^AHb^S$ | Half the red blood cells are round and carry oxygen well; the other half carry oxygen less well and collapse in low oxygen concentration (the sickle cells in Figure 2.19). |
| $Hb^SHb^S$ | All the red blood cells are sickle cells |

You might expect from Table 2.7 that individuals with either of the genotypes $Hb^AHb^S$ or $Hb^SHb^S$ would suffer anaemia. This is the case – they both suffer from a disease called **sickle-cell anaemia,** which is particularly severe in individuals with the genotype $Hb^SHb^S$. You might then use the scheme in Table 2.6 to describe how the $Hb^S$ allele would be selected against and how its frequency in human populations would decline. This would be true but for one important fact – possession of the $Hb^S$ allele protects a person from malaria. Sickle cells are not attacked by the malarial parasite as successfully as round, red blood cells

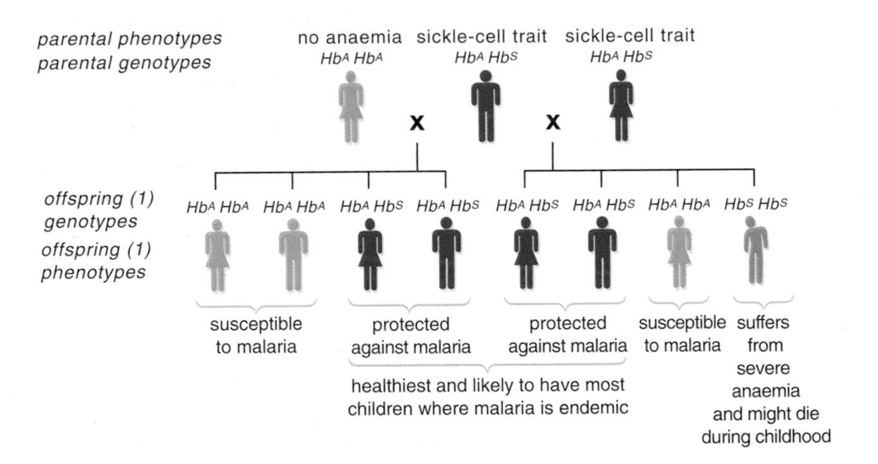


Figure 2.20
Natural selection acting on sickle-cell anaemia in humans. Whether the 'normal' or the sickle cell allele of the gene for haemoglobin-A is most common in a population is related to the incidence of malaria in the environment

Figure 2.20 represents families with the sickle-cell condition. If one parent is $Hb^AHb^A$ and the other parent is $Hb^AHb^S$, there is a 50% chance that any of the offspring will be susceptible to malaria or resistant to malaria. If these individuals live in a part of the world where malaria is common, the susceptible offspring might have fewer offspring because of the effect that malaria has on them. If both parents have the $Hb^AHb^S$ genotype, you will see in Figure 2.20 that there is a 75% chance that any of the offspring will be resistant to malaria. However, one of the genotypes ($Hb^SHb^S$) will be so severely anaemic that she/he is unlikely to be healthy enough to have children. In this example of natural selection, it is the heterozygotes that are at an advantage in parts of the world where malaria is common.

**Q 13 Why would you expect natural selection to reduce the frequency of the sickle-cell allele in malaria-free parts of the world?**

## Summary

- When taking samples, errors can occur through chance or through human bias. We can reduce these errors by taking more than one sample and by using random sampling techniques.
- Data collected by sampling from a population can be represented as a frequency distribution. A normal frequency distribution is a special type of distribution in which the mean, median and mode have the same value.
- The standard deviation is a measure of the way that data are spread out from the mean. The more the data are spread out from the mean, the greater the standard deviation.
- The standard error (SE) is calculated from the standard deviation of a sample and is used to make allowances for the difference between samples and the total population from which they were taken. 95% of data lie within a range ± 1.96 SE of the mean.
- An organism's phenotype results from an interaction between its genotype and factors in its environment. Whereas some characteristics are influenced by a single gene, others are influenced by many genes. This is called polygenic inheritance and gives rise to continuous variation.
- Variation exists between the members of any species. This is caused by processes that occur during meiosis and by gene mutation.
- Within a population, the frequency of the different alleles of a gene will stay the same from generation to generation provided the population is large, and there is no mutation, migration or natural selection.
- The frequency of alleles in a population, and the genotypes and phenotypes resulting from them, can be calculated using the Hardy-Weinberg equation. This is given as $p^2 + 2pq + q^2 = 1$, where $p$ is the frequency of the dominant allele of a gene and $q$ is the frequency of the recessive allele of the same gene.
- Natural selection occurs on populations when different genotypes have different fertilities. Fit organisms have more offspring than less fit organisms, so that succeeding generations inherit the advantageous alleles of the genes that made them successful. As a result, the frequency of the alleles of these genes changes.
- The relationship between areas where malaria is common and the frequency of sickle cell anaemia in humans, the evolution of pesticide resistance in rats and the evolution of resistance to antibiotics in bacteria are examples of natural selection in action.

# Assignment

## Rats and super-rats

This chapter contains a number of examples of how selection can lead to a change in the frequency of particular alleles in a population. One of the examples included was the evolution of warfarin resistance in a population of rats. In this assignment we will look at this example in a little more detail. You will need to use material from various parts of the specification as well as from this chapter to answer the questions. Before you start, it would be worth reading about proteins and protein structure in your notes or textbook.

When you cut yourself, blood clots rapidly and seals the wound. The mechanism governing blood clotting is complex and involves a number of substances called clotting factors. Some of these factors are formed in the liver and contain an unusual amino acid residue called γ-carboxyglutamate. Vitamins are substances which are needed in small amounts in the diet. They are needed because the body is unable to synthesise them. One of these vitamins is called vitamin K (It is called this because it was discovered in Germany and named the Koagulation-vitamin). Vitamin K is necessary for the synthesis of γ-carboxyglutamate. Figure 2.21 summarises the biochemical pathway involved. Look at this diagram carefully and then answer questions 1 and 2.

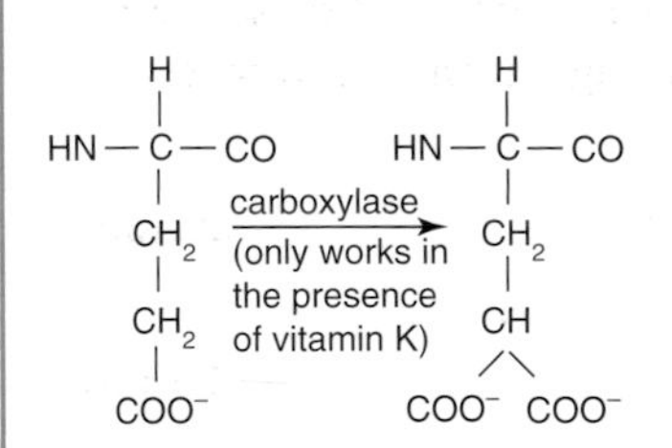


Figure 2.21
The biochemical reaction in which γ-carboxyglutamate is produced in the body. This reaction requires an enzyme, carboxylase, which will only work in the presence of vitamin K

1 Amino acids join together to form a polypeptide chain. When an amino acid has joined to another in this way it is called an amino acid residue.

   (a) Draw a diagram to show the structure of a glutamate molecule.
   *(1 mark)*

   (b) Explain why the structure of a glutamate residue differs from the structure of a glutamate molecule.
   *(2 marks)*

   (c) Explain why the enzyme controlling this reaction is known as a carboxylase.
   *(1 mark)*

2 The R-group of the γ-carboxyglutamate residue binds readily to phospholipids. Suggest why this might be useful in blood clotting.
   *(2 marks)*

Figure 2.22
Rats are serious pests on farms. With a lot of grain and animal food available large populations build up rapidly. Warfarin is a very effective poison and still widely used to control rats

Warfarin is a very effective rat poison. It is a competitive inhibitor of vitamin K.

3 (a) Look at Figure 2.22. The rat in this photograph has recently eaten warfarin. Use the information given so far in this assignment to suggest an explanation for the blood on the rat's face.
   *(4 marks)*

(b) If humans are accidentally poisoned by warfarin, they are treated with large doses of vitamin K. Use your knowledge of competitive inhibition to explain why this treatment is successful.

*(1 mark)*

In 1959, rats resistant to warfarin were found in an area in central Wales. It was found that this resistance was determined by a single gene with two codominant alleles, $W^R$ and $W^S$. The three different genotypes have the following characteristics.

- $W^SW^S$ Rats with this genotype are susceptible to warfarin. They only require small amounts of vitamin K in their diet.
- $W^SW^R$ These rats are resistant to warfarin. They require more vitamin K in their diet than rats with the genotype $W^SW^S$ .
- $W^RW^R$ Rats with this genotype are also resistant to warfarin but as they need extremely large amounts of vitamin K in their diets, they cannot survive under natural conditions.

4 Giving a reason for your answer in each case, which of these three genotypes would be at the greatest advantage in an area where:

(a) no warfarin had ever been used;

*(2 marks)*

(b) warfarin was commonly used as a rat poison.

*(2 marks)*

5 Use genetic diagrams to show the genotypes of the expected offspring of crosses between rats with the following genotypes:

(a) $W^SW^R$ and $W^SW^R$
(b) $W^SW^R$ and $W^SW^S$.

*(3 marks)*

6 The percentage of rats resistant to warfarin in central Wales rose rapidly at first. It then levelled out. Use the results of your answer to question 5 to explain this.

*(4 marks)*

Thrombosis is the formation of a blood clot inside a blood vessel. It can be very serious because the clot might block the flow of blood to the heart or the brain and cause the death of the patient. Patients with thrombosis are treated with anticoagulant drugs. One commonly used anticoagulant is warfarin.

7 A gene has been described which controls warfarin resistance in humans. In this gene, the allele for warfarin resistance is dominant. Suggest why this allele is rare in the human population.

*(3 marks)*

# Examination questions

1 The diagram shows stages of meiosis in a human testis. Each circle represents a cell.

(a) In each empty circle, write the number of chromosomes that would be found in the cell.

*(1 mark)*

(b) Describe **two** ways in which meiosis contributes to genetic variation.

*(2 marks)*

(c) Explain the importance of genetic variation in the process of evolution.

*(1 mark)*

2 In the flour beetle, the allele for red body colour (**R**) is dominant to the allele for black body colour (**r**).

(a) Complete the genetic diagram to show the result of a cross between a heterozygous red beetle and a black beetle.

| | | |
|---|---|---|
| *Parental phenotypes* | **Red** | **Black** |
| *Parental genotypes* | | |
| *Gametes* | | |
| *Offspring genotypes* | | |
| *Offspring phenotypes* | | |

*(2 marks)*

(b) A mixed culture of red beetles and black beetles was kept in a container in the laboratory under optimal breeding conditions. After one year, there were 149 red beetles and 84 black beetles in the container.

(i) Use the Hardy-Weinberg equation to calculate the expected percentage of heterozygous red beetles in this population.

*(2 marks)*

(ii) Several assumptions are made when using the Hardy-Weinberg equation. Give **two** of these.

*(2 marks)*

3 The figure shows the length of cobs in two breeding varieties of maize plant and the $F_1$ and $F_2$ generations derived from a cross between them.

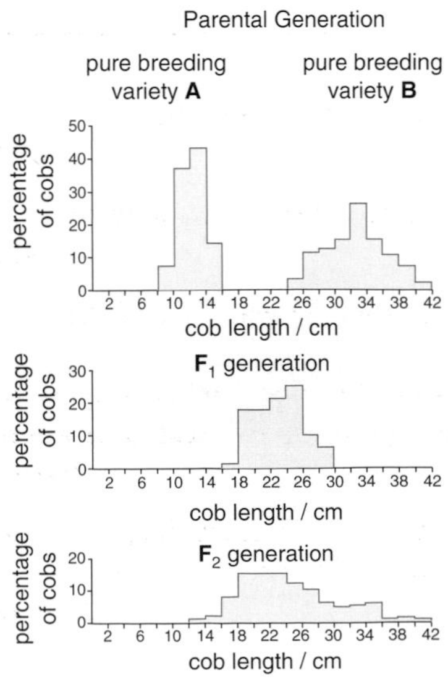


(a) What information does the term *pure breeding* give about genotype?

*(1 mark)*

(b) Give the modal class of the $F_1$ generation.

(c) (i) Explain what is meant by polygenic inheritance.

*(1 mark)*

(ii) Explain the evidence from the figure which suggests that the inheritance of cob length is polygenic.

*(2 marks)*

(d) What is the evidence that differences in cob length in the parental generation are partly due to:

(i) genetic difference;

*(1 mark)*

(ii) environmental differences?

*(1 mark)*

In maize a single gene with two alleles controls the type of carbohydrate stored in the cells of the plant. Starchy varieties of maize have starch grains which stain blue-black with iodine solution; waxy varieties have starch grains which stain red. The allele for starch, **W**, is dominant to that for waxy, **w**.

(e) Explain what is meant by:

(i) a gene;

*(1 mark)*

(ii) an allele.

*(1 mark)*

(f) Pollen from a single maize plant was dusted on a microscope slide and stained with iodine solution. The results are shown in the table.

| Pollen grains stained blue-black with iodine solution | Pollen grains stained red with iodine solution |
|---|---|
| 58 | 64 |

What is the genotype of:

(i) the pollen grains stained red with iodine solution;

*(1 mark)*

(ii) the parent plant from which these pollen grains were taken?

*(1 mark)*

(g) In a field of maize the frequency of allele **W** was 0.7 and the frequency of allele **w** was 0.3.

(i) If the maize plants were randomly fertilised, what frequencies of these two alleles would be expected in the next generation?

*(1 mark)*

(ii) Use the Hardy-Weinberg equation to calculate the percentage of heterozygous plants in the field of maize.

*(3 marks)*

# 3 The Evolution and Classification of Species

Figure 3.1
This bandro lives in the papyrus reed beds of Lake Alaotra in Madagascar. Biologists dispute whether bandros belong to an individual species or whether they are a subgroup of another species of lemur

**The animal in Figure 3.1 is a type of lemur, called the bandro. It is found in only one habitat in the world – the papyrus reed beds around Lake Alaotra in Madagascar – where no other lemurs live.**

**Bandros live in small family groups. The head and body length of an adult is about 30 cm and an adult weighs about 900 g. Bandros are active during the day. They spend most of this time feeding on papyrus. Like the animal in Figure 3.1, bandros use their hands, feet and tail to hold on to the papyrus reeds. Unfortunately, bandros are classed as a critically endangered species because their habitat is being destroyed. They are also hunted for food and for use as pets.**

**The bandro is a confusing animal. Firstly, it has several common names – the bandro, the Lake Alaotra bamboo lemur and the grey gentle lemur. It even has two biological names – *Hapalemur alaotrensis* and *Hapalemur griseus alaotrensis*. Secondly, biologists dispute whether the bandro is a species or whether it is a localised subgroup of another species. The biological names reflect this dispute. The name *Hapalemur alaotrensis* gives bandros the status of a species whereas the name *Hapalemur griseus alaotrensis* suggests that bandros are a subspecies of the lesser bamboo lemur (*Hapalemur griseus*).**

**In this chapter, you will learn how biologists recognise individual species and begin to understand why there is a dispute about the status of the bandro. You will also learn how species are thought to have evolved, how they are classified and the origin of their biological names.**

## The concept of a species

Humans seem to have a need to classify objects or organisms in the world around them. If you watch a young child putting things into its mouth, you can see it classify objects as pleasant to taste or distasteful. You continue to do this as an adult – some foods you like to eat and others you do not (Figure 3.2). You probably classify animals as dangerous or harmless, berries and fungi as poisonous or edible, people as helpful or unfriendly, and so on. When we make a formal study of different organisms, we need to agree a classification system that is universal.

Figure 3.2
Karela is the unripe fruit of *Momordica charantia*. It is also known around the world as bitter gourd, bitter melon, balsam pear or foogwa. It is a popular vegetable in northern India, where it is believed to be good for the heart, and has been recommended by the Department of Health in the Philippines as a remedy for diabetes mellitus. However, many people find the bitter taste of karela, especially strong in its seeds, makes it unpleasant to eat

**Figure 3.3**
The horse (*Equus equus*) and donkey (*Equus hemionus*) belong to different species. There are clear differences in their external appearance and, although they can interbreed, their offspring (mules) are always sterile

The **species** is the basic unit of biological classification. We can define the meaning of species in two ways.

1. **Variation** – Members of a species are similar to each other but different from members of other species. These similarities might be physical (e.g. two trees have a similar branching pattern), biochemical (e.g. the haemoglobin structure of two animals is identical), immunological (e.g. an antibody made against antigens from one organism is equally effective against those from a second organism), developmental (e.g. growth of the embryo is similar in two organisms), or ecological (e.g. organisms occupy an identical ecological niche). In Chapter 2, we learned about variation within a species and you can see this variation if you look around at your fellow students or at the members of your family. Consequently, we need to be careful about the importance of the differences that we use to identify different species.

2. **Potential for breeding** – Members of a species breed together in their own environment and produce offspring that are fertile. Look at the horse and donkey in Figure 3.3. You can see the differences between horses and donkeys that led to them being classed as different species. In captivity, they can mate and produce offspring, called mules. However, mules are always sterile. In this case, the potential for breeding confirms our classification of horses and donkeys that was based on observable differences.

These two definitions of the term species are not always helpful to biologists. Some species seldom, if ever, reproduce sexually. Other species are known only from their fossil record. Finally, some groups of similar organisms live so far from each other that they are unlikely ever to meet. In these cases, it is difficult to use our second definition to decide whether these organisms belong to a separate species.

**Q** 1 **Use the information above to suggest why biologists are undecided about the biological status of the bandro in Figure 3.1.**

## Speciation

Speciation is the formation of new species from existing species. It is often called evolution. The evolution of new species is a controversial idea, which many groups of people reject. Most biologists believe that the species we see around us have evolved from other species and this idea is central to biological classification.

New species are thought to arise in one of two ways. One of these involves:

- **hybridisation** – the production of offspring from parents of two different species
- **polyploidy** – an increase in the number of sets of chromosomes.

We saw one example of hybridisation above. Horses and donkeys can mate to produce mules, but these mules are always sterile. Their sterility is explained by the chromosomes that mules inherit from their parents. The parental horse chromosomes and the parental donkey chromosomes are so different that they cannot form homologous pairs. Although this has no effect on mitosis, it prevents meiosis, which is essential for egg and sperm formation in animals.

**Q 2 At which stage of meiosis do homologous chromosomes pair together?**

The sterility of mules could theoretically be overcome if the mule were to double its diploid chromosome number, i.e. become **tetraploid** (4n). There would now be two copies of every individual chromosome and these could form homologous pairs during meiosis. This process has not happened in mules and rarely happens at all in animals. It is, however, common in plants. In fact, over 50% of flowering plants are polyploid. A few examples of familiar, polyploid plants are shown in Table 3.1.

| Wild species of plant | Cultivated species of plant |
|---|---|
| Wild cotton (2n = 26) | Cultivated cotton (2n = 52) |
| Wild dahlia (2n = 32) | Garden dahlia (2n = 64) |
| Wild potato (2n = 24) | Cultivated potato (2n = 48) |
| Wild rose (2n = 14) | Garden rose (2n = 42) |
| Wild tobacco (2n = 24) | Cultivated tobacco (2n = 48) |

Table 3.1
Like over 50% of flowering plants, these cultivated plants are polyploid. The diploid number of chromosomes is shown as 2n

## Extension box 1

### Polyploidy in wheat

Wheat is a crop plant that is grown all over the world. It is used to make bread and to make pasta. Modern wheat is thought to have arisen thousands of years ago by the accidental crossing of different species of grass to produce hybrid offspring that became polyploid. Figure 3.4 shows how the evolution of modern wheat is thought to have occurred.

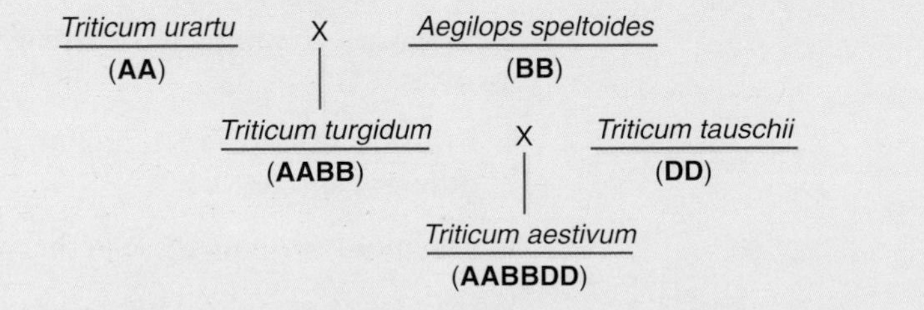


Figure 3.4
Modern wheat is thought to have arisen following chance hybridisation and polyploidy on at least two occasions in the past four thousand years

*Triticum urartu* was an ancestral wheat species that accidentally crossed with a wild grass, called *Aegilops speltoides*. The hybrid offspring of this cross would have been sterile except that a chance mutation caused it to become polyploid. This is represented in Figure 3.4, which shows the chromosome combination of the parents as **AA** and **BB**, where **A** and **B** represent one haploid set of parental chromosomes. Instead of being **AB**, the polyploid hybrid is **AABB**, i.e. it has two sets of each of the parent's haploid chromosomes. This new species is *Triticum turgidum*, commonly called durum wheat, and is used to make pasta.

A second hybridisation is then thought to have occurred in which *Triticum turgidum* formed a hybrid with *Triticum tauschii* (chromosome combination **DD**). Chance mutation of the hybrid formed a polyploid, *Triticum aestivum*. This is commonly called bread wheat because its flour is used to make bread.

Hybridisation and polyploidy can explain speciation in plants but not in animals. We need another explanation of the way in which new species are formed. This method of speciation involves:

- the separation of two groups of organisms of one species
- reproductive isolation between the two separated groups, i.e. members of the two groups cannot breed with each other, so that they are genetically isolated
- changes in the frequency of alleles that is different in the two reproductively isolated groups that are so great that the two groups become new species

Table 3.2 (see page 58) gives some examples of the barriers that can lead to reproductive isolation. It is important to realise that, whilst reproductive isolation is likely to result in genetic differences between the isolated groups, it will not always lead to new species. For example, humans in the Australian subcontinent were reproductively isolated from humans in Asia and Europe for thousands of years. As a result, modern Australian aborigines have features that distinguish them from humans inhabiting Asia and Europe. However, breeding amongst all these human groups occurs and results in fertile offspring. All these humans still belong to the same species, *Homo sapiens*.

## Allopatric speciation

It is easiest to imagine populations diverging from each other if barriers that they cannot cross physically separate them. In their different environments, the populations will be subjected to different natural selection pressures, which will cause their allele frequencies to change in different ways. When speciation occurs as a result of such barriers, we call this **allopatric speciation**.

The Hawaiian Islands were formed thousands of years ago by the eruption of volcanoes on the floor of the Pacific Ocean. As the volcanoes cooled, they became suitable for plant and animal life. The islands were

| Stage of life cycle at which barrier is effective | Barrier to interbreeding | Explanation |
|---|---|---|
| Before mating | geographical separation | populations inhabit different continents, different islands or different sides of a large canyon |
| | habitat isolation | populations inhabit different local habitats within one environment |
| | temporal isolation | populations use the same environment but are reproductively active at different times |
| After mating | incompatibility of gametes | female might kill sperm of wrong genotype or pollen of wrong genotype fails to germinate |
| | hybrid not viable | hybrid fails to develop to maturity |
| | hybrid sterility | hybrid grows to maturity but is sterile (Figure 3.3) |

Table 3.2
Examples of barriers that lead to reproductive isolation of two populations of a single species. If sufficient genetic differences occur in the isolated populations, new species are formed

probably colonised by organisms from the neighbouring mainland, in a random way. The only group of flies that colonised the Hawaiian Islands was the fruit fly, *Drosophila*. There are now hundreds of species of fruit fly in these islands. They are thought to have evolved from an ancestral population through allopatric speciation.

**Q 3 Name the barrier(s) to reproduction that might have led to the evolution of different species of fruit fly in the islands of the Hawaiian archipelago.**

## Sympatric speciation

The formation of new species without geographical isolation is called **sympatric speciation**. Where this occurs, strong forces of natural selection cause genetic differences between two populations that can, at least in theory, interbreed.

The waste from copper mines forms unsightly tips in many parts of Britain. One species of grass, called *Agrostis tenuis* (common bent grass), is able to grow on these tips because some plants possess an allele of a gene that makes them tolerant to high concentrations of copper ions in the soil. These plants can grow on polluted soil that is inhospitable to susceptible *A. tenuis* plants. Sometimes, the copper-tolerant populations grow very close to populations of *A. tenuis* that are susceptible to poisoning by high concentrations of copper ions. However, copper-tolerant plants have an earlier flowering time than the copper-susceptible plants. Although these two varieties of *A. tenuis* belong to the same species, there is a potential for reproductive isolation that might lead to sympatric speciation.

**Q** 4 **Distinguish between allopatric speciation and sympatric speciation.**

## The five-kingdom classification of organisms

In our biological classification system, all species are placed in larger groups, called genera (singular: genus). A single **genus** contains species that, although different from each other, have similarities that make them distinct from other genera. In turn, genera are placed into larger groups, and so on. Figure 3.5 shows the names of these biological groups, called **categories**. Notice that organisms in the categories at the bottom of Figure 3.5 show a greater degree of similarity than those in categories at the top of the diagram. Each set of organisms within a category is called a **taxon**. Taxonomy is the study of biological classification.

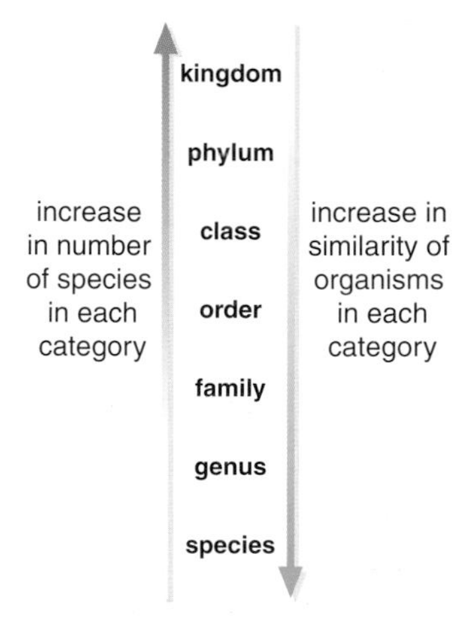


**Figure 3.5**
Species are grouped into biological categories of decreasing similarity. The largest category is the biological kingdom, of which there are five

Table 3.3 shows how three animals previously named in this chapter are classified. It also shows the classification of a familiar plant, the tomato. Each name in the body of the table is a taxon. The names are in Latin, which no one uses today and so the names cannot be corrupted. The taxa (plural of taxon) shared by humans, horses and fruit flies are shown in green. Notice that the biological name of an organism is the **binomial** formed from the names of its genus and species. Through international convention, the name of the genus is always written with an upper case and the name of the species is always written with a lower case. The binomial is always printed in italics (e.g. *Homo sapiens*) or, if written by hand, is underlined (e.g. Homo sapiens).

| Category | Human | Horse | Fruit fly | Tomato |
|---|---|---|---|---|
| kingdom | Animalia | Animalia | Animalia | Plantae |
| phylum | Chordata | Chordata | Arthropoda | Angiospermophyta |
| class | Mammalia | Mammalia | Insecta | Dicotyledoneae |
| order | Primata | Perissodactyla | Diptera | Solanales |
| family | Hominidae | Equidae | Drosophilidae | Solanaceae |
| genus | *Homo* | *Equus* | *Drosophila* | *Lycopersicon* |
| species | *sapiens* | *equus* | *melanogaster* | *esculentum* |

**Table 3.3**
The taxa of three animals, mentioned earlier in this chapter, and of a familiar plant, demonstrate the major categories used in classification. Humans, horses and fruit flies share the taxa shown in green

The first three organisms in Table 3.3 belong to the same kingdom – Animalia. This is one of only five kingdoms into which all organisms are grouped. The five kingdoms are:

- **Prokaryotae** (prokaryotes, which are all bacteria)
- **Protoctista** (protoctists)
- **Fungi** (fungi)
- **Plantae** (plants)
- **Animalia** (animals)

We will now look at the distinguishing characteristics of each kingdom, which are summarised in Table 3.4. In a Unit Test, you will not be expected to recall the classification of any specific organism.

**Q 5 What is the binomial for the tomato plant?**

Table 3.4
A simple comparison of the five kingdoms. Unicellular organisms consist of only a single cell, colonial organisms are groups of cells and multicellular organisms have many cells arranged in tissues. Autotrophic organisms use external energy sources to form organic compounds from inorganic compounds whereas heterotrophic organisms digest organic molecules into smaller products, which they then absorb into their own bodies

| Kingdom | Cell structure | Cell wall | Nutrition | Notes |
|---|---|---|---|---|
| Prokaryotae | prokaryotic; unicellular or colonial | present (peptidoglycans) | autotrophic (photosynthesis and chemosynthesis) | contains only the bacteria |
| Protoctista | eukaryotic; includes unicellular, colonial and multicellular forms | sometimes present (polysaccharide) | some autotrophic, some heterotrophic, some both | organisms are classed in this kingdom if they cannot be placed in any other |
| Fungi | eukaryotic; can be single-celled (e.g. yeast) but most are multicellular | present (chitin) | heterotrophic | most fungi are made of a mass (mycelium) of thread-like filaments, called hyphae; they lack cilia and flagella at all stages of their life cycle; they reproduce by forming resistant spores, which they produce by mitosis |
| Plantae | eukaryotic and multicellular | present (cellulose) | autotrophic (photosynthesis using chloroplasts) | plants develop from multicellular embryos, which are formed from zygotes and nourished by the maternal plant; they have a complex life cycle that involves two different types of adult – a haploid, sexually reproducing gametophyte and a diploid, asexually reproducing sporophyte |
| Animalia | eukaryotic and multicellular | absent | heterotrophic, involving a digestive cavity | animals develop from embryos that, at some stage, form a hollow ball of cells – the blastula; they have nervous and hormonal control systems |

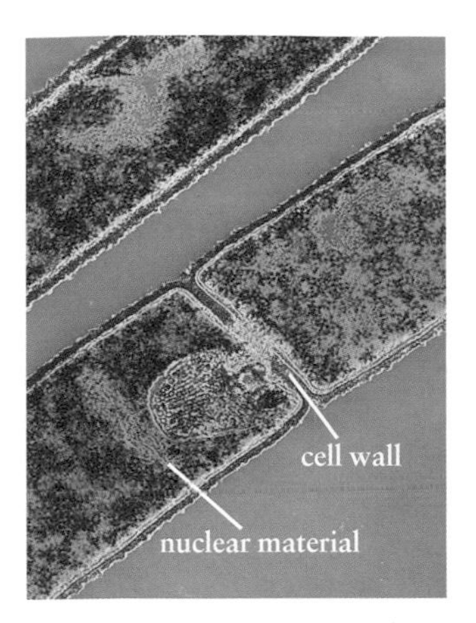


## Prokaryotae

In your AS course, you learned about prokaryotic and eukaryotic cells (e.g. refer back to pages 14 and 15 of the AS book in this series). Table 3.5 summarises the differences between prokaryotic cells and eukaryotic cells. All organisms that have prokaryotic cells are classed in the kingdom Prokaryotae. They are all bacteria, such as the bacterium shown in Figure 3.6. There are about 10 000 known species of bacteria.

Figure 3.6
An electron micrograph of a rod-shaped bacterium (bacillus)

| Feature | Prokaryotic cells | Eukaryotic cells |
|---|---|---|
| size | relatively small, of the order of 1 to 10 μm | relatively large, of the order of 10 to 100 μm |
| cell structure | cell capsule sometimes present | cell capsule never present |
| | cell wall usually present and made of polymers of sugars and amino acids (peptidoglycans) | cell wall present in some protoctists (cellulose), fungi (chitin) and plants (cellulose) |
| | cytoplasm lacks membrane-bound organelles | cytoplasm contains membrane-bound organelles, such as mitochondria |
| | ribosomes relatively small (70S*) | ribosomes relatively large (80S*) |
| genetic material | DNA not coated by proteins | DNA coated by proteins (called histones) |
| | DNA forms a single circular genophore | DNA forms one, or more, linear chromosomes |
| | DNA free in cytoplasm as a nucleoid | chromosomes held within a membrane-bound nucleus |
| cell division | neither mitosis nor meiosis is involved | division is by mitosis or meiosis |
| | cytoplasm divides into two (binary fission) and replicated genophore is separated without a spindle of microtubules | cytoplasm divides into two and replicate chromosomes are separated by a system of microtubules, which form a spindle |
| cell metabolism | great variation in metabolic pathways | similar metabolic pathways |
| | some photosynthesise, forming sulphur, sulphates or oxygen as a waste product | some protoctists and most plants photosynthesise and form oxygen as a waste product |
| | include obligate aerobes (need free oxygen for respiration), obligate anaerobes (need absence of free oxygen for respiration) and facultative aerobes (can adapt their respiration to both aerobic and anaerobic conditions) | most are obligate aerobes, though some fungi are facultative aerobes |

Table 3.5
A comparison of prokaryotic and eukaryotic cells. All bacteria have prokaryotic cells and belong to the kingdom Prokaryotae.
(*) The symbol S stands for Svedberg unit and is a measure of sedimentation rate during centrifugation. A 70S ribosome has a slower sedimentation rate than an 80S ribosome because it is smaller

## Protoctista

This kingdom contains about thirty different phyla (plural of phylum), whose members are all simple-bodied eukaryotes. There are about 100 000 known species of protoctists. The kingdom Protoctista is best described by exclusion, i.e. if an organism is not a prokaryote, a fungus, a plant or an animal, then it is a proctoctist. Among the protoctists are:

- organisms that consist of only one cell (**unicellular**), organisms that form filaments of cells, organisms that form ball-like colonies of cells and organisms that are made of many cells organised into tissues (**multicellular**)
- organisms that behave like animals (e.g., *Plasmodium*, which causes malaria) and organisms that behave like plants (e.g. seaweeds)
- slime moulds, protozoa and algae

Figure 3.7 shows examples of all these types of protoctists. Please note that the terms protozoa (unicellular, 'first animals') and algae (unicellular or very simple multicellular photosynthetic organisms), though commonly used, do not refer to taxa.

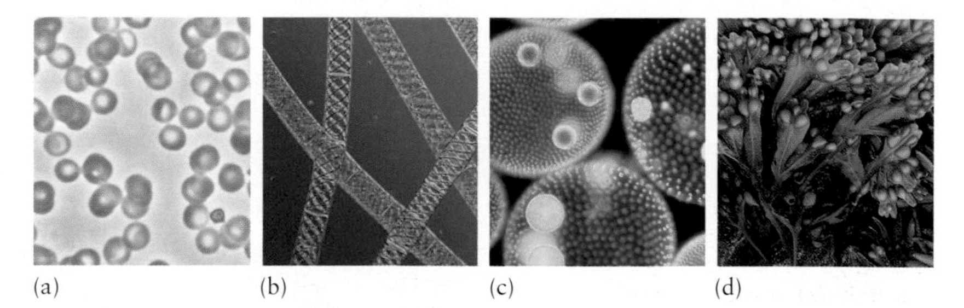


Figure 3.7
Examples of protoctists
(a) *Plasmodium* is an animal-like, single-celled organism that causes malaria in humans (×400)
(b) *Spirogyra* is a filament of photosynthetic cells (×400)
(c) Volvox is a ball-like colony of photosynthetic cells (×40)
(d) The seaweed wrack, *Fucus*, is a photosynthetic, multicellular protoctist

### Extension box 2

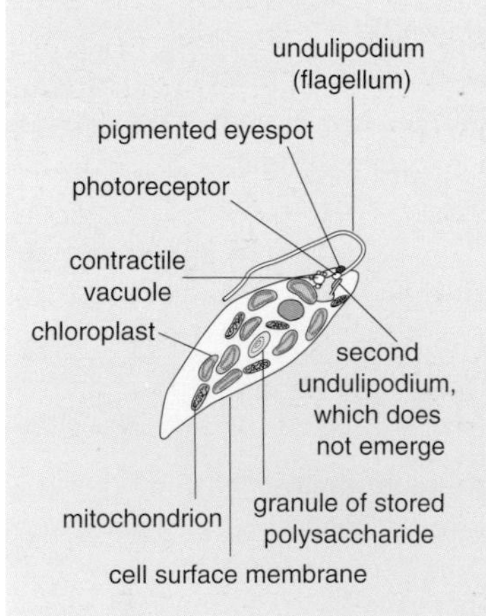


Figure 3.8
*Euglena* is a protoctist that has some plant-like qualities and some animal-like qualities

### *Euglena* – a protoctist with both animal and plant-like qualities

Figure 3.8 shows a protoctist called *Euglena*, which is common in freshwater ponds. You can see that it is a unicellular eukaryote with many chloroplasts. When a single *Euglena* is in light, these chloroplasts photosynthesise in just the same way as those in plants. For this reason, you might be tempted to classify *Euglena* as a plant. However, you can see that *Euglena* also has properties that are not shared by plants. For example, it has no cellulose cell wall. As a result:

- it is not able to prevent the inward movement of water by osmosis. The contractile vacuole that you can see in the diagram uses energy from ATP to collect this water and pump it out of the cell, preventing osmotic lysis of the cell.
- it is able to change its shape.

Unlike a plant, *Euglena* is able to move about. Waves of contractions pass down its flagellum, in a whip-like action that pulls the cell forwards and makes it spin at the same time. Contractile fibrils in its outer surface change the shape of the cell and can also move it forwards, in as process called euglenoid movement. The photoreceptor, which you can see at the base of the flagellum in the diagram, helps *Euglena* to detect light. It moves towards light of moderate intensity (positive phototaxis) but away from very bright light (negative phototaxis).

Although *Euglena* can photosynthesise, in darkness it can feed like fungi and many bacteria, i.e. by secreting enzymes onto organic molecules in its surroundings and absorbing the digested products.

Our inability to classify *Euglena* clearly into one of the other kingdoms results in it being classified in the kingdom Protoctista.

**Q 6 How can you tell from Figure 3.8 that *Euglena* has a eukaryotic cell?**

## Fungi

Figure 3.9
Like all fungi, this puffball produces fruiting bodies, which is usually all we see of a fungus. The fruiting bodies are made of masses of thread-like hyphae and contain the reproductive spores that all fungi produce. You can see some of the spores of this puffball being released in a smoke-like cloud

Fungi are commonly called moulds because we see them most often on objects that are decaying. If you have kept fruit or bread for too long, you will have seen it go 'mouldy'. The mould that you see is mainly the fruiting bodies of the fungi. These fruiting bodies form spores, which is the way in which fungi disperse. If you look at Figure 3.9, you will see spores being released into the air by the fruiting body of a puffball fungus. Like mushrooms, many puffballs are fungal fruiting bodies that are eaten by humans.

There are about 100 000 known species of fungi. They all have eukaryotic cells, which are surrounded by a cell wall made of chitin. Some fungi are unicellular (e.g. yeast), but most have bodies that are made of thread-like filaments, called hyphae (singular, **hypha**). Once a fungal spore lands on a suitable environment, it germinates to form a single hypha. Figure 3.10 shows how, by continued growth and branching, a network of hyphae, called a **mycelium**, is formed. This is the body form of the great majority of fungi. Although you cannot see them with the naked eye, vast fungal mycelial networks surround us in soil and anywhere where there is decaying organic matter.

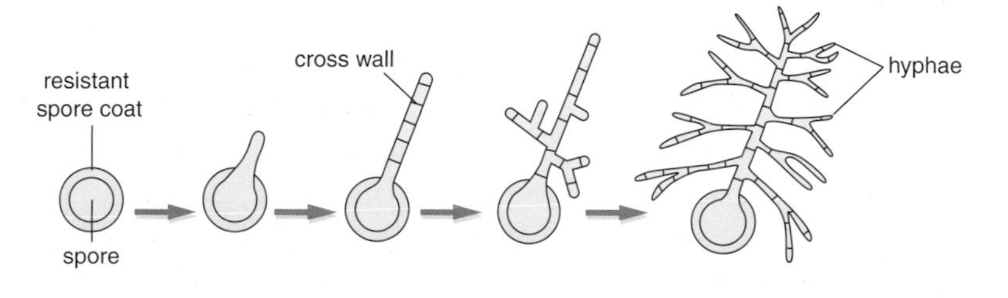


Figure 3.10
Fungi reproduce asexually and sexually to form spores. On a suitable environment, a single spore germinates to form a fungal hypha. Continued growth and branching forms a network of hyphae, or mycelium

Figure 3.11a shows a section through a hypha. In it, you can see the cell wall. This is made of chitin and, in some fungi, separates the hypha into cells. A hypha is most active at its tip. It is here that:

- growth of the hypha occurs
- most digestive enzymes are released onto the surroundings
- most absorption of digested food back into the hypha occurs.

In contrast, Figure 3.11b shows a section through a yeast cell. This unicellular fungus is used in industry to make beer and wine and to make bread rise. Cultures of yeast are commonly used in school and college laboratories to investigate respiration.

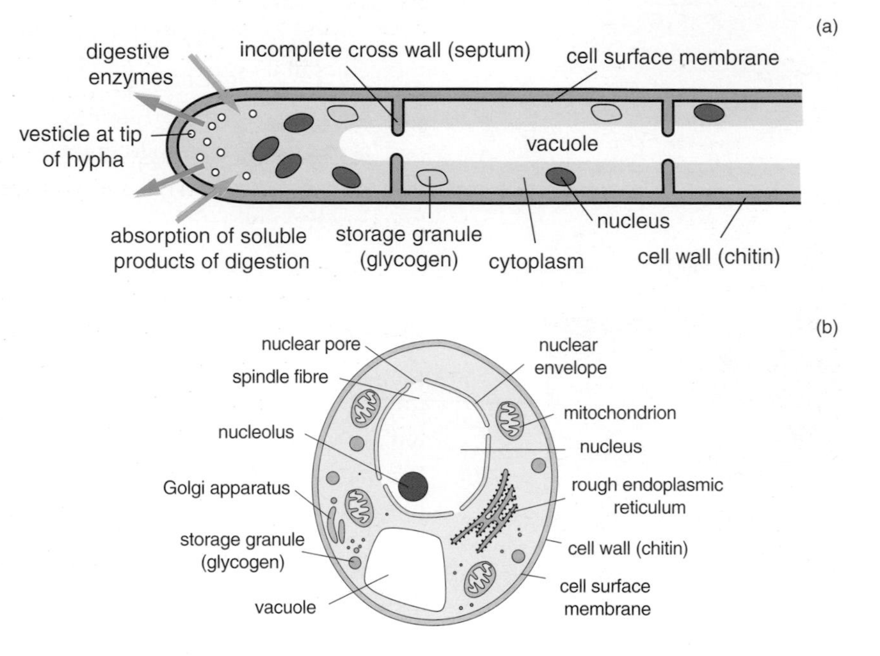


**Figure 3.11**
(a) The hypha shown in this diagrammatic section has walls across its length, which are not present in all fungi (b) Yeast is a unicellular fungus. During aerobic respiration, cultures of yeast produce large amounts of carbon dioxide gas: this causes bread to rise in the baking industry. During anaerobic respiration, cultures of yeast produce ethanol: this is the basis of the beer and wine industries

**Q 7 Suggest why there are usually many nuclei concentrated in the cytoplasm at the tip of a fungal hypha.**

The mould that you might have seen on old bread was mentioned above. One of the fungi that is most likely to have caused this mould is a species of *Penicillium*. Figure 3.12 shows the generalised structure of fungi in the genus *Penicillium*. In this diagram, you can see most of the features that characterise fungi. You can see:

- a mycelium, formed by hyphae that are separated into cells by cross walls
- special feeding hyphae, called haustoria, that secrete digestive enzymes onto the surroundings and absorb the products of digestion
- erect hyphae that grow upwards from the mycelium
- chains of spores on the erect hyphae. It is the colour of these spores that you would notice on mouldy bread. It is also the brush-like appearance of these chains of spores that gave the genus *Penicillium* its name, from the Latin penicillus, or little brush.

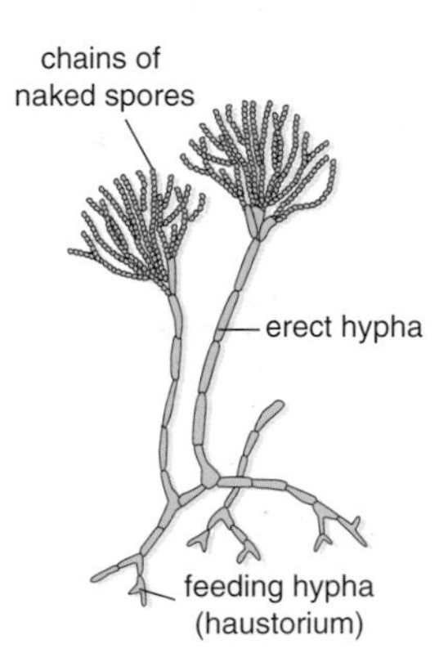


**Figure 3.12**
The general features of the genus *Penicillium*. The body of this fungus is a mass of hyphae which grow horizontally through the medium on which it lives. Erect hyphae grow from this mycelium and hold spores, which are dispersed by touch or air currents. The spores are blue, green or yellow and give rise to the coloured mould you might see on stored food

*Penicillium* can be a nuisance. Apart from causing stored food to go mouldy, it also causes spoilage of stored textiles. However, *Penicillium* does have its uses. Many species of *Penicillium* live in soil, where they contribute to nutrient cycles by breaking down dead organic matter. *P. roqueforti* and *P. camemberti* are used to ripen cheeses. *P. notatum* is the species in which Sir Alexander Fleming first discovered the antibiotic penicillin.

**Figure 3.13**
These organisms are representative of five different plant phyla (a) Liverwort (b) Moss (c) Fern (d) Conifer (e) Flowering plant

## Plantae

There about 350 000 known species of plants. All plants are multicellular organisms, with eukaryotic cells containing large vacuoles and surrounded by cellulose cell walls. They usually grow into a branched body form and grow from restricted patches of actively dividing cells, called meristems. Most plants photosynthesise, using chloroplasts, and produce oxygen as a waste product of their photosynthesis. Figure 3.13 shows a variety of plants. The liverwort and moss are relatively small, so that you might not notice them during a normal day. Ferns, conifers and flowering plants are large and you are likely to see examples of these plants every day.

Plants are adapted to living on land and many of their features are adaptations to life on land. They have strong supporting tissues, their leaves allow gas exchange in the air and they are fairly well waterproofed to prevent drying out. The life cycle of plants is complex and includes two different types of adults. Figure 3.14 summarises this life cycle, known as alternation of generations. The liverwort and moss plants shown in Figure 3.13 are the gamete-producing adults (gametophytes). The fern, conifer and flowering plants shown in Figure 3.13 are spore-producing adults (sporophytes). Their gamete-producing adults are hidden within them and you would not see them without an optical microscope.

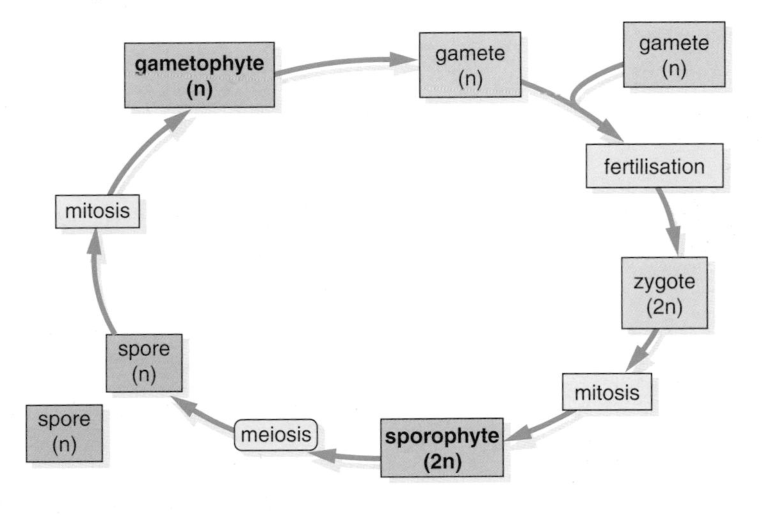


**Figure 3.14**
Plants have a complex life cycle, known as alternation of generations because it involves two separate types of adult. Sexual reproduction by gametophyte adults forms multicellular zygotes that develop into sporophyte adults. These sporophytes reproduce asexually to form spores that germinate into gametophytes

### Extension box 3

### Alternation of generations in flowering plants (phylum, Angiospermaphyta)

Alternation of generations is a feature of the plant kingdom. In its life cycle, every plant has two different types of adult: a sexually reproducing gametophyte and an asexually reproducing sporophyte. These are difficult to distinguish in flowering plants because the gametophytes are kept within the flowers produced by the sporophyte.

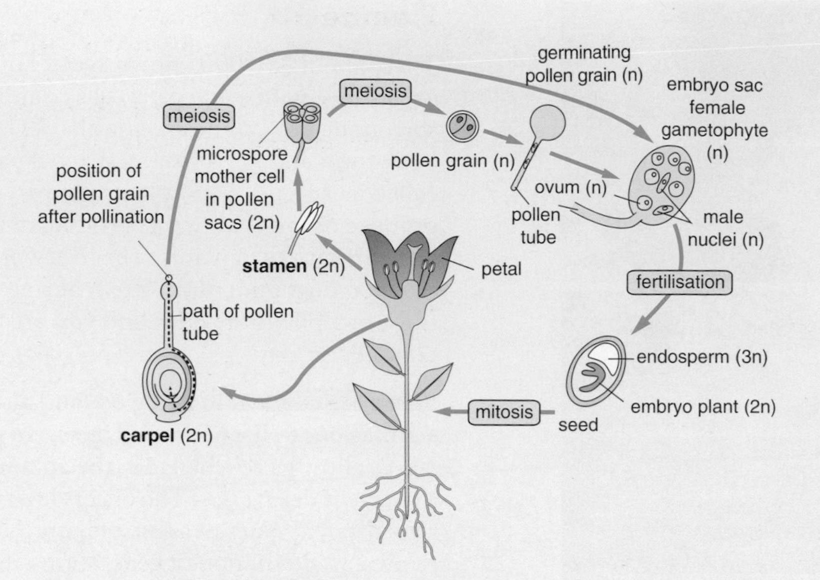


**Figure 3.15**
Alternation of generations in a flowering plant. The sexually reproducing gametophytes are haploid and are kept within the body of the diploid, asexually reproducing sporophyte

Figure 3.15 represents the life cycle of a flowering plant. What we would normally call the plant is actually the diploid sporophyte generation. In its flowers, it forms two types of haploid gametophyte, a male (the germinated pollen grain) and a female (the embryo sac). The embryo sac contains a number of haploid nuclei. One of these is the egg cell. It is fertilised by one haploid nucleus from the germinated pollen grain, which travels down the pollen tube that grows from the pollen grain to the embryo sac within the flower. This fertilisation forms a zygote, which divides by mitosis to form an embryo plant. A second haploid nucleus in the pollen tube fuses with two of the other haploid nuclei within the embryo sac to form a triploid nucleus (3n). This divides by mitosis to form the endosperm that nourishes the growing embryo plant. The embryo sac and endosperm are enclosed with a protective seed.

Plants are incredibly useful to humans. They provide us with food, building materials, rubber and many drugs that are used in the pharmaceutical industry.

## Animalia

Animals are multicellular organisms, whose cells are eukaryotic and always lack cell walls. Animals are unable to photosynthesise and so are always heterotrophic. They have a variety of ways of obtaining nourishment from ready-made organic sources (food) and a variety of body cavities for digesting and absorbing inorganic nutrients from their food. You will learn more about the structure of the digestive system of mammals and about the way that insects deal with food during different stages of their life cycles in Chapter 12. Animals develop from a multicellular embryo that, at some stage, forms a hollow ball of cells, called a **blastula**.

The animal kingdom contains over 1 million species, which are organised into many different phyla. They include sponges, jellyfish, corals, worms, crabs, insects, snails, squid, sea urchins, fish, amphibians, reptiles, birds and mammals. Figure 3.16 shows examples of some of these phyla. As we saw in Table 3.3, we are animals, belonging to the phylum Chordata.

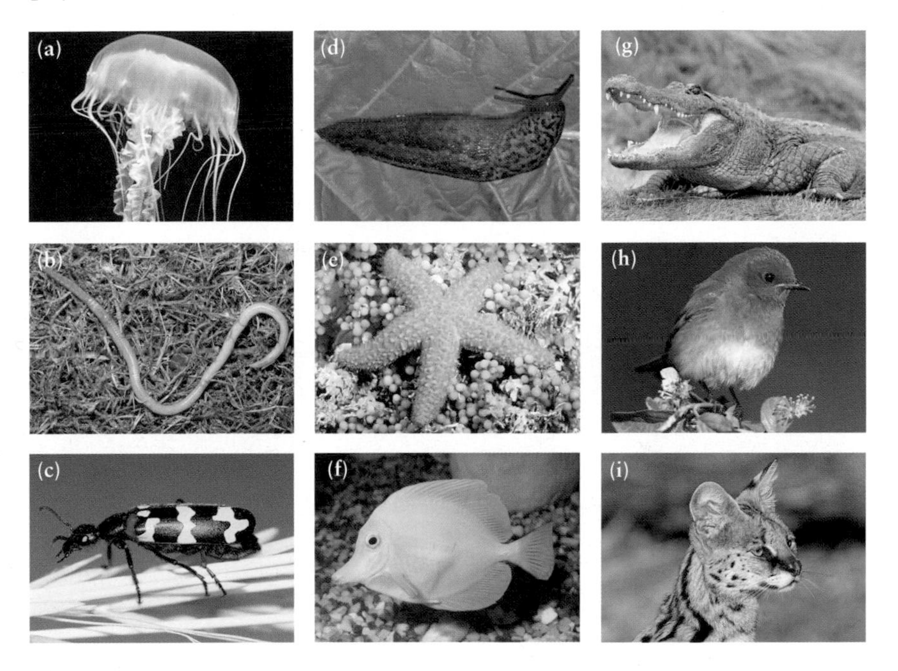


**Figure 3.16**
The animal kingdom contains a huge variety of organisms. The distinctive feature of them all is that they developed from a multicellular embryo that formed a hollow ball of cells, called a blastula (a) Jellyfish (b) Earthworm (c) Insect (d) Slug (e) Starfish (f) Fish (g) Reptile (h) Bird (i) Mammal

**Q** 8 **An organism has cells, which contain nuclei and are surrounded by cell walls. To which of the five biological kingdoms might this organism belong?**

## Summary

- A species is the smallest unit of biological classification.
- Species can be described in two ways.

  1. Variation – members of the same species are very similar to each other but different from members of other species.

  2. Potential for breeding – members of the same species can breed together naturally and produce fertile offspring.
- Existing species are thought to have evolved from other species. One method of speciation involves the formation of a hybrid, the offspring from two different species, which then increases the number of sets of chromosomes in its cells (polyploidy). This method of speciation is thought to be responsible for the evolution of most species of flowering plants, but not of animals.

- Another method of speciation involves the reproductive isolation of two or more groups of a single species. Different pressures of natural selection on the isolated groups results in different frequencies of alleles in the groups. If these changes in allele frequency are great enough to prevent interbreeding, new species are formed.
- In the latter method of speciation, allopatric speciation occurs when the barrier to reproduction is geographical separation. Sympatric speciation occurs if the groups are reproductively isolated without geographical separation.
- Species are organised into biological categories in which the members show decreasing similarity to each other. Species are grouped into genera, which are then grouped into families. Families are grouped into orders, orders into classes, classes into phyla and phyla into kingdoms.
- The biological name of any species is a binomial, formed from the name of the genus and the name of the species. Through international convention, the name of the genus is written with a capital initial letter and the name of the species is written with a small initial letter. These names are printed in italics or, if hand-written, are underlined. The biological name for humans is *Homo sapiens*.
- There are five biological kingdoms: Prokaryotae, Protoctista, Fungi, Plantae and Animalia.
- The kingdom Prokaryotae contains all the bacteria. These organisms have prokaryotic cells (no true nucleus) and cell walls made of peptidoglycans. Some prokaryotes are heterotrophic and others are autotrophic, obtaining their nutrition by photosynthesis or by chemosynthesis.
- Organisms are classed in the kingdom Protoctista if they cannot be classed into one of the other four kingdoms. Protoctists have eukaryotic cells. They include organisms that resemble single-celled animals, organisms that resemble single-celled, filamentous or colonial plants and multicellular organisms, such as seaweeds.
- Fungi have eukaryotic cells, which are surrounded by cell walls made of chitin. They are heterotrophic, with most releasing enzymes onto their surroundings and absorbing the digested food materials into their cells. Most fungi are made of thread-like hyphae, which grow horizontally in the food source, but have erect hyphae, which carry their reproductive spores. The pigment in the spores produces the colour that we see on mouldy food. Yeast is a unicellular fungus that is used extensively in industry.
- Members of the phylum Plantae have eukaryotic cells surrounded by a cellulose cell wall. The phylum includes liverworts, mosses, ferns, conifers and flowering plants. The life cycle of plants is complex, containing two different types of adult. The haploid gametophyte adult reproduces sexually to form a zygote. The diploid adult that

develops from a zygote is the sporophyte. It reproduces asexually to produce spores that germinate into gametophytes again.

- All members of the animal kingdom (Animalia) have eukaryotic cells that lack a cell wall. Animals always develop from a multicellular zygote that forms a hollow ball, called a blastula, at some stage. Humans belong to this kingdom.

# Assignment

## When is a primrose not a primrose?

In this chapter you have seen that it is not always easy to decide whether two organisms belong to the same or to different species. We will now consider this problem in a little more detail by looking at a plant which is common throughout much of Britain – the primrose. Some of the material in this assignment is about plant reproduction. Although this will probably be unfamiliar to you, you should be able to understand the principles involved from the information you have been given.

**Figure 3.17**
Primrose flowers are usually pale yellow but occasionally plants with pink flowers are found. Pink flowered primroses are quite common in some parts of Wales

1 Figure 3.17 shows a clump of primroses. Use the information in this chapter to describe how you could show that the yellow-flowered plants and the pink-flowered plants belong to the same species.

*(3 marks)*

In many parts of Britain another species of *Primula* may be found growing near primroses. This is the cowslip. Figure 3.18 shows primroses, cowslips and the hybrids which sometimes form between these plants.

**Figure 3.18**
(a) Primroses (the plants with larger paler flowers) are growing together with cowslips on this bank. (b) A hybrid plant growing nearby. It has some characteristics of its primrose parent and some characteristics of its cowslip parent

Even in areas where primroses and cowslips are common, hybrids between them are relatively rare. This may be the result of slightly different ecological requirements – cowslips are generally found in more open areas. It might also be due to cowslips generally flowering later than primroses.

In experimental conditions, primroses and cowslips readily produce hybrids but this only happens when the female parent is a cowslip and the pollen comes from a primrose. The reason for this is associated with the way in which flowering plants reproduce. When fertilisation takes place in a flowering plant, the pollen grain provides two male gametes. One of these male gametes fuses with the egg nucleus to become the embryo. The other fuses with the two endosperm nuclei. The endosperm nucleus divides by mitosis and gives rise to endosperm. This is a tissue which provides a food source for the developing embryo.

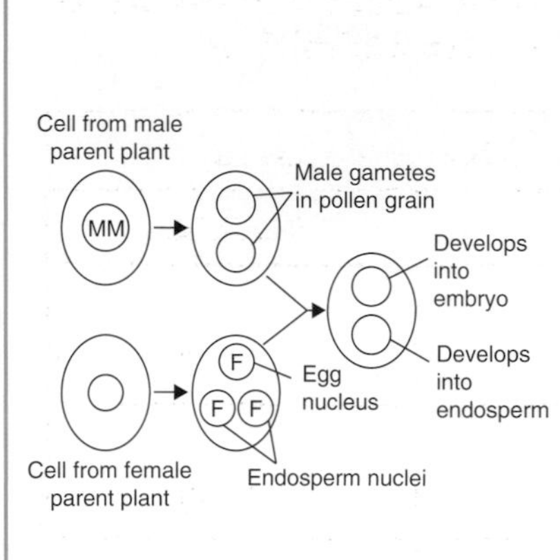


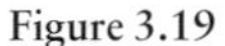
**Figure 3.19**

2 Copy the diagram in Figure 3.19. Use the information in the paragraph above to complete this diagram to show the chromosomes in each of the nuclei. Use **M** to represent one set of chromosomes in or from the male parent and **F** to represent one set of chromosomes in or from the female parent. Some stages have been put in to help you.

*(4 marks)*

Now we will explain why we only get hybrids if the cowslip acts as the female parent. Primroses and cowslips have different rates of seed development. Primrose seeds develop more slowly than cowslip seeds. Not only does the whole seed develop more slowly but so does the embryo and the endosperm. Look at Table 3.6. This table shows the sets of cowslip chromosomes (**C**) and the sets of primrose chromosomes (**P**) in possible hybrids.

**Table 3.6**
The number of sets of cowslip chromosomes and primrose chromosomes in the endosperm and in the embryo will depend on whether the female parent is a cowslip or a primrose.

| Female parent | Number of sets of cowslip chromosomes (C) and primrose chromosomes (P) in a nucleus from | |
|---|---|---|
| | endosperm | embryo |
| Cowslip | CCP | CP |
| Primrose | CPP | CP |

3 (a) The endosperm formed when the female parent is a primrose develops more slowly than the endosperm formed when the cowslip is the female. Use the information in the table to explain this difference.

*(2 marks)*

(b) Experiments have shown that when the female parent of a primrose-cowslip cross is a primrose only about 1% of the seeds formed have embryos present. When the female parent of this cross is a cowslip, between 80 and 100% of the seeds contain embryos. Explain these findings.

*(2 marks)*

4 Do you think that primroses and cowslips are the same or different species? Use evidence from this assignment to support your answer.

*(4 marks)*

## Examination questions

1 The Hawaiian islands are a chain of volcanic islands in the middle of the Pacific Ocean. **Table 3.7** compares the number of species of some different groups of insect in Hawaii and in the British Isles.

| | Number of native species | |
|---|---|---|
| | Hawaii | British Isles |
| Flies belonging to the genus *Drosophila* | 800 | 37 |
| Total number of species of fly | 800 | 5 950 |
| Total number of species of insect | 6 500 | 21 833 |

Table 3.7

Suggest an evolutionary explanation for each of the following observations.

(a) All the flies native to Hawaii belong to the same genus.

*(1 mark)*

(b) When the volcano erupted, lava poured out. This left isolated patches of vegetation surrounded by large areas of bare lava. There are slight differences between the populations of flies in neighbouring patches of vegetation even though they belong to the same species.

*(3 marks)*

(c) There are more species of *Drosophila* in Hawaii than there are in the British Isles.

*(2 marks)*

2 The diagram shows how four species of pig are classified.

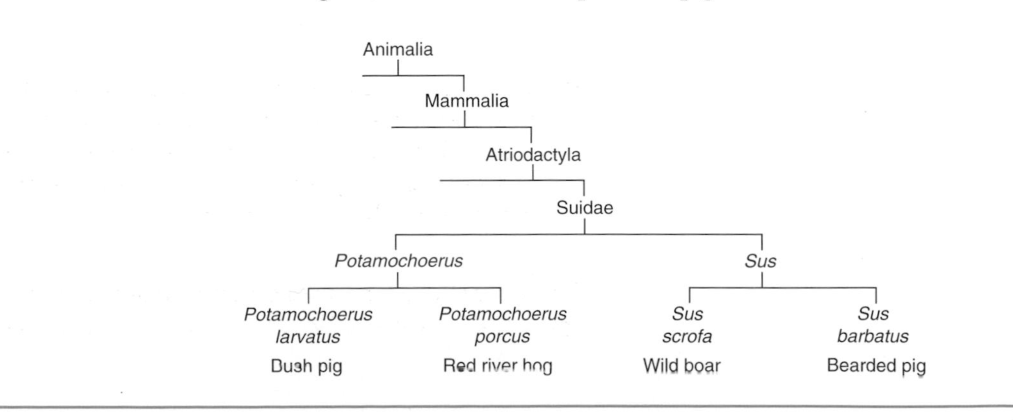

(a) (i) To which family does the red river hog belong?

*(1 mark)*

(ii) To which genus does the bearded pig belong?

*(1 mark)*

(b) Some biologists think bush pigs and red river hogs belong to the same species. The list below summarises some features of the biology of bush pigs and red river hogs.

- The bush pig has a body length of 100–175 cm and a mass of 45–150 kg. The red river hog has a body length of 100–145 cm and a mass of 45–115 kg.
- The red river hog is found in West Africa. The bush pig is found in East Africa.
- Both animals are omnivorous but feed mainly on a variety of underground roots and tubers.
- The ranges of these animals overlap in Uganda. In this area populations of animals which have all characteristics intermediate between those of bush pigs and red river hogs have existed for many years.

Do you think that bush pigs and red river hogs belong to the same or to different species? Explain how the information above supports your answer.

*(3 marks)*

# 4 Numbers and Diversity

Figure 4.1
This tree has recently been felled. It has left a space where light can penetrate to the forest floor. Plants such as macaranga rapidly become established in areas like this

Yanamamo is an area of tropical rain forest in the Peruvian Amazon. It probably has more species of animals and plants than anywhere else on Earth. A single hectare of the forest contains almost 300 different species of tree. Compare that with the United Kingdom where there are only 33 native tree species! These trees support huge numbers of animals. A biologist working in this area found a single tree, for example, which had 43 different species of ant living on it – that's about the same number as in the whole of the United Kingdom.

Not only are there enormous numbers of different species in a tropical rain forest but the relationships between them are often very complex. Macaranga is a fast-growing tree. It springs up wherever taller trees fall and allow light to reach the forest floor (Figure 4.1). Macaranga has hollow stems in which a particular species of ant lives. These ants 'farm' scale insects. The scale insects feed on the tree sap and produce a sugary liquid which is readily consumed by the ants. Macaranga trees with ants living in them are usually free of climbers and other plants living on them because the ants bite off any foreign shoot or leaf coming into contact with their host tree. In addition, experimental work has shown that once ants are removed, macarangas are attacked by many species of leaf-eating insect.

When humans clear patches of forest, species are lost and this may have far reaching effects. Jackfruit (Figure 4.2) is a tree which is pollinated by nectar-feeding bats. These bats need other species of tree on which to feed for those months of the year when jackfruit is not in flower. If you fell the forest, there will be no bats and no jackfruit.

Figure 4.2
Jackfruit, one of many tropical plants which are pollinated by bats. A good crop of jackfruit depends on a good population of bats

The huge number of different species and the complex biological relationships between them make tropical rain forests very difficult places to study. We find, however, that there are a number of basic principles which apply just as readily to a rain forest as they do to a seashore or a pond in Britain. In the next three chapters we shall look at these ecological principles in a little more detail.

An understanding of the ecology of a particular area – seashore, pond or rain forest – requires us first of all to gain some idea of the numbers of organisms found there and how they are distributed. Then we can begin to use our understanding of ecological principles to explain these observations. In this chapter we will start by looking at some of the different methods biologists use to find out about numbers and distribution. We will then consider two basic ecological ideas that help us to explain differences in distribution – diversity and succession.

## Organisms and their environment

Nettles are common plants. You can find them growing in woodland, by the side of ponds and streams and in areas of waste land in towns and cities. Look at Figure 4.3. It shows a clump of nettles and some of the animals which can be found living on them.

**Figure 4.3**
Some of the common organisms associated with a clump of nettles

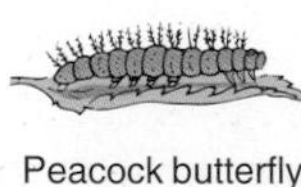
Peacock butterfly (larva)

Peacock butterfly

Nettle aphid

Dark bush cricket

7-spot ladybird

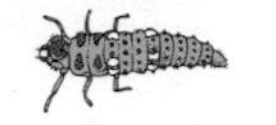
7-spot ladybird (larva)

2-spot ladybird

The **environment** of the organisms in this clump of nettles is the set of conditions which surrounds them. It consists of an abiotic component, and a biotic component. **Abiotic factors** make up the non-living part of the environment. Nettles grow particularly well, for example, where there is a high concentration of phosphate in the soil. Warm, humid conditions result in large populations of aphids on the leaves. The concentration of mineral ions such as phosphate, temperature and humidity are all abiotic factors. **Biotic factors** are those relating to the living part of the environment. The number of aphids on these nettle plants will be affected by the number of ladybirds present since ladybirds feed on aphids. Predation by ladybirds is a biotic factor.

A **population** is a group of organisms belonging to the same species. The members of a population are able to breed with each other, so the two-spot ladybird population consists of all the two-spot ladybirds found in a particular area at the same time. The two-spot ladybirds found on our clump of nettles will form a different population from the ladybirds found on another clump two or three hundred metres away. A **community** is the term used to describe all the populations of different organisms living in the same place at the same time. When we talk about the community of organisms living in this nettle patch we mean the nettles, the aphids, the ladybirds and all the other organisms that we have not mentioned, such as the fungi and the bacteria which live in the soil round the roots of the nettles.

An **ecosystem** is made up of a community of living organisms and the abiotic factors which affect them. It is sometimes convenient to think of ecosystems as being separate from one another but they seldom are. A small bird for example, such as a blue tit, may feed on insects in a nettle patch but it might nest in a nearby wood and, in winter, visit bird tables on a housing estate.

**Q** 1 **Suggest how increased use of pesticides on farmland in North America could result in an increase in the concentration of pesticides in the tissues of penguins living in the Antarctic.**

Figure 4.4
Mute swans are conspicuous birds and they are confined to aquatic habitats. It is possible to estimate the population of mute swans fairly accurately by counting all the birds in a particular area. There are likely to be errors, however, resulting from birds moving round or being present in areas where access is difficult

There are two more terms we often use to explain ecological ideas. The **habitat** of an organism is the place where it lives. Its **niche** describes not only where it is found but what it does there. It is a description of how an organism fits into its environment. We can describe the niche of the two-spot ladybird, for example, in terms of the abiotic features of the habitat in which it lives, such as the temperature range it can tolerate and the position on the nettle plant where it is found. We can also bring into our description some idea of its feeding habits, referring to the size and species of aphids that it eats.

## Counting and estimating

It is not often that we can find the size of a population by counting all the organisms of a particular species. It can sometimes be done if the animal or plant is comparatively rare, very conspicuous or lives in a small area (Figure 4.4).

If it is not possible to count every single organism of a particular species, we need to take samples. Ecologists usually base their estimates of populations on samples. Whatever method is used, however, we must make sure that our samples are representative of the population as a whole. In order to be sure of this, they must be large enough and must be taken at random.

### Sample sizes

The larger the size of the sample the more reliable the results, but a very large sample may be too large to study accurately in the time available. Small samples, on the other hand, can be quite different from the rest of the population. We have to strike a balance. Look at the graph in Figure 4.5. It shows that a very small sample has very few species in it. As the sample size increases, the number of species it contains increases. There comes a time, however, when the number of species does not increase significantly, however much bigger the sample. This sample size is obviously representative of the population as a whole.

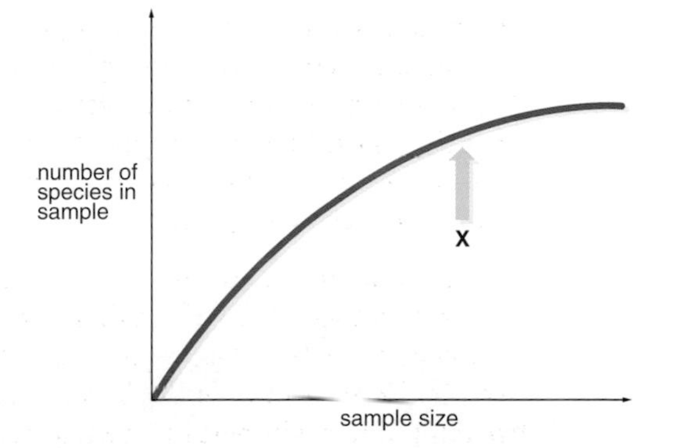


Figure 4.5
The number of species present in samples of different sizes. At the point marked **X**, the number of species present does not increase much more. This sample size is representative of the population as a whole. A larger sample size would simply mean more work.

**Figure 4.6**
A quadrat frame being used to sample organisms on the seashore. In this investigation, a random sample of limpets is being measured

## Random sampling

If we do not take samples at random, results will obviously be biased. Quadrat frames (Figure 4.6) are often used to sample vegetation or sedentary animals living on a seashore. If we use a quadrat frame for sampling, it is important that we use a method (Figure 4.7) that will result in it being placed genuinely at random in our study area.

**Q 2 Which of the methods below would allow samples to be taken at random?**
**A Closing your eyes, turning on the spot and throwing a quadrat frame over your shoulder**
**B Picking numbers out of a hat to give coordinates on a grid**
**C Placing quadrat frames at 5 metre intervals**

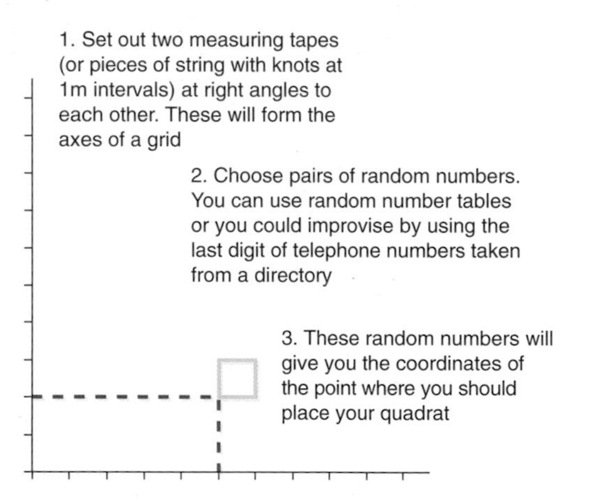


**Figure 4.7**
Placing quadrats at random. Where possible, a method like this should be used. Throwing the quadrat frame over your shoulder will not result in genuinely random sampling

## Quadrats and transects

Two important ways of sampling involve the use of quadrats and transects (Table 4.1) Both of these are generally used for plants but they can be used for other organisms which do not move about much – such as many of those that live on the seashore. There are three measures commonly used to describe the distribution of organisms once the quadrat or the transect is put in position. These are described in Figure 4.8.

(a) population density

This quadrat measures 0.5m × 0.5m. It contains 6 dandelion plants. The population density of dandelions would be 24 plants per m$^2$. To get an accurate figure you would need to collect the results from a large number of quadrats.

**Figure 4.8**
Three measures commonly used to describe the distribution of plants and other organisms. These are (a) population density, (b) frequency and (c) percentage cover

(b) frequency

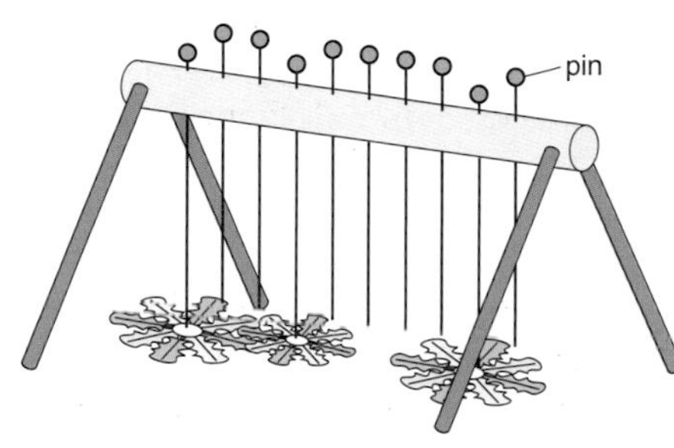


This point quadrat frame is being used to measure frequency. The pins of the frame are lowered. Suppose 3 out of 10 pins hit a dandelion plant. The frequency of dandelion plants will be 3 out of 10 or 30%.

(c) percentage cover

Percentage cover measures the proportion of the ground in a quadrat occupied by a particular species. The percentage cover of the dandelions in this quadrat is approximately 30%.

| Method | What it is used for | How it is used |
|---|---|---|
| **Frame quadrat**<br>A sample area marked out in order to study the organisms it contains. | Usually used to study the distribution of plants in a fairly uniform area. | Placed at random and used to find population density, frequency or percentage cover. In some investigations, such as those involving succession or the effects of grazing, permanent quadrats may be used. They remain in place for many years. |
| **Transect**<br>A line through a study area along which samples are taken | Used where the environment gradually changes and species vary | Placed so that it follows the gradient, for example up a seashore or from full sunlight into dense shade. Quadrats may be used at regular intervals along the transect to sample in more detail. |

**Table 4.1**

Sampling methods – what for and how

## Mark–release–recapture

Most animals move around so it would not be possible to estimate their population size using a quadrat. We can use the mark–release–recapture method. This relies on capturing a number of animals and marking them so that they can be recognised again. They are then released. Some time later a second sample is trapped and the numbers of marked and unmarked animals recorded. From the data collected, the size of the population can be estimated. The calculation is very straightforward and relies on the fact that the proportion of marked animals in the second sample will be the same as the proportion of marked animals in the whole population (Extension box 1).

**Extension box 1**

### Using the mark–release–recapture method to estimate the size of a population of grasshoppers

| | |
|---|---|
| Number of grasshoppers caught, marked and released | 56 |
| Number of marked grasshoppers in second sample | 16 |
| Total number of grasshoppers in second sample (marked + unmarked) | 48 |

Proportion of marked grasshoppers in sample = Proportion of marked grasshoppers in population

$$\frac{\text{Number of marked grasshoppers in sample}}{\text{Total number of grasshoppers in sample}} = \frac{\text{Number of marked grasshoppers in population}}{\text{Total number of grasshoppers in population}}$$

$$\frac{16}{48} = \frac{56}{\text{Total population}}$$

$$\text{Total population} = \frac{48 \times 56}{16}$$

$$= 168$$

**Q** 3 **A sample of 40 trout in a fish pond were netted and each fish was marked on one of the fins. They were then released back into the pond. One week later a second sample was netted. It contained 17 marked trout and 41 unmarked trout. Estimate the size of the trout population in the pond.**

The mark–release–recapture method involves making a number of assumptions. These are:

- The number of animals in the population does not change between marking and releasing animals trapped in the first sample, and capturing the second sample. It will not work if lots of animals are born or hatch from eggs, or if large numbers die in the time between the two samples. It will not work either if the animal concerned is migrating and large numbers are either entering or leaving the study area.

- When the marked animals are released, they must mix thoroughly in the population.
- Marking must not affect the animal in any way. A large blob of white paint put on the back of a grasshopper, for example, will make it conspicuous and more likely to be eaten by a predator.

**Q 4 Blue tits breed during the spring. During the breeding season, pairs of birds establish territories which they defend from other blue tits. Give two reasons why the mark–release–recapture technique would not be suitable for estimating the population of blue tits in a wood in springtime.**

## Diversity

Biodiversity is a word frequently used by politicians and, like all words used by politicians, its meaning is not always clear! A biologist uses the word 'diversity' and means something very specific. Diversity is a way of describing a community in terms of the numbers of species and the number of organisms present. Different communities differ in their diversity. In the introduction to this chapter you read about some of the organisms present in a tropical rain forest. Tropical rain forests have a very high diversity. On the other hand, many Arctic communities (Figure 4.9) have a low diversity.

**Figure 4.9**
This area, high in the mountains of Norway, has a very low diversity of living organisms. It is characterised by a small number of species, although many of these are very numerous

As scientists, we try to describe diversity in numerical terms. In this way we can compare different communities. We will now look at how we can do this by calculating the **index of diversity, d**, from the formula:

$$d = \frac{N(N-1)}{\sum n(n-1)}$$

In this formula, **N** is the total number of organisms of all the species present in the community and **n** is the total number of organisms of each individual species. The Greek letter Σ (sigma) means 'the sum of ...' so, in the bottom line of this formula we have to work out **n(n −1)** for each species and then add all these separate figures together. Extension box 2 shows a worked example.

## Extension box 2

## Index of diversity – worked example

The figures in Table 4.2 show the birds visiting a bird table on a cold dry day and again, two days later, when it was raining heavily.

| Species | Maximum number of birds around bird table in one hour period | |
|---|---|---|
| | on dry day | on wet day |
| Blue tit | 8 | 1 |
| Coal tit | 1 | 1 |
| Great tit | 3 | 1 |
| Robin | 1 | 1 |
| House sparrow | 5 | 2 |
| Greenfinch | 3 | 1 |
| Starling | 7 | 7 |
| All species | 28 | 14 |
| Index of diversity, d | 5.8 | 4.1 |

**Table 4.2**
Number of birds visiting a bird table on a cold dry day and on a rainy day

We will look at the way in which we calculate the figures in the first column. The formula is:

$$d = \frac{N(N-1)}{\sum n(n-1)}$$

We can substitute the figures for N and n from the information in Table 4.2

$$d = \frac{28(28-1)}{8(8-1)+1(1-1)+3(3-1)+1(1-1)+5(5-1)+3(3-1)+7(7-1)}$$

$$d = \frac{28 \times 27}{(8 \times 7)+(1 \times 0)+(3 \times 2)+(1 \times 0)+(5 \times 4)+(3 \times 2)+(7 \times 6)}$$

$$d = \frac{756}{130} = 5.8$$

You might have thought 'Why do we not simply compare the number of species present? Surely, this would give us an idea about diversity and would save the need for a lot of tedious calculation.' The answer is that it does provide information in some cases, such as in comparing the number of species in a tropical rain forest with the number in an Arctic ecosystem. But, it does not take into account the fact that many species will be rare and only likely to be encountered in very small numbers. Look at Table 4.2 again. You have the same number of species in both columns. In the column representing the wet day, however, 5 out of the 7 species were encountered once only. An index of diversity gives us a much better idea of differences between communities in situations like this.

**Figure 4.10**
The seaweeds growing on the seashore and the animals living in them form distinct zones. There are fewer species living in the harsher conditions on the upper shore than in the less severe conditions on the lower shore

**Figure 4.11**
Abiotic conditions vary on this seashore. On the lower shore, organisms are covered by sea water for most of the time. On the upper shore, they are exposed to the air. Few marine species can tolerate these harsh conditions

On its own, the index of diversity tells us very little. It is important, however, in allowing us to make comparisons. In the example in Extension box 2, we can begin to see a pattern. With more observations, we might be able to say convincingly that the wetter the weather, the lower the diversity of birds visiting a garden bird table. By looking at a lot of different communities, ecologists have established a very important principle.

- The harsher the environment, the fewer the species present and the lower the diversity. In harsh environments it is generally abiotic factors which determine the species present.
- The less harsh the environment, the more species there are and the greater the diversity. In these environments it is often biotic factors such as competition which determines whether particular species can survive.

We will illustrate this principle by looking at the distribution of seaweeds on a rocky seashore such as that shown in Figure 4.10.

Seaweeds found on the upper shore are only covered in sea water for the small part of the day around high tide. For the rest of the day, they are exposed to the air. In these conditions the temperature undergoes considerable variation, evaporation of salt water and rain may bring about large changes in salt concentration, and long, hot summer days are likely to have a considerable drying effect. These are harsh conditions and very few seaweeds can survive them (Figure 4.11). The index of diversity for these seaweeds will be very low and it will be their tolerance of abiotic conditions which will determine the height up the shore to which they are found.

**Q 5 What data would it be necessary to collect to calculate an index of diversity for seaweeds growing on the upper shore?**

Now we will look at the low-tide level. Seaweeds growing here will be covered in sea water for most of the day. Conditions are much more stable and not subject to large fluctuations in temperature and salt concentration. There is a much higher diversity of seaweeds. The distance they extend down the shore is likely to be due to biotic factors such as competition with other seaweeds. We will explore this principle further in Extension box 3 and in the last section in this chapter.

## Extension box 3

## Diversity and freshwater pollution

In this chapter we have been looking at an important ecological principle – the more extreme an environment, the fewer the species of organism found and the lower the diversity. We can apply this principle to pollution. Polluted habitats are extreme environments in which the diversity of organisms is low. In the assignment in Chapter 5 we will look at the way in which air pollution by sulphur dioxide affects the distribution of lichens. Here, we will consider how diversity can be used to study the amount of organic pollution in freshwater streams and rivers.

We can monitor organic pollution by measuring abiotic factors such as the concentration of dissolved oxygen in the water. With a suitable meter this can be done rapidly, but it only gives us a picture of the pattern of pollution at a particular time. If we want to get an overall picture, we need to take a lot of measurements, or we have to use another technique.

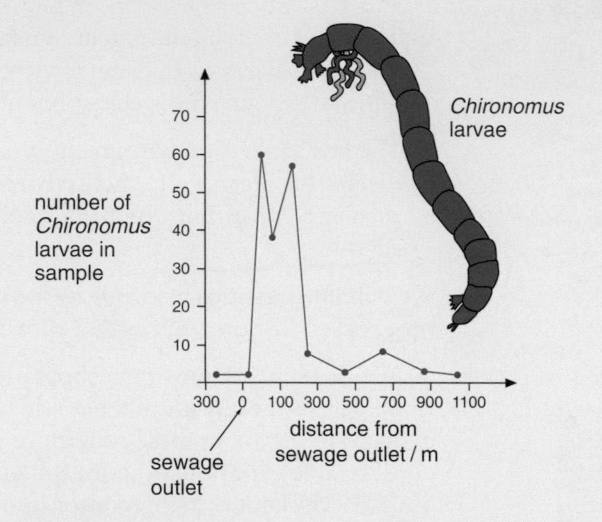


**Figure 4.12**
*Chironomus* is an insect whose larvae are very tolerant of organic pollution. The graph shows that there are very large numbers of *Chironomus* larvae immediately downstream of the sewage outlet

Organic pollution affects living organisms. We can use these organisms as indicators of the amount of pollution (Figure 4.12). The results obtained by looking at the distribution of a single species are not very reliable. In the example shown in Figure 4.12, there could be other factors which influence the distribution of *Chironomus* other than organic pollution. We get a better picture if we look at all the organisms present. There are various ways of doing this. The simplest is to collect a sample of living organisms and use it to calculate the sequential comparison index (SCI). This is given by the formula:

$$\text{SCI} = \frac{\text{number of sequences of organisms of the same species}}{\text{total number of organisms}}$$

Let us look at how it works. At an extremely polluted sampling point we may catch twenty individual organisms, all the same species. The SCI will therefore be 1/20 or 0.05 since there is only one sequence of organisms of the same species and twenty organisms in total. Suppose we sample at a less polluted point. We again catch a total of twenty organisms but they are of different species. We remove them from the net in the order:

**A A A | B | A A | B | C C C | D | E | F | A A | B | F F | C C**

Look at this list and you will see that there are twelve sequences of the same species – they are marked by vertical lines. The SCI will therefore be 12/20 or 0.6.

This method has a big advantage over other methods. It only requires a person to be able to recognise that organisms differ from each other. It does not rely on identification so it can be used by a non-biologist. The disadvantage is that it is rather crude and unreliable. In addition, it is influenced by other sorts of pollution. Because of this biologists have produced other measures called biotic indices which are more closely related to organic pollution, but they still rely on measuring the diversity of living organisms (Figure 4.13.)

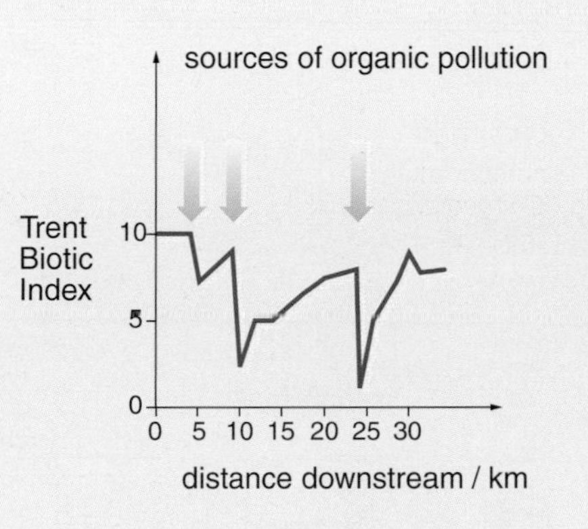


**Figure 4.13**
The Trent Biotic Index, an index based on the diversity of living organisms, at various points along the river North Esk in Scotland. The three main dips in the curve indicate sources of organic pollution

## Succession

Some of the organisms living in a community are gradually replaced by others. This is **succession**. It is an ecological process resulting from the activities of the organisms themselves. Over a period of time they modify their environment (Figure 4.14) and these modifications produce conditions better suited to the growth of other species.

**Figure 4.14**
Plants like bog bean grow in shallow water and trap mud around their roots. This eventually leads to drier conditions in which other species become established

In this book, we cannot describe all the examples of succession that you might encounter on a field course, so what we will do is to choose one particular example and use this to illustrate the principles which apply to them all. Sand dunes occur in many coastal areas and they show very clearly the process of succession. Obviously it is not possible to sit in one place and observe successional changes as they take place. The process is far too slow for that. What we can do, however, is to walk up from the sea shore into the sand dunes. As we do, we will pass through different areas which represent different stages in succession (Figure 4.15)

In understanding what has happened, it is useful to think again about the principles we considered when we discussed diversity. We will start by considering the mobile dunes. They get this name because they are continually changing shape as the wind scours the surface and blows the sand. This is a very harsh environment. Look at Table 4.3. It compares abiotic factors in these mobile dunes with those from the fixed dunes later in the succession.

| Abiotic factor | Mobile dunes | Fixed dunes |
|---|---|---|
| Mean wind velocity 5 cm above dune surface/km $hour^{-1}$ | 12.1 | 2.4 |
| Organic matter/% | 0.3 | 1.0 |
| $Na^+$/ppm | 8.5 | 4.2 |
| $Ca^{2+}$/ppm | 637.0 | 297.0 |
| $NO_3^-$/ppm | 48.0 | 380.0 |

**Table 4.3**
Abiotic factors in mobile and fixed dunes

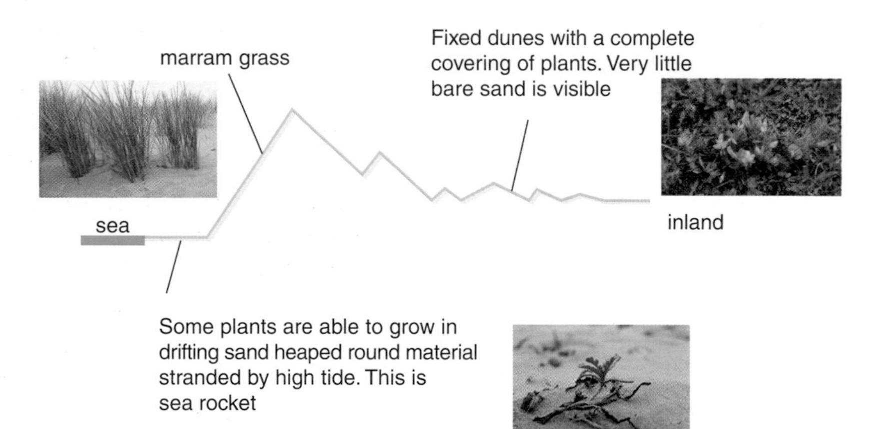


**Figure 4.15**
Succession in sand dunes. As we walk up the shore we pass from an area of isolated plants growing in the drifting sand heaped round material left stranded by high tide, through an area dominated by marram grass to an area of fixed dunes with a complete covering of plants

The high wind speed has two important effects. It will pile sand on top of any plants growing there and it will lead to a high rate of water loss by transpiration (Chapter 11). There is very little organic matter, so the sand does not retain water very well. It will dry out rapidly after rain. Important soil nutrients such as nitrates are in short supply, while sea spray results in high concentrations of sodium and calcium ions. Very few plants can grow here, but one that does is marram grass, illustrated in Figure 4.15. Because it is one of the first plants to colonise the area, we refer to it as a **pioneer species**. Marram grass has a number of adaptations. Like many grasses, the leaves grow from an underground stem, but this stem does not grow horizontally, it grows vertically upwards. The more sand that is deposited on top, the more vigorously it grows.

**Q** 6 **How is the vertically growing underground stem of marram grass an adaptation for conditions found in mobile dunes?**

Marram grass also has many structural features that enable it to grow in dry conditions. You can find out more about these in Chapter 11.

We have here a good example of the principle we encountered when we discussed diversity.

- The harsher the environment, the fewer the species present and the lower the diversity. In harsh environments it is generally abiotic factors which determine the species present.

**Q 7 What is the value of the index of diversity in mobile sand dunes if the only organism present is marram grass?**

Now let us go inland to the area of fixed dunes. The first thing you notice is that there are many more species of plants growing here. Table 4.3 shows us that abiotic factors have changed and many of these changes are due to the activity of the living organisms growing on the mobile dunes. The roots and leaves of the grass form a windbreak. The wind velocity is lower so less sand is being blown around. Dead material falls from the marram grass and is broken down by the soil bacteria. The amount of humus is higher so the developing soil can retain moisture rather better and the concentrations of important nutrients such as nitrates rises. Rain is also beginning the process of leaching the soluble sodium and potassium ions from the surface layers of the soil. Other plants begin to appear and gradually replace marram grass as the dominant vegetation. We are beginning to get to a situation where

- The environment is less harsh, so there are more species and a greater diversity. In this environment it is biotic factors such as competition which determines whether particular species can survive.

The process of the organisms in the community affecting their environment in such a way that they are replaced by other species continues. The fixed dune community that we have been looking at is gradually colonised by bushes and then trees. Ultimately we reach a stage when no further change takes place. This is the **climax community** and, in Britain, it would generally be woodland of some sort.

We have described the process of succession as it takes place in sand dunes but there are other examples such as the edges of ponds and streams and bare rock where the process may be studied. The details may be different in each case but the principles are very similar. They are summarised in Figure 4.16.

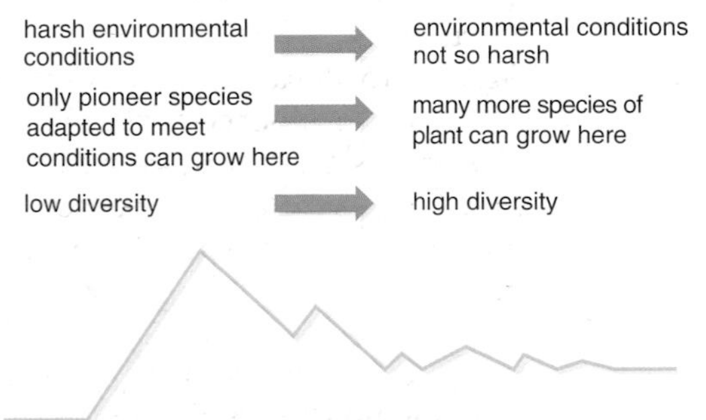


**Figure 4.16**
A summary of the processes involved in succession

Succession does not always proceed all the way to a climax community however. It can be stopped by various factors such as human activity. In a series of experiments, control plots were compared with plots in which the plant cover had been partly destroyed by allowing motorcycles to ride over them. The results are shown in Figure 4.17.

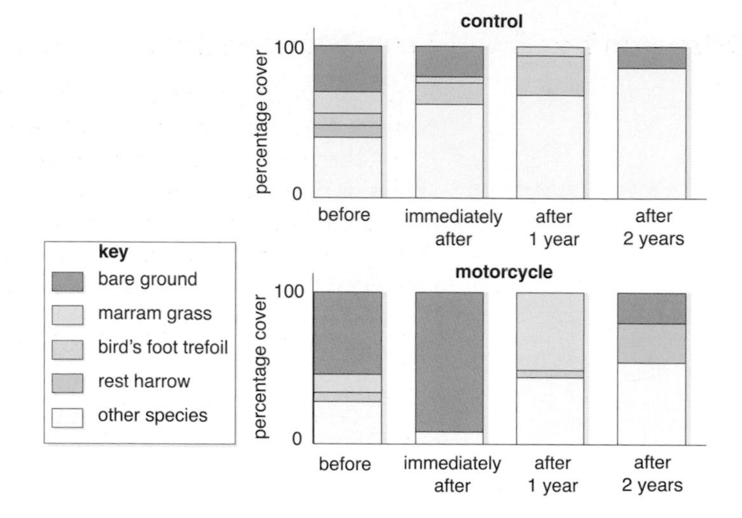


**Figure 4.17**
Bar charts showing the effect of damage by motorcycles to the plants growing on sand dunes

In the control plot, you can see that the process of succession has continued. Early colonisers such as marram grass and rest harrow have decreased in amount while the mean percentage of the plots covered with other species has increased. In the experimental plots, the initial effect of the motorcycles was to remove much of the plant cover. This allowed the spread of rest harrow.

Experiments like this show us that when we halt the process of succession, it rarely returns to where it began. In this example, it is quite easy to see why. The areas in which the vegetation has been partly removed by motorcycles are already quite different from those in which the marram grass became established. The sand contains more humus and is not so likely to dry out, it has a higher concentration of nitrate and lower concentrations of calcium and sodium ions, and there is less wind-blown sand. These are conditions in which rest harrow grows better than marram grass. Human activity, then, can alter the path of succession.

Another example where we see this happening is when we clear an area of tropical rain forest. If we leave this area, more forest grows in its place. This secondary forest, however, is quite different from the original forest. It tends to be denser and it contains different species.

Ecologists sometimes distinguish two types of succession. **Primary succession** occurs in places where previously there were no living organisms. The colonisation of bare sand dunes by marram grass and the changes which follow this are an example of primary succession. **Secondary succession** takes place in areas where there has already been a community of living organisms. An example of secondary succession is provided by the changes which follow when human activity damages the plant cover of sand dunes.

**Q 8** **Breeding elephant seals live on islands on which tussock grass grows. Their activity results in the death of this grass. When the seals leave at the end of the breeding season other plants grow in place of the tussock grass. Explain whether this is primary or secondary succession.**

## Extension box 4

## How old is a hedge?

Fields enclosed by hedges are a characteristic feature of the landscape in many parts of Britain. Some of these hedges are very old and may have been planted in Roman times. Some have been planted much more recently. How can we find the age of a particular hedge? We have seen that, during the process of succession, many changes take place in a community. Is there any way we can use our knowledge of these changes to suggest when a hedge was originally planted?

Look at Figure 4.18. This is a scatter diagram showing the number of different species of shrubs in random thirty metre lengths of hedge. The hedges in this study were all dated from historical records, so we have been able to plot the number of species against the exact age of the hedge

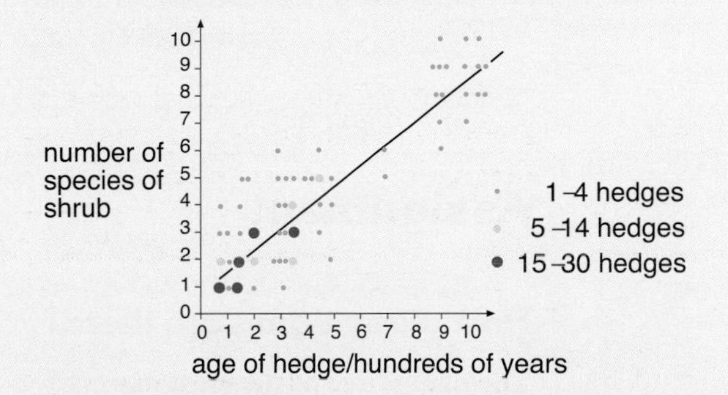


**Figure 4.18**
This scatter diagram shows a positive correlation between the age of a hedge and the number of species of shrubs it contains

You can see that there is an obvious trend. The older the hedge, the more species it contains. In approximate terms, there is one additional species every hundred years. So, a hedge containing only hawthorn is likely to be less than 100 years old; a hedge with three different species may have been planted at the beginning of the nineteenth century; and a hedge with six species probably dates back to somewhere around 1500.

Of course we have to be careful how we use information like this. The scatter diagram also shows us that there is a lot of variation in the number of species of shrubs in hedges of the same age. There are many reasons for this variation. More than one species might have been planted in the first place; variation in the soil and climatic conditions may have a considerable effect on the species which become established during the process of succession; and different hedges are different distances away from woods and other sources of seeds. Because of all these reasons, the best we can probably do is suggest a date but to remember that for a particular hedge we could be as much as 200 years out.

**Q 9** **How would you expect the diversity of shrubs in an old hedge to differ from that in a recently planted hedge?**

## Summary

- Random sampling with quadrats and counting along transects may be used to obtain quantitative data about the distribution of organisms.
- The number of organisms that move around can be estimated with the mark–release–recapture technique.
- Diversity is a measure of the number of species and number of individuals in a community.
- The harsher the environment, the fewer the species present and the lower the diversity. In harsh environments it is generally abiotic factors which determine the species present.
- The less harsh the environment, the more species there are and the greater the diversity. In these environments it is often biotic factors such as competition which determine whether particular species can survive.
- Succession involves colonisation by pioneer species. Subsequent changes lead to the establishment of a climax community.

## Assignment

### How many tigers are there?

The tiger is one of the most impressive of all wild mammals. Unfortunately, it is also one of the rarest. Poaching for skins and for body parts used in traditional Chinese medicine, and the inevitable conflict with ever-increasing human populations, have led to a dramatic decline in wild tigers. Whether the tiger will survive as a wild mammal long into this century is open to considerable doubt.

In trying to conserve animals such as tigers, one of the first things biologists must do is to monitor the population as accurately as possible. In this assignment we will look at a study carried out in the Way Kambas National Park in Sumatra (Figure 4.19).

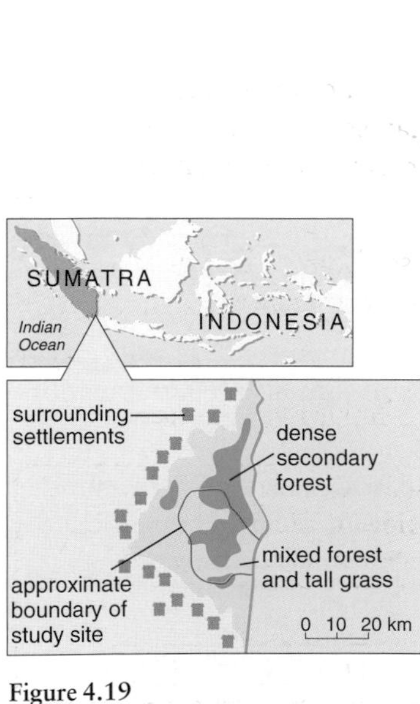


**Figure 4.19**
A map of the Way Kambas National Park on the Indonesian island of Sumatra

1 Apart from poaching, suggest why tigers are particularly threatened in this national park.

*(4 marks)*

2 The map shows that this national park is a mixture of dense secondary forest and tall grass. Suggest why it would be difficult to produce an accurate count of the tigers living in the park.

*(2 marks)*

An ecological study was started in August 1995 using camera 'traps'. Automatic cameras were set up at selected points along trails used by tigers in the study area. These cameras took a photograph each time an animal passed through an infra-red beam crossing the trail. The cameras had built-in flash and were spread evenly through the study area.

3 Why was it necessary for the cameras to have built-in flash in order to get accurate results ?

*(1 mark)*

The pattern of stripes on a tiger is unique. By analysing the photographs taken by the cameras, it proved possible to identify individual animals. The graph in Figure 4.20 shows some of the results obtained from this study. Curve **A** shows the cumulative number of tigers photographed. Curve **B** shows the cumulative number of tigers photographed which were thought to be resident in the study area.

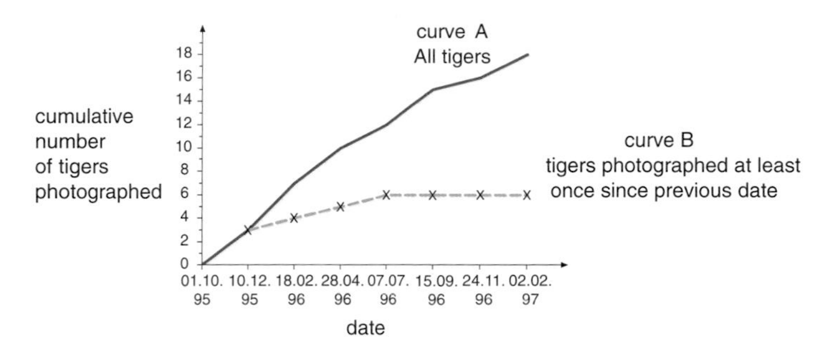


Figure 4.20

Tigers are territorial animals. They spend most of their time in a relatively small area which they defend against other tigers. The territory of a male tiger is considerably larger than the territory of a female tiger.

4 How many tigers had territories in the study area? Explain the evidence from the graph which supports your answer.

*(3 marks)*

**Table 4.4** below shows some more data from this investigation.

| Tiger | Number of times photographed (P) | Number of camera locations at which animal photographed (L) | $\frac{P}{L}$ |
|---|---|---|---|
| A | 13 | 2 | |
| B | 43 | 11 | |
| C | 15 | 4 | |
| D | 23 | 8 | |
| E | 13 | 5 | |
| F | 19 | 11 | |

5 Copy and complete the table by calculating the ratio of the number of times a tiger was photographed to the number of camera locations at which it was photographed.

*(1 mark)*

6 (a) What does this ratio tell us about the size of a particular tiger's territory?

*(1 mark)*

(b) Two of the animals in the table were adult males. Use the information in this assignment to suggest which two

*(1 mark)*

7 The study area was approximately 12% of the total park area. A suggestion was made that the total number of tigers in the park could be calculated from the following formula:

$$\text{Total number of tigers} = \frac{\text{Number of tigers with territories in the study area} \times 100}{12}$$

What assumptions does the use of this formula make? Suggest why using the formula would give an inaccurate estimate of the number of tigers in the park.

*(4 marks)*

## Examination questions

1 (a) Suggest **two** ways in which modern farming might affect the diversity of living organisms.

*(2 marks)*

(b) "Set-aside" is the common name given to a European policy under which farmers receive a subsidy for land taken out of cultivation. A study was carried out to investigate how the amount of time a set-aside field was left uncultivated would affect the species of birds feeding there.

**Table 1**

| Species | Number of birds of that species feeding in the field |
|---|---|
| Greenfinch | 12 |
| Goldfinch | 8 |
| Wood pigeon | 3 |
| Pheasant | 1 |

Table 1 on page 90 shows the number of birds of different species feeding in one field which had been left uncultivated for a year.

(i) Use the formula $d = \frac{N(N-1)}{\sum n(n-1)}$

where $d$ = index of diversity
$N$ = total number of organisms of all species
and $n$ = total number of organisms of a particular species

to calculate the index of diversity for the birds feeding in the field. Show your working.

*(2 marks)*

(ii) Explain why it is more useful in a study of this sort to record diversity rather than the number of species present.

*(2 marks)*

(c) **Figure 1** is a graph showing the relationship between bird species diversity and plant species diversity in this study.
**Figure 2** is a graph showing the relationship between bird species diversity and plant structural diversity for the same study.
Structural diversity refers to the different forms of plants such as herbs, shrubs and trees.

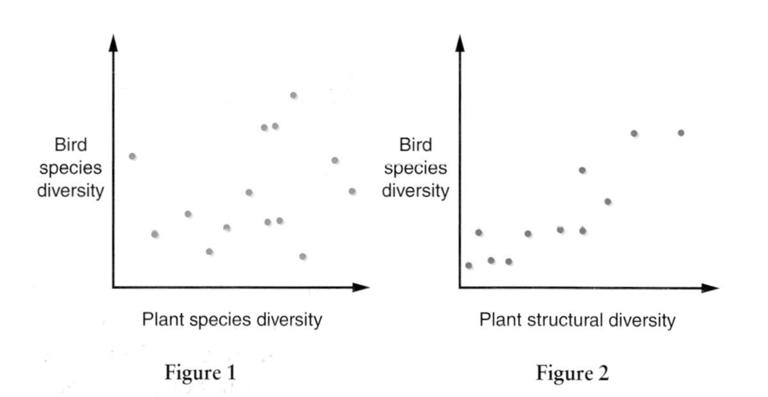


**Figure 1** **Figure 2**

(i) Explain briefly how you could obtain the data that would enable you to calculate the diversity index for the species of plants growing on a set-aside field.

*(3 marks)*

(ii) Describe the difference in the relationships shown in **Figures 1** and **Figure 2.**

*(2 marks)*

(iii) Suggest an explanation for the relationship between bird species diversity and plant structural diversity and plant structural diversity shown in **Figure 2.**

*(2 marks)*

(d) In another study of fields taken out of cultivation, the figures shown in **Table 2** were obtained.

| Value of index of diversity for bird species | Time in years since cultivaton stopped |
|---|---|
| 2.1 | 5 |
| 3.2 | 15 |
| 5.6 | 20 |
| 4.1 | 25 |
| 4.8 | 40 |
| 9.4 | 60 |

Table 2

(i) Plot these data as a suitable graph.

*(4 marks)*

(ii) Predict what might happen to the bird species diversity in the study summarised in **Table 2** over the next 100 years. Explain how you arrived at your answer.

*(3 marks)*

**2** The diagram shows a number of stages in an ecological succession in a lake.

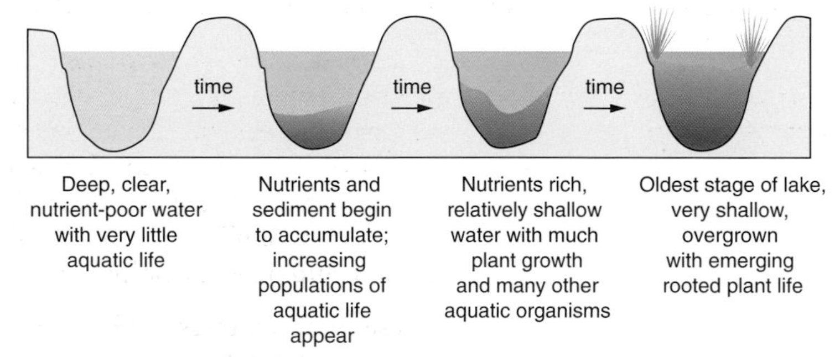


(a) Use information in this diagram to help explain what is meant by an ecological succession.

*(2 marks)*

(b) Give two general features which this succession has in common with other ecological successions.

*(2 marks)*

(c) A number of small rivers normally flow into this lake. These rivers flow through forested areas. Explain how deforestation of the area might affect the process of succession in the lake.

*(2 marks)*

# Energy transfer

Figure 5.1
This deep-sea angler fish lives at great depths. It is part of a community of organisms that is able to survive because of a constant rain of organic matter drifting down from the sunlit water at the surface. Breakdown of this organic matter provides the energy necessary to sustain a whole community of organisms

**The bottom of the Pacific Ocean is an inhospitable place. It is pitch dark; no light ever penetrates to its depths. And it is cold; the water remains all year round just above freezing. In a few areas, volcanic vents bubble out a mixture of sulphur-rich gases. The ocean floor is a place where you might think life could not possibly exist. But you would be wrong.**

**Around volcanic vents, bacteria are found. They make use of the rich source of sulphur-containing substances bubbling from the vents. They obtain energy from chemical reactions involving these substances and use it to build up the complex organic molecules that make up their cells. Various single-celled organisms feed on the bacteria and these, in turn, support a community of large worms and other invertebrate animals.**

**Far from being lifeless, the ocean depths contain many living organisms. These organisms depend on energy transferred from chemical reactions involving substances such as the sulphur-containing compounds emerging from volcanic vents. Or they may depend on energy contained in large organic molecules, which form part of dead organisms falling to the ocean floor (Figure 5.1).**

**This energy is then used to build up small molecules into the larger ones that make up cells and tissues. Communities of living organisms exist because chemical potential energy in large molecules is transferred from one organism to another along food chains and through food webs.**

In this chapter we shall look more closely at energy transfer. As we will not be looking at ocean depths, we shall be concerned mainly with another energy source, sunlight. As you can see from Figure 5.2, plants and other chlorophyll-containing organisms photosynthesise, using carbon dioxide and water to produce carbohydrates. In this process energy from sunlight is transferred to chemical potential energy in the glucose and other carbohydrates that are formed.

These carbohydrates can be used directly by the plants or they can be transferred to other organisms which either feed on plant material or break down dead plant tissue. They can also be used for respiration. During respiration, chemical potential energy in glucose is transferred to another molecule, ATP. ATP provides an immediate source of energy for various forms of biological work such as movement, the synthesis of large organic molecules and active transport.

Finally we need to remember that none of these processes is totally efficient. During respiration, for example, we cannot transfer all the chemical potential energy from a molecule of glucose into molecules of ATP. Some of the energy in the glucose molecule will inevitably be lost as heat. This is obviously important when we come to look at the transfer of energy from one organism to another along a food chain or through a food web.

**Q 1 Some animals are cannibals and eat their own young. Use your knowledge of energy transfer to explain why there are no animals that only feed on their own young.**

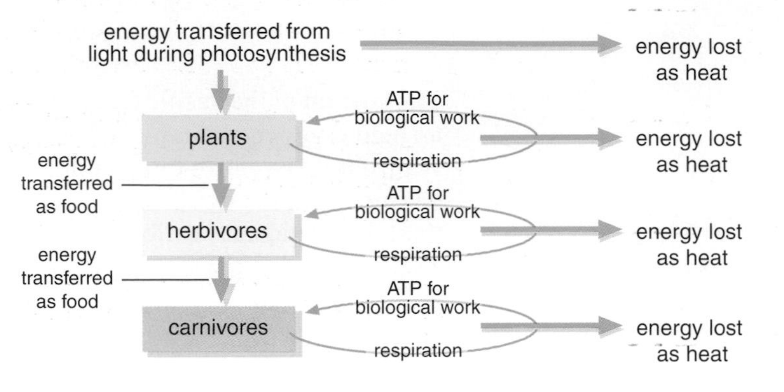


**Figure 5.2**
A summary of the way in which energy is transferred within and between organisms. Note that every time a transfer occurs, some energy is lost as heat

In this chapter we shall look at four main topics relating to this general theme of energy transfer.

- ATP as an energy source
- Photosynthesis
- Respiration
- Energy transfer and ecosystems

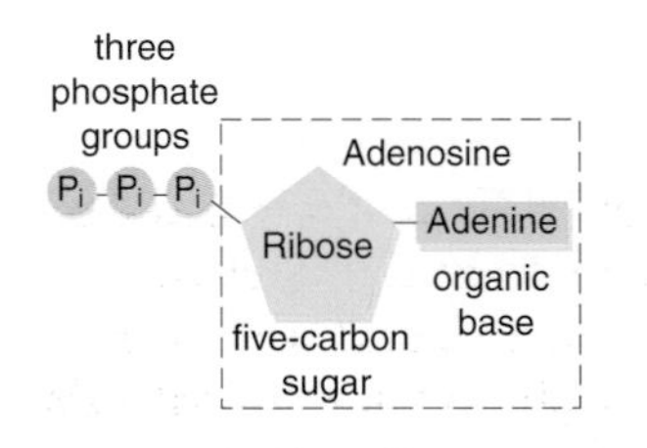


**Figure 5.3**
This simplified diagram of an ATP molecule shows the arrangement of its main components

## ATP as an energy source

Figure 5.3 shows a simplified diagram of an ATP molecule. You will see it has several components. **Ribose**, a five-carbon sugar, and **adenine**, an organic base, combine to give **adenosine**. Adenosine is joined in turn to three **phosphate groups**. When a single phosphate group is joined we have a molecule of adenosine monophosphate (AMP); two phosphate groups give us **adenosine diphosphate** (ADP); and three phosphate groups produce **adenosine triphosphate** (ATP).

In a living cell, ATP is normally produced by adding a third phosphate group to a molecule of ADP. This reaction requires a considerable amount of energy and this comes from chemical reactions such as those that occur in respiration and photosynthesis.

ATP breaks down to form ADP and a phosphate ion (written as $P_i$ for inorganic phosphate). This reaction involves hydrolysis of the ATP molecule and it releases a large amount of energy which can be used for various energy-requiring reactions. Since more energy is released than is used, some heat is always lost. ATP is a very convenient short-term store of chemical potential energy in living organisms. We can summarise the reactions involved in its formation and breakdown in the simple equation in Figure 5.4.

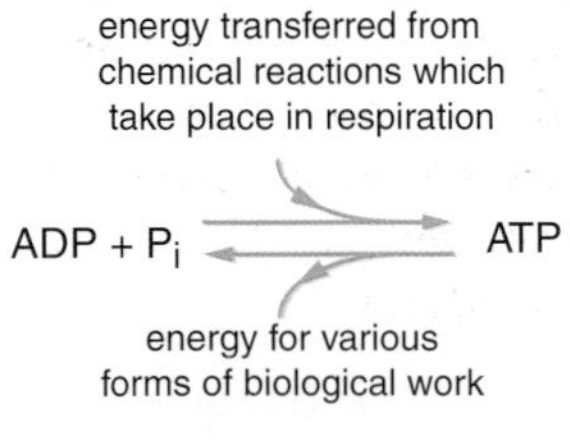

energy transferred from chemical reactions which take place in respiration

$$ADP + P_i \rightleftharpoons ATP$$

energy for various forms of biological work

**Figure 5.4**
The formation and breakdown of ATP allows energy released in reactions such as those which take place in respiration to be made available for processes such as the synthesis of large molecules and active transport

Why do we need to transfer chemical potential energy from glucose molecules to ATP molecules? ATP is more useful than glucose as an immediate source of energy because:

- the breakdown of ATP to ADP and phosphate is a single reaction. It makes energy instantly available. The breakdown of glucose to carbon dioxide and water is, as we shall see when we look at respiration in more detail, a complex process involving many stages. It would take much longer to make energy available from glucose than from ATP.
- the breakdown of a molecule of ATP releases a small amount of energy, ideal for fuelling the energy requiring reactions which take place in the body. The breakdown of a molecule of glucose would produce more energy than is required.

**Q** 2 **It is sometimes suggested that a molecule of ATP can move around a cell more easily than a molecule of glucose because it is smaller. True or false?**

## Photosynthesis

### What is it all about?

Humans must have a supply of complex organic molecules to provide the chemical potential energy they need to build up new cells and tissues. These organic molecules come from our food – sometimes from other animals but ultimately from plants. It is only by photosynthesis that light energy can be converted into chemical potential energy and simple inorganic molecules such as carbon dioxide and water can be built up into organic ones. Ultimately we all depend on photosynthesis.

Photosynthesis is a complex process involving a number of separate reactions. It is useful to get an idea of the overall process before we look at any detail. There are two basic steps (Figure 5.5). In the **light-dependent reactions**, two substances are produced. Light energy is captured by chlorophyll and is transferred to chemical potential energy in ATP. The second substance, which we will look at in more detail in the next part of this chapter, is reduced NADP. In order to produce these substances, a molecule of water is split and oxygen is given off as a waste product. In the **light-independent reactions** ATP and reduced NADP are used in the conversion of carbon dioxide to carbohydrate.

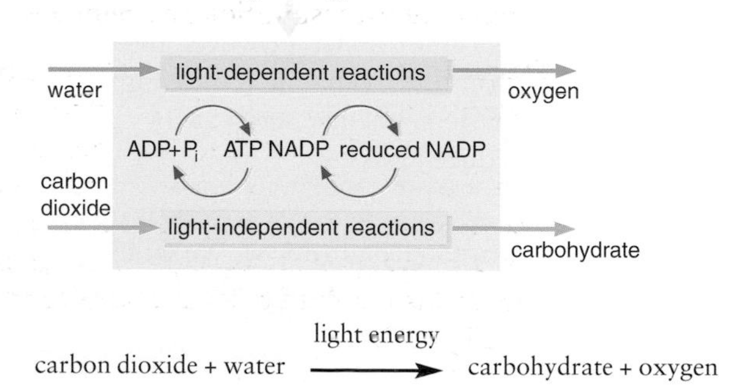


**Figure 5.5**
The main steps in photosynthesis. The substances entering and leaving the main box can be arranged to give you the basic equation for photosynthesis

$$\text{carbon dioxide} + \text{water} \xrightarrow{\text{light energy}} \text{carbohydrate} + \text{oxygen}$$

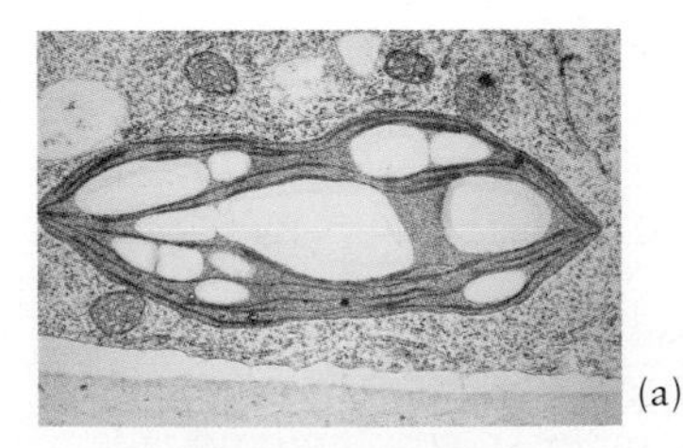


**Figure 5.6**
(a) A transmission electron micrograph of a chloroplast. This chloroplast is approximately 5 μm in diameter. (b) The chloroplast has a three-dimensional structure.

A simple model of the way in which the membranes inside a chloroplast are arranged. You will need some coins – 15 to 20 2p pieces will do fine and about 5 pieces of paper cut out like this:

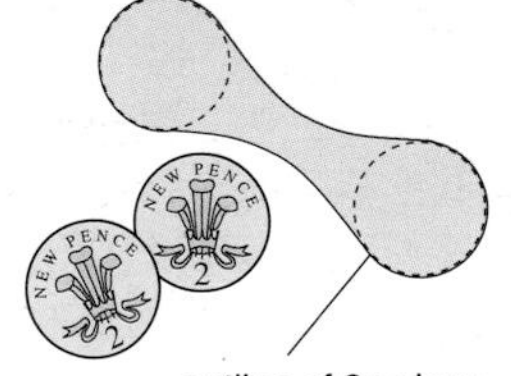


Now arrange your coins in stacks. Vary the number of coins in a stack. Link them to each other with the pieces of paper you have cut out, like this:

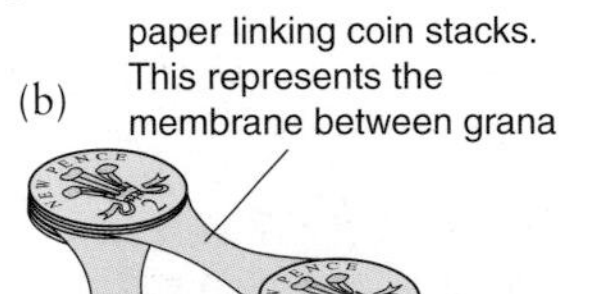


## Chloroplasts, chlorophyll and light energy

You should remember from your AS course that chloroplasts are the site of photosynthesis. Each chloroplast (Figure 5.6) is surrounded by two plasma membranes. A system of membranes is also found inside the organelle. These membranes form a series of flattened sacs called **thylakoids.** In some places the **thylakoids** are arranged in stacks called **grana.** More membranes join the grana to one another.

The membranes which form the grana provide a very large surface for chlorophyll and other light-absorbing pigments. These pigments form clusters called photosystems (Figure 5.7). Each photosystem contains a chlorophyll molecule called a **primary pigment molecule.** Situated around this are several hundred **accessory pigment molecules.** These accessory pigment molecules include various different forms of chlorophyll and other light-absorbing pigments. The whole photosystem acts as a light-harvesting system. Light energy is captured and passed from one accessory pigment molecule to another before it finally reaches the primary pigment molecule.

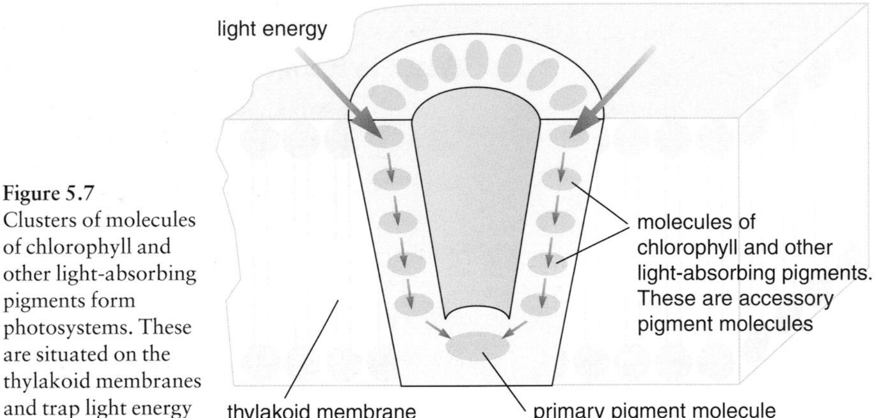


**Figure 5.7**
Clusters of molecules of chlorophyll and other light-absorbing pigments form photosystems. These are situated on the thylakoid membranes and trap light energy

## The light-dependent reactions

If we make a solution of chlorophyll and shine a bright light on it, it fluoresces. Instead of appearing green, it looks red in colour. What happens is that light energy raises the energy level of some of the electrons in the chlorophyll. These electrons leave the chlorophyll molecule. They lose most of this energy as light of a slightly different wavelength as they fall back into their place in the chlorophyll molecules. In a chloroplast, however, the electrons do not return to the chlorophyll molecule from which they came. They pass down a series of electron carriers, losing energy as they go. This energy is used to produce ATP and reduced NADP.

A whole series of reactions is involved which we can represent as a diagram (Figure 5.8). We will look at these reactions in a little more detail starting with light striking photosystem II.

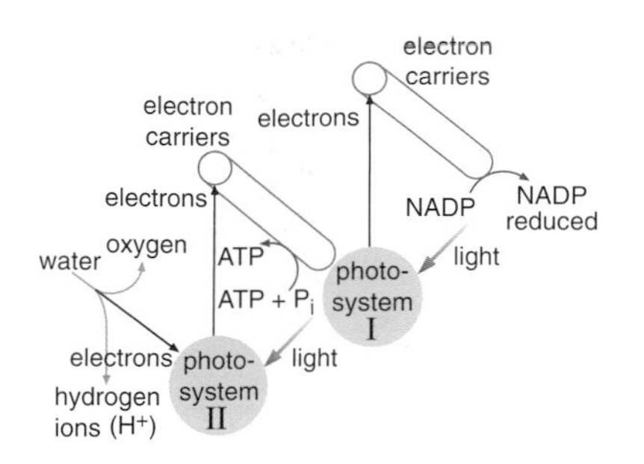


Figure 5.8
A summary of the light-dependent reactions of photosynthesis. The bullet points in the text should help you to understand this diagram

- Light strikes photosystem II. The energy levels of two of the electrons in the chlorophyll molecule which acts as the primary pigment molecule are raised. These electrons leave the chlorophyll molecule and pass to an electron acceptor. This results in the chlorophyll molecule now being positively charged.
- Photosystem II contains an enzyme which catalyses the breakdown of water in a process called **photolysis**.

$$H_2O \rightarrow 2H^+ + 2e^- + \tfrac{1}{2}O_2$$

The oxygen is given off as a waste product and the electrons replace those lost from photosystem II. We will look at what happens to the hydrogen ions later.

- The electrons which have been passed to the electron acceptor now pass down a series of electron carriers losing energy as they go. This energy is used to produce ATP.
- Light also strikes photosystem I and a similar sequence of events occurs to that which took place when light struck photosystem II. Electrons again leave the chlorophyll molecule which acts as the primary pigment molecule. They pass to an electron acceptor, leaving the chlorophyll molecule positively charged.
- The electrons which have passed down the electron carriers from photosystem II replace those lost from photosystem I.
- The electrons from photosystem I also pass down a series of electron carriers. They are used, together with the hydrogen ions from the photolysis of water to produce reduced NADP.

$$\text{NADP} + 2H^+ + 2e^- \rightarrow \text{reduced NADP}$$

**Q 3 What happens to the electrons which come from:**
**(a) the water molecule**
**(b) photosystem I**
**(c) photosystem II?**

## The light-independent reactions

In the light-independent reactions, ATP and reduced NADP are used to convert carbon dioxide into carbohydrate. These reactions get their name because they are independent of light. However, they cannot take place without products produced with light energy. Because of this, the light-independent reactions cannot continue for long in the dark.

**Q4 The light-independent reactions of photosynthesis are sometimes called the 'dark reactions'. Explain why this term is misleading.**

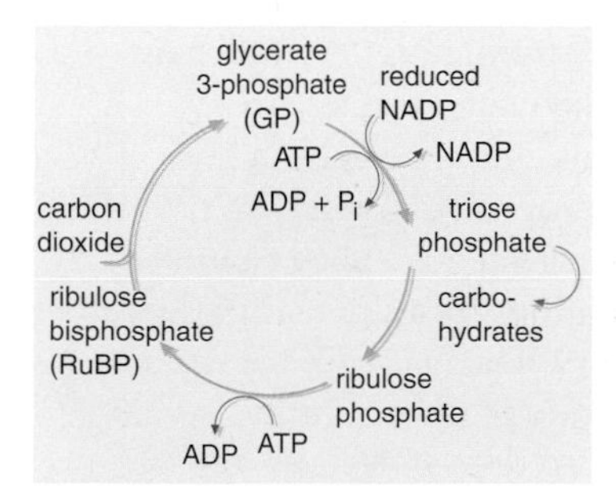


Figure 5.9
The light-independent reactions of photosynthesis form a cycle of reactions called the Calvin cycle. The diagram summarises the main steps in this cycle

In a cycle of reactions, carbon dioxide combines with a five-carbon sugar called **ribulose bisphosphate** (RuBP). Two molecules of a three-carbon compound, glycerate 3-phosphate (GP) are formed (Figure 5.9). The next step in the cycle is the conversion of glycerate 3-phosphate to triose phosphate. This reaction involves reduction and requires ATP and reduced NADP. Triose phosphate is a carbohydrate. Some of it is built

**Figure 5.10**
A potato crop like that shown here is able to convert around 2% of the energy in visible light into chemical potential energy in organic molecules

up into other carbohydrates, such as glucose and starch, and into amino acids and lipids. The rest of it is used to make more ribulose bisphosphate so the cycle can continue.

**Q 5 ATP supplies energy for the conversion of GP to triose phosphate. Use Figure 5.9 to describe one other function of ATP in the light-independent reactions of photosynthesis.**

## How efficient is photosynthesis?

A lot of light energy falls on the Earth's surface. Only a small part of this is actually use by plants and converted into chemical potential energy. Look at the potato crop shown in Figure 5.10. About half of all the visible-light energy falling on the plants in this crop is absorbed and converted into heat energy. Much of this heat energy evaporates water from the surface of the soil and from plant leaves during transpiration.

Of the remaining energy, just over 15% is reflected; and 32%, almost a third, is transmitted. It passes directly through the plant without striking any chlorophyll molecules on the way. As you can see, only a very small proportion of the incoming light energy ends up as chemical potential energy in substances produced by photosynthesis. For a crop grown in the UK, this is about 2%. We should also remember that not all of this energy goes into substances which form new tissues and cells. Quite a lot is used for respiration.

**Q 6 In mid-summer the amount of light energy falling on an acre of farmland in the United Kingdom is approximately 18 800 kJ $m^{-2}$ $day^{-1}$. Use information in the paragraph above to calculate the amount of energy converted to chemical potential energy in the tissues of the crop.**

## Extension box 1

## Any advance on 2%?

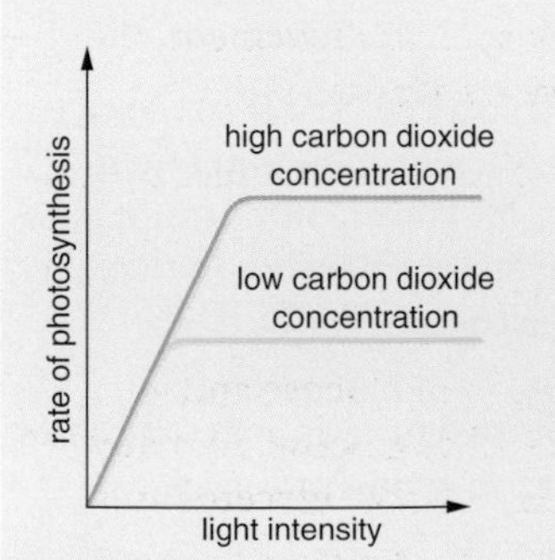


**Figure 5.11**
The effect of light intensity and carbon dioxide concentration on the rate of photosynthesis

You may not be very impressed with the fact that a typical British crop plant only converts about 2% of the light energy falling on it into chemical potential energy. The figure of 2% though, is very much an approximation. Some organisms, such as the lichens which are described in the assignment at the end of this chapter are even less efficient than this; others, like the tropical crop, sugar cane, are rather better. In order to understand why the figures vary, we need to look at the graph in Figure 5.11. It shows us that at low light intensities the rate of photosynthesis is limited by light intensity. The evidence for this is that as you increase the light intensity, the rate of photosynthesis increases. At high light intensities, however, it does not make any difference if you increase the light intensity further. The rate of photosynthesis does not get any faster. At high light intensities it is something else, such as carbon dioxide concentration, which is limiting.

A crop of sugar cane (Figure 5.12) is able to convert 7 to 8% of the light energy it receives into chemical potential energy. It is probably the most productive of all plants. Why is it so efficient?

Figure 5.12
Sugar cane is a crop plant and like all crop plants it has been selected over many generations for particular characteristics. One of these is the efficiency with which it converts light energy into chemical potential energy in useful substances

- Light intensity is generally much higher in the tropics than in temperate regions. During the daylight hours, it is unlikely to limit the rate of photosynthesis. Tropical temperatures are also higher than those in temperate regions. The enzyme-controlled reactions involved with synthesis of organic molecules and growth are therefore likely to be faster in a tropical crop.
- Sugar cane is a grass. It therefore has tall upright leaves. This arrangement is very efficient at intercepting light; much more efficient than the horizontal arrangement found in many other plants.
- It is a cultivated plant grown in plantations. It is frequently irrigated and it is provided with fertiliser, so a lack of water and mineral ions are unlikely to limit its growth. In addition, weeds are controlled so the light energy goes to produce chemical potential energy in the sugar cane plants, not in other plants.
- You may remember from your AS course that tropical plants such as maize and sugar cane have a particular type of photosynthesis that is based on a special biochemical pathway, the C4 pathway. This pathway involves a series of biochemical reactions in which carbon dioxide is concentrated. This makes it much more efficient in hot, sunny conditions such as are found where sugar cane grows.

## Respiration

### What's it all about?

Respiration takes place in all living cells. It is a biochemical process in which organic substances are used as fuel. They are broken down in a series of stages and the chemical potential energy they contain is transferred to another molecule, ATP. This provides an immediate source of energy for various forms of biological work such as movement, the synthesis of large organic molecules and active transport.

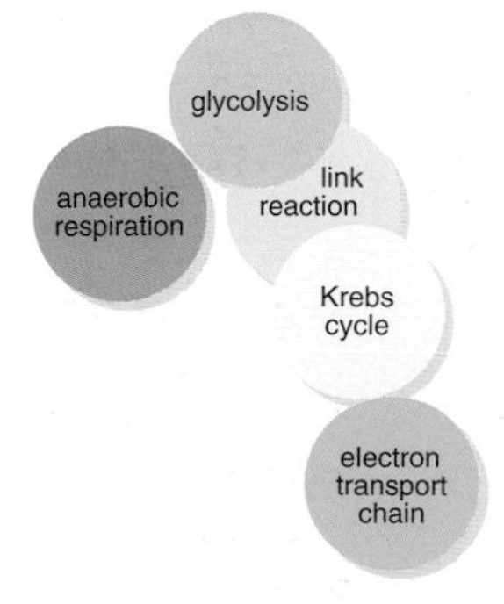


Figure 5.13
The breakdown of glucose in respiration. Aerobic respiration takes place in the presence of oxygen. Respiration can continue when there is no oxygen present. This is anaerobic respiration

You may be familiar with the simple equation that we sometimes use to summarise the process of respiration.

$$C_6H_{12}O_6 + 6O_2 \rightarrow 6CO_2 + 6H_2O + \text{Energy}$$

It is really an equation for the complete oxidation of glucose and is misleading in a number of ways. It shows the fuel as glucose. In many living cells, although the main fuel is glucose fatty acids, glycerol and amino acids are also respiratory substrates and can be used for respiration. The equation also shows that oxygen is required. Respiration can, however, take place anaerobically, that is without the presence of free oxygen. Finally the equation fails to show that respiration involves a number of reactions in which the respiratory substrate is broken down in a series of steps, releasing a small amount of energy each time.

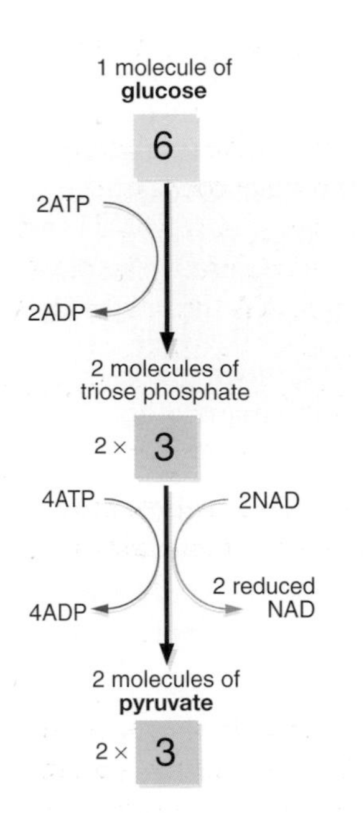


**Figure 5.14**
A summary of glycolysis. In this first step in respiration, a molecule of glucose is broken down to give two pyruvate groups. The boxes show the number of carbon atoms in some of the molecules and ions

We will look in more detail at glucose breakdown. It involves the four steps shown in Figure 5.13.

## Glycolysis

Glycolysis (Figure 5.14) is the first step in the biochemical pathway of respiration. A molecule of glucose is broken down into two pyruvate groups, each of which contains three carbon atoms.

Glucose contains a lot of chemical potential energy but it is not a very reactive substance. In order to release this energy, we need to use some energy from ATP. In the first stage of glycolysis, a molecule of glucose is converted into two molecules of triose phosphate. This requires two molecules of ATP.

Triose phosphate is then converted to pyruvate. This process produces four molecules of ATP, two for each triose phosphate. The conversion of triose phosphate to pyruvate is an oxidation and involves the transfer of hydrogen to a carrier molecule. This carrier molecule or **coenzyme** is known as NAD. Adding hydrogen reduces NAD to reduced NAD.

Summarising glycolysis, for each glucose molecule we get two pyruvate groups, a net gain of two molecules of ATP and two molecules of reduced NAD. We will look at reduced NAD again in the section on the electron transport chain and see how it is used to produce more ATP.

## The link reaction

Pyruvate still contains a lot of chemical potential energy. When oxygen is available, this energy can be released in a cycle of reactions known as the **Krebs cycle.** The link reaction (Figure 5.15) is a term used to describe the reaction linking glycolysis and the Krebs cycle.

Pyruvate combines with coenzyme A to produce acetylcoenzyme A. This involves the loss of a molecule of carbon dioxide. It is also an oxidation reaction and results in the formation of another molecule of reduced NAD.

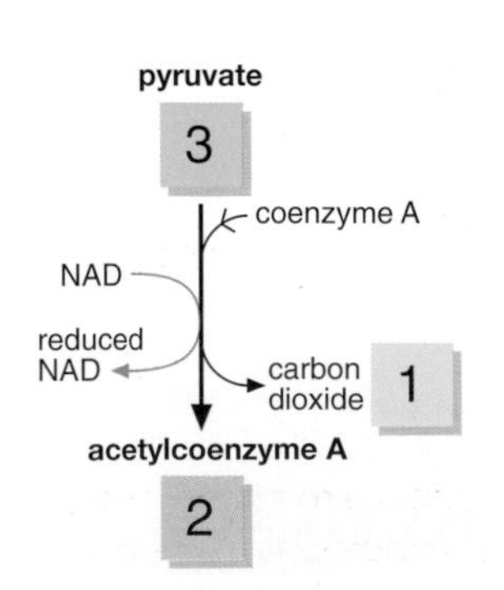


**Figure 5.15**
The link reaction. The boxes show the number of carbon atoms which are directly involved in the respiratory pathway. Acetylcoenzyme A contains more than two carbon atoms but only these two enter the next stage of respiration, the Krebs cycle

**Q 7 Complete the following equation which summarises the link reaction**

**pyruvate + coenzyme A + NAD →**

## The Krebs cycle

The Krebs cycle (Figure 5.16) is a cycle of reactions which involves a number of intermediate steps. We will only concern ourselves here with its main features.

- Acetylcoenzyme A produced in the link reaction is fed into the cycle. It combines with a 4-carbon compound (oxaloacetate) to produce a 6-carbon compound (citrate).
- In a series of reactions, the citrate is converted back to oxaloacetate. This involves the production of two molecules of carbon dioxide which is given off as a waste gas.

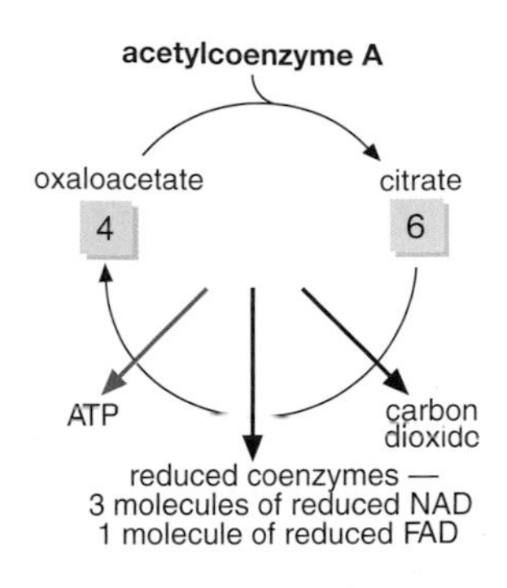


Figure 5.16
The Krebs cycle plays a very important part in respiration. It is the main source of the reduced coenzymes which are used to produce ATP in the electron transport chain

- For each complete turn of the Krebs cycle, one molecule of ATP is produced.
- The reactions which form the Krebs cycle again involve oxidation. The hydrogens which are given off are used to reduce coenzymes. Each turn of the cycle produces three molecules of reduced NAD and one molecule of another reduced coenzyme, reduced FAD. The most important function of the Krebs cycle in respiration is the production of reduced coenzymes. These are passed to the electron transport chain where the chemical potential energy these molecules contain is used to produce ATP.

**Q 8 One molecule of glucose produces two pyruvate groups. How many molecules of reduced NAD are produced for each molecule of glucose respired?**

## The electron transport chain

The reduced NAD and reduced FAD produced as a result of oxidation now pass to chains of molecules situated on the internal membranes in the mitochondria. These are the electron transport chains (Figure 5.17). The hydrogen is removed from the two reduced coenzymes and is split into hydrogen ions ($H^+$) and electrons.

The electrons pass from one molecule to the next along the electron transport chain. At each transfer, a small amount of energy is released. This is used to pump the hydrogen ions out through the inner mitochondrial membrane into the space between the inner and the outer membrane. When the hydrogen ions return into the matrix of the mitochondrion they release their potential energy. This is used to produce ATP. The last molecule in the chain is oxygen. Oxygen combines with hydrogen ions and electrons to produce water.

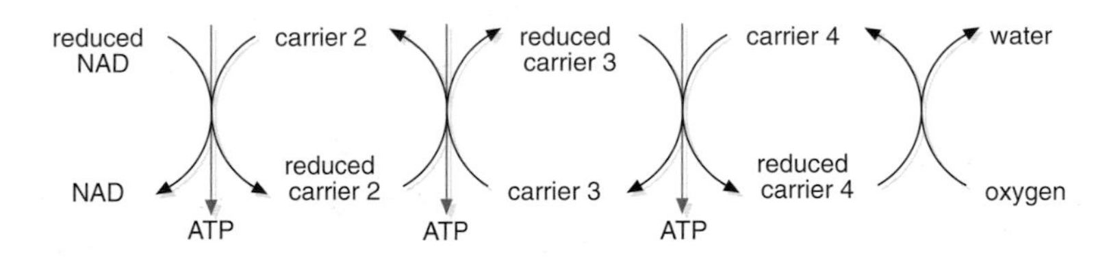


Figure 5.17
As electrons pass down the electron transport chain, energy is released. Some of this energy is lost as heat but a lot of it goes to produce ATP. Each molecule of NAD entering the chain produces three molecules of ATP. A molecule of reduced FAD produces two molecules of ATP

## Anaerobic respiration

Sometimes there is not enough oxygen for an organism to respire using the pathway we have just described. Under these conditions it has to produce ATP by respiring anaerobically. The only stage in the anaerobic pathway which produces ATP is glycolysis, so it is not as efficient as aerobic respiration. The chemical potential energy from a single molecule of glucose can be used to produce 38 molecules of ATP in aerobic respiration. It can only produce 2 molecules of ATP in anaerobic respiration.

Look back at Figure 5.14. You will see that during glycolysis NAD is reduced. Reduced NAD is normally converted back to NAD when its hydrogen is transferred in the electron transport chain. This will only happen when oxygen is present. Obviously, if all the NAD in a cell was converted to reduced NAD, the process of respiration would stop. In anaerobic respiration in animals, pyruvate is converted to lactate. In plants and microorganisms such as yeast, it is converted to ethanol and carbon dioxide. Both of these pathways, summarised in Figure 5.18, involve the conversion of reduced NAD to NAD and allow glycolysis to continue.

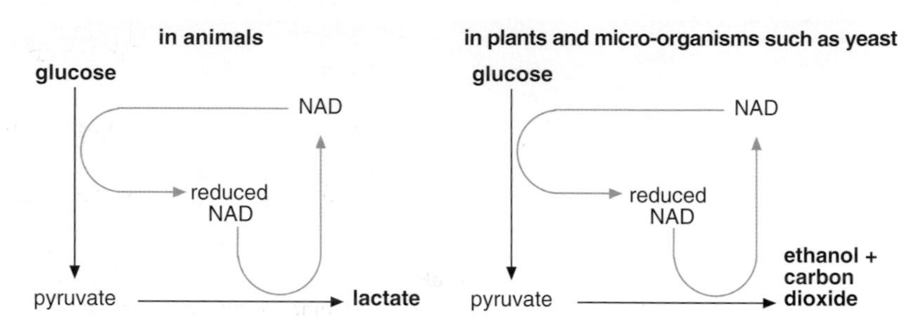


Figure 5.18
Anaerobic respiration allows organisms to produce ATP even in the absence of oxygen

## Using different respiratory substrates

A **respiratory substrate** is the organic substance which forms the starting point for respiration. So far we have only looked at the respiratory pathway involving glucose as a respiratory substrate but other substances can also be respired (Figure 5.19)

Figure 5.19
Insects such as this locust are able to fly long distances. When they start flying, their main respiratory substrate is glucose. After they have been in flight for 15 minutes or so, it is triglyceride

One way in which it is possible to get some idea of an organism's respiratory substrate is to calculate its **respiratory quotient** (RQ). RQ can be calculated by dividing the amount of carbon dioxide produced by the amount of oxygen consumed in the same time. This ratio is different for different respiratory substrates so it can be used to indicate the substance which is being respired. The calculation in Extension box 2 shows how we can calculate the expected respiratory quotient for a triglyceride.

### Extension box 2

### RQ calculations

The chemical equation below represents the respiration of a triglyceride

$$2C_{51}H_{98}O_6 + 145O_2 \rightarrow 102CO_2 + 98H_2O$$

The equation for respiratory quotient is:

$$RQ = \frac{\text{Amount of carbon dioxide produced}}{\text{Amount of oxygen consumed}}$$

Using the chemical equation we can enter the figures for the amounts of carbon dioxide produced and the amount of oxygen consumed.

$$RQ = \frac{102}{145}$$

$$= 0.70$$

We can carry out similar calculations to that shown in Extension box 2 to find the expected respiratory quotients for carbohydrates and proteins. These are shown in Table 5.1. We have to be careful how we interpret these figures. An RQ of 0.9, for example could mean that an organism is using protein as its respiratory substrate, but it could also be respiring a mixture of triglycerides and carbohydrates.

| RQ | Respiratory substrate |
|---|---|
| 0.7 | Triglycerides |
| 0.9 | Amino acids and proteins |
| 1.0 | Carbohydrates |

Table 5.1
The RQs for some important respiratory substrates

**Q 9 An organism is using protein as a respiratory substrate. In one minute it consumes 5.4 $cm^3$ of oxygen. How much carbon dioxide would you expect it to produce?**

**Q 10 How would you expect the RQ of the locust shown in Figure 5.19 to change over the first fifteen minutes of a flight?**

## Extension box 3

Figure 5.20
When the animal is hibernating, the body temperature of a dormouse falls from its normal level of about 35 °C to just above 0 °C. Brown fat plays a very important role in allowing a dormouse to regain its normal body temperature rapidly when it comes out of hibernation

## Brown fat

White fat plays an important part in insulating the body of a mammal, and helps to limit heat loss. Brown fat is very different. For one thing, brown fat cells are packed with mitochondria. This provides a clue as to its function. Another clue is provided by looking at those organisms which have large quantities of brown fat. It is found only in mammals, in particular, in very young mammals such as human babies, and in those which hibernate during the coldest part of the year (Figure 5.20).

Brown fat is the only tissue in the body of a mammal whose sole function is to produce heat. The internal membranes of its mitochondria have a special adaptation. In mitochondria in other tissues, hydrogen ions pass back from the space between the two mitochondrial membranes into the matrix through pores. These pores are associated with the enzyme ATP synthetase. The electron transport chain is therefore linked to ATP production and most of the potential energy in the hydrogen ions is transferred to ATP. In the mitochondria of brown fat, however, this mechanism can be over-ridden. The hydrogen ions can flow back through channels which are not associated with ATP synthetase so the energy they release will not be used to form ATP. It will all be released as heat.

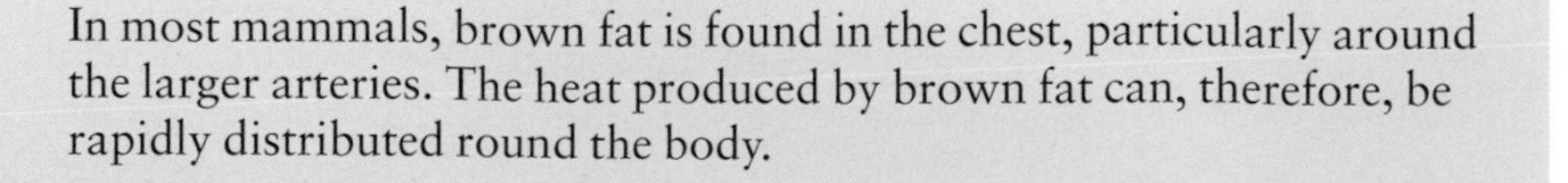
In most mammals, brown fat is found in the chest, particularly around the larger arteries. The heat produced by brown fat can, therefore, be rapidly distributed round the body.

## Energy transfer and ecosystems

### Food chains and food webs

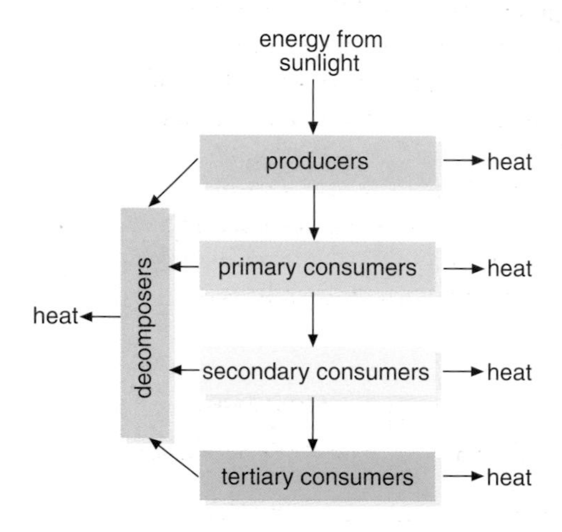


**Figure 5.21**
In this diagram, which shows the transfer of energy in an ecosystem, the boxes represent trophic levels. The arrows show the direction in which energy is transferred

In any ecosystem different organisms gain their nutrients in different ways. Green plants are the **producers**. They are able to produce organic molecules from carbon dioxide and water. They rely on photosynthesis to transfer energy from sunlight to chemical potential energy in organic molecules. The other organisms which make up the community rely either directly or indirectly on organic molecules produced by the plants. **Primary consumers** feed on plants; **secondary consumers** feed on primary consumers; and **tertiary consumers** feed on secondary consumers. Organisms which are not eaten eventually die. Another group of organisms, the **decomposers**, break dead tissues down and use the organic molecules which make up these tissues as a source of potential chemical energy. We can summarise all this as a simple diagram (Figure 5.21).

We often talk about food chains suggesting perhaps that we frequently encounter situations where animal **B** feeds only on plant **A**. In turn, animal **C** eats animal **B**, and animal **D** eats animal **C**. This hardly ever happens in nature. Food chains are generally linked with each other to form complex food webs. We will now look at a specific ecosystem – the patch of nettles we mentioned at the beginning of the previous chapter. Look at Figure 5.22 but bear in mind we have kept the food web that this represents as simple as possible. We have left out a number of different organisms and completely omitted the decomposers. Note also that:

- some organisms, such as the dark green bush cricket, feed at different trophic levels. They feed on nettle leaves and they also feed on various species of insect.
- some organisms feed on different foods when they are larvae and when they are adult. The larva of the peacock butterfly, for example, feeds on nettle leaves. The adult feeds on the nectar produced by various flowers.

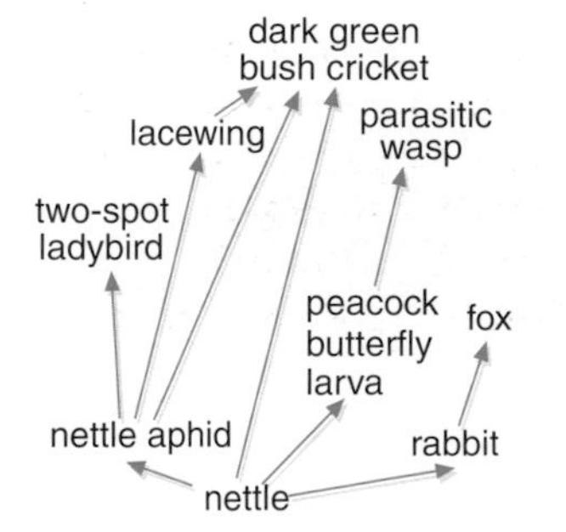


**Figure 5.22**
A simplified food web showing some of the organisms which feed on nettles

**Q 11 Sundew is an insectivorous plant. It has sticky hairs on its green leaves which trap and digest small insects such as aphids. What trophic levels does sundew occupy?**

### Ecological pyramids

Diagrams such as Figure 5.22 only provide qualitative information. They only show what food different organisms eat. Biologists are interested in quantitative information. This is where ecological pyramids are particularly useful.

### Pyramids of numbers

The simplest way to compare the different trophic levels in an ecosystem, is to count the organisms present. We can represent this information as a pyramid of numbers. Three typical pyramids of number are shown in Figure 5.23. If we take the pyramid representing the food chain:

nettle → rabbit → fox

you will see that it is obviously pyramid shaped. If you count all the nettle plants, there are a lot more than the total number of rabbits. In turn there will be more rabbits than there are foxes There are two important exceptions to this pyramid shape when we are talking about numbers of organisms. The pyramid will be upside-down or inverted if a lot of small animals are feeding off a large plant. This is the case in the second example where the primary consumers are aphids. A second example where we get an inverted pyramid is where an organism is supporting a large number of small parasites such as humans and head lice or, in our example in Figure 5.23, parasitic wasps feeding on the caterpillars of peacock butterflies.

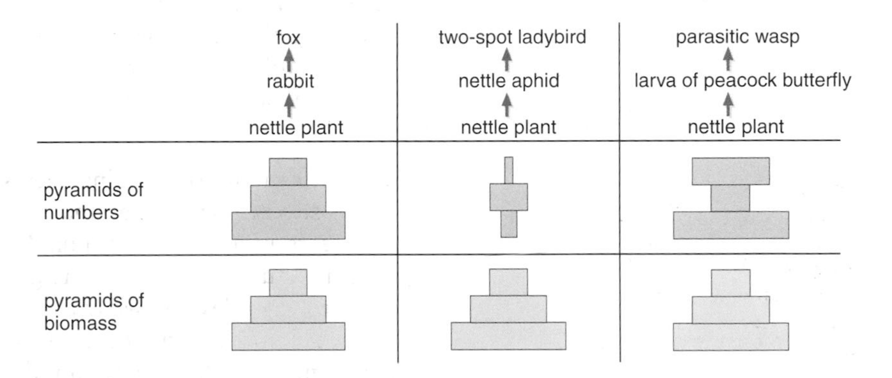


Figure 5.23
These examples of pyramids of numbers and pyramids of biomass all refer to food chains based on nettle plants

### Pyramids of biomass

Biomass is a measure of the total amount of living material present. If we shook all the aphids from a clump of nettle plants and weighed them we would have the biomass of the aphids on that clump of nettles. A pyramid of biomass allows us to compare the mass of organisms present at each trophic level of a particular food chain or food web. Pyramids of biomass overcome the problem of trying to compare numbers when the organisms are of very different size. Because of this, all the pyramids of biomass shown in the table have a similar pyramid shape. There is one important exception to this – organisms such as the phytoplankton shown in Figure 5.24 which multiply very rapidly. At any one time, their total biomass may be less than the biomass of the primary consumers which feed on them.

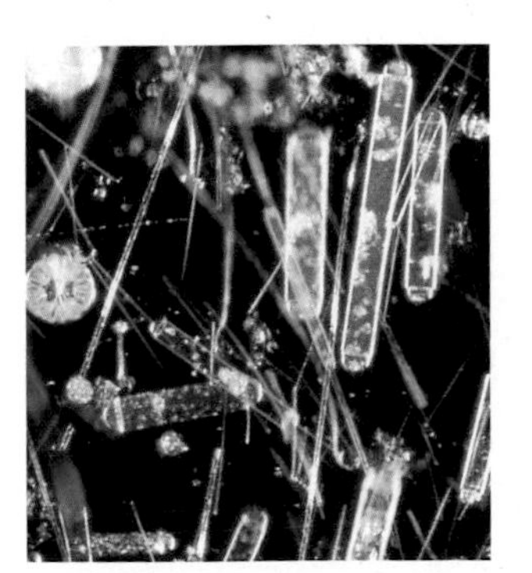

Figure 5.24
Phytoplankton are tiny single-celled algae. They are producers and float in the surface waters of lakes and oceans. Under favourable conditions they multiply very rapidly. Because of this, their total biomass may be less than that of the primary consumers. However, if we looked at their biomass over a period of say a month, then it would be greater than that of the primary consumers

## The transfer of energy

Earlier in this chapter we considered the efficiency with which energy is transferred to plants in photosynthesis. In this section we will look at the efficiency with which energy is transferred to the consumers in an ecosystem. Look at Figure 5.25. It shows the percentage of energy transferred between different trophic levels. We can look at this in another way. For every 10 000 kJ of light energy absorbed by the producers, 200 kJ will be incorporated into their tissues; 20 kJ will be transferred to the tissues of the primary consumers; and 2 kJ will end up as chemical potential energy in the tissues of secondary consumers. The numbers in this diagram are only generalisations. There are many factors which influence exactly how much energy is transferred at each stage. Some of these are considered in Extension box 4.

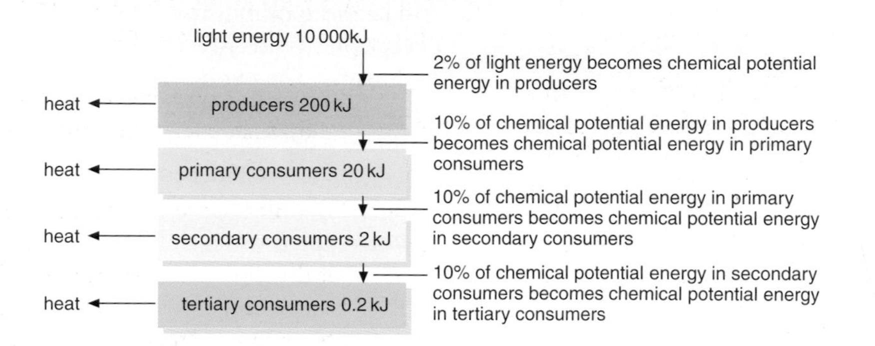


**Figure 5.25**
Only a small proportion of energy is transferred from one trophic level to another. The rest is used to make ATP in the process of respiration and will ultimately be lost as heat. Transferring only a small proportion limits the number of trophic levels in food webs. In practice it is rare for there to be more than five

**Q 12 What percentage of the chemical potential energy in the tissues of a primary consumer would you expect to be transferred to the tissues of a tertiary consumer?**

We can construct a third type of ecological pyramid which allows us to compare the amount of energy passing through the different levels in an ecosystem over a given period of time. This is called a pyramid of energy. Pyramids of energy are always pyramid-shaped. There are no exceptions.

## Extension box 4

## Energy transfer and efficiency

The figure we gave for the transfer of energy from producers to primary consumers and between primary consumers and secondary consumers was 10%. This figure, however, is only an average value. The actual proportion transferred varies from food chain to food chain. In this box we will look at some of the factors which help to determine the efficiency of energy transfer.

Mammals are homeothermic. This means that they are able to keep their body temperatures more or less constant at a value somewhere between about 35 °C and 40 °C, depending on the species. This high temperature is a result of heat produced during metabolism. Monitor lizards like that shown in Figure 5.26 are found in many parts of the tropics. They rely on their environment to maintain a high body temperature. More of the food they eat can therefore be converted into new cells and tissues and less chemical potential energy will be lost in maintaining body temperature.

**Figure 5.26**
Monitor lizard

Figure 5.27
Palm squirrel

Figure 5.28
Lion

You should remember from your AS course that the surface area to volume ratio of a small mammal such as the palm squirrel shown in Figure 5.27 is much bigger than that of a large mammal. Small mammals therefore lose a lot more heat relative to their size and cannot convert as much of the food they eat into new cells and tissues.

In general carnivores, like the lion in Figure 5.28, convert the food they eat into new tissue more efficiently than do herbivores. Herbivores feed on plant material and plants contain a lot of substances such as cellulose and lignin which are difficult to digest. A much higher proportion of the food that a herbivore eats passes through the gut and is lost as faeces.

**Q 13 Which do you think would be more efficient at transferring chemical potential energy in food into chemical potential energy in its tissues – a young animal or a mature animal? Explain your answer.**

So which animal converts food into new tissue most efficiently? The answer is probably trout living in fish farms. They can convert as much as 20% of the food they eat into new tissue. The information in this box will help to explain why. Trout are poikilotherms so do not need to use much of their food to maintain body temperature; they are carnivores and there is little waste from the food pellets on which they are fed; and they are harvested while they are still young and growing rapidly.

## Summary

- Photosynthesis uses energy from sunlight to synthesise organic molecules from carbon dioxide and water. The biochemical reactions involved can be divided into the light-dependent and light-independent reactions.
- Respiration is a biochemical process in which organic substances, called respiratory substrates, are broken down in a series of stages and the chemical potential energy they contain is transferred to ATP.
- ATP is an immediate source of energy for various forms of biological work such as movement, the synthesis of large organic molecules and active transport.
- Energy is transferred from one trophic level to another through food chains and food webs in a community.

**Figure 5.29**
A selection of different lichens. Each lichen consists of a fungus and an alga. The lichen fungi are unable to live without their algal partners. Many of the algae, however, can be found free-living

# Assignment

## Sharing a life

A lichen, such as one of those shown in Figure 5.29, is made up of two entirely different organisms. It is a partnership of an alga and a fungus. When a section through a lichen is examined with a microscope, green algal cells are clearly seen among thread-like fungal hyphae.

Many lichens are able to withstand extreme conditions. Some species colonise and grow on bare rock and must be able to survive for long periods without water. Others live in the Arctic or in deserts. Despite this, they are extremely sensitive to pollution. In this assignment we will look at the way in which lichens are affected by sulphur dioxide in the atmosphere. In order to answer the questions, you will need to draw on your knowledge of photosynthesis as well as your understanding of ecology.

Read the following passage about the relationship between the algae and the fungus which make up a lichen.

> Nutrition in lichens has been investigated with carbon dioxide
> containing the radioactive isotope of carbon, $^{14}C$. The lichen is
> incubated with radioactive carbon dioxide. The total amount of
> carbon dioxide fixed by the algal cells and the total amount
> 5 transferred to the fungus are then calculated. The results show that
> the total amount transferred is between 60 and 80%. These results
> may be misleading, however. First, it is very difficult to separate the
> algal and fungal partners. Second, some of the photosynthate
> transferred to the fungus is used as a respiratory substrate.
>
> 10 The substance transferred from the algae varies from one lichen to
> another. Once in the fungus, however, it is used for respiration or to
> form other substances. A lot of the photosynthate taken up by fungi
> is converted to substances called polyols. Polyols are soluble and
> have a marked effect on the water potential of the fungal hyphae.
> 15 This enables them to absorb water rapidly.

1 Explain the meaning of the following terms used in the passage:

(a) fixed (line 4)
(b) photosynthate (line 8)

*(2 marks)*

2 Explain how each of the following contributes to the results of these investigations being unreliable:

(a) the difficulty of separating the algal and fungal partners (lines 7–8);

*(2 marks)*

(b) lichens using photosynthate as a respiratory substrate (lines 9–10).

*(2 marks)*

3 Lichens have no means of controlling water loss. They dry out readily and remain inert until they are wetted again. Explain, in terms of water potential, how the presence of polyols enable lichens to absorb water rapidly when they are wetted.

*(2 marks)*

Lichens are very sensitive to the presence of sulphur dioxide. The main effect of this pollutant is on photosynthesis in the algal partner. Increasing the concentration of sulphur dioxide produces a series of increasingly severe effects.

These are:

- a temporary reduction in the rate of photosynthesis which recovers as soon as sulphur dioxide concentration is reduced.
- a permanent reduction in the rate of photosynthesis. This is probably the result of damage to chloroplast membranes inhibiting ATP production.
- a permanent reduction in the rate of photosynthesis related to the breakdown of chlorophyll.

4 (a) Explain why damage to chloroplast membranes may inhibit ATP production.

*(2 marks)*

(b) Explain how a lack of ATP will affect the light-independent reaction of photosynthesis.

*(3 marks)*

5 Describe briefly how you could use chromatography to investigate whether chlorophyll had been broken down in lichens which had been exposed to very high concentrations of sulphur dioxide.

*(4 marks)*

It is thought that lichens are very susceptible to sulphur dioxide pollution because they have a low proportion of chlorophyll-containing tissue. Experimental work has shown that different species of lichen are affected to different extents by sulphur dioxide. Table 5.2 shows the effect of different concentrations of sulphur dioxide on the rate of uptake of carbon dioxide in three species of lichen:

| Sulphur dioxide concentration/ arbitrary units | Rate of uptake of carbon dioxide by lichen/arbitrary units | | |
|---|---|---|---|
| | *Usnea subfloridana* | *Hypogymnia physodes* | *Lecanora conizaeiodes* |
| 0 | 300 | 330 | 280 |
| 0.1 | 320 | 440 | |
| 0.2 | 210 | 340 | 320 |
| 0.3 | 30 | 390 | |
| 0.4 | 0 | 0 | 280 |
| 0.6 | | | 40 |
| 0.8 | | | 10 |

Table 5.2
Effect of different concentrations of sulphur dioxide on the rate of uptake of carbon dioxide in three species of lichen

6 Plot these data as a suitable graph on a single pair of axes.

*(4 marks)*

7 Suggest how sulphur dioxide may affect the distribution of these three species of lichen.

*(2 marks)*

## Examination questions

1 The diagram shows the flow of energy through a marine ecosystem. The units are kJ $m^{-2}$ year $^{-1}$.

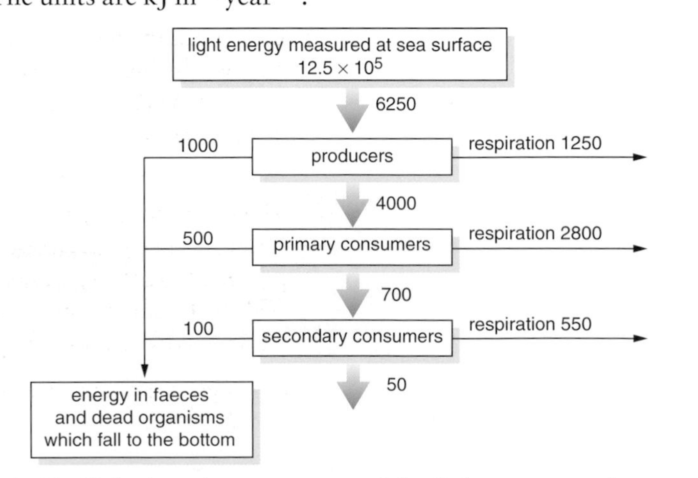


(a) (i) Calculate the percentage of the light energy at the sea surface which is converted into chemical potential energy in the producers. Show your working.

*(2 marks)*

(ii) The percentage of the light energy at the sea surface which is converted into chemical potential energy in the producers is very small. Give **two** reasons for this.

*(2 marks)*

(b) Use the information in the diagram to explain why marine ecosystems such as this rarely have more than five trophic levels.

*(2 marks)*

(c) What happens to the energy in faeces and dead organisms which fall to the bottom of the sea?

*(2 marks)*

(d) Light energy is important in the light-dependent reaction of photosynthesis. The energy changes which take place in the light-dependent reaction are shown in the diagram.

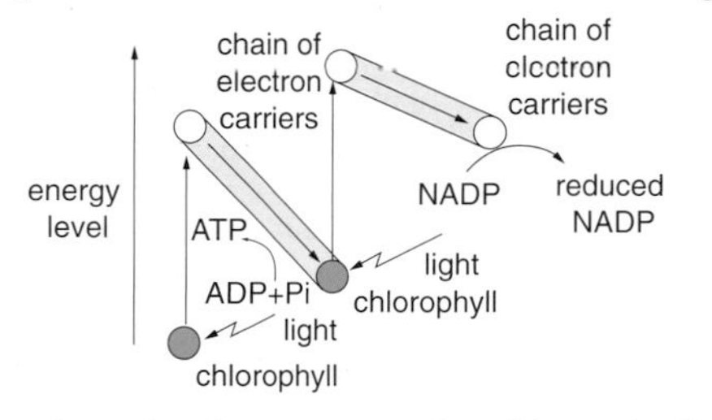


(i) Describe what happens to the chlorophyll when it is struck by light.

*(2 marks)*

(ii) The weedkiller DCMU blocks the flow of electrons along the chains of electron carriers. Describe and explain the effect this will have on the production of triose phosphate in the light -independent reaction.

*(3 marks)*

(e) Living organisms release energy from organic molecules such as glucose during respiration. Much of this energy is used to produce ATP. Explain why ATP is better than glucose as an immediate energy source for cell metabolism.

*(2 marks)*

(f) The production of ATP is said to be coupled to the transport of electrons along the carrier chain. Normally, electrons are only passed along the electron chain if ADP is being converted to ATP at the same time. When the amount of ADP in a cell is low, electrons do not flow from the reduced coenzyme to oxygen.

(i) Suggest how the rate of respiration is linked to the needs of the cell.

*(3 marks)*

(ii) DNP is a substance which allows electron transport to take place without the production of ATP. When DNP is given to rats, their body temperatures rise. Explain why.

*(2 marks)*

2 A suspension of pure chloroplasts was prepared from plant cells by cell fractionation and centrifugation. Some of these chloroplasts were broken up and used to produce a suspension of chloroplast membranes. The table shows the results of a number of investigations with the intact chloroplasts and with the chloroplast membranes.

| | Intact chloroplasts | | Chloroplast membranes | |
|---|---|---|---|---|
| | light | dark | light | dark |
| Volume of oxygen produced/$mm^3$ $hour^{-1}$ | 480 | 0 | 400 | 0 |
| Amount of carbon dioxide taken up/arbitrary units | 52 000 | 238 | 40 | 37 |
| Amount of inorganic phosphate taken up/ arbitrary units | 4 250 | 950 | 1 100 | 300 |

(a) (i) Describe how you could obtain a preparation of chloroplasts which was not contaminated with cell nuclei.

*(2 marks)*

(b) (i) Describe what happens to the inorganic phosphate taken up by the chloroplasts.

*(1 mark)*

(ii) Explain the evidence from this table that the light-independent reaction takes place in the stroma of the chloroplasts.

*(2 marks)*

3 (a) The inner membrane of a mitochondrion is folded to form cristae. Suggest how this folding is an adaptation to the function of a mitochondrion.

*(2 marks)*

(b) The equation represents oxidation of a lipid.

$$C_{57}H_{104}O_6 + 80O_2 \rightarrow 57CO_2 + 52H_2O + \text{Energy}$$

Use the equation to calculate the Respiratory Quotient (RQ) of this lipid. Show your working.

*(1 mark)*

(c) Suggest an explanation for each of the following:

(i) the RQ of germinating maize grains is 1;

*(1 mark)*

(ii) the RQ of a normal healthy person varies over a 24-hour period.

*(1 mark)*

# 6 Decomposition and Recycling

Figure 6.1
A reed bed

All too often summer holidays in Cornwall are spoilt by rain. And when it rains, tourists need somewhere to go. One popular place is the Elizabethan manor house at Trerice, now owned by the National Trust. Several hundred people may visit the house and its teashop in a single day and most of them will use the toilets and washrooms. Unfortunately neither the house nor the teashop is connected to mains drainage.

Waste runs first of all to a septic tank. Here, the solid material separates out to form sediment at the bottom of the tank. The liquid is then pumped into a reed bed (Figure 6.1) where it is purified further. It flows through the sandy soil in which the reeds are growing. There are huge numbers of microorganisms around the roots of the reeds and these act on the suspended organic substances in the incoming water. They break these substances down and release soluble nitrates and phosphates as a result. Some nitrate and phosphate is contained in the water leaving the reedbed but a lot of it is taken up by the plants. In their cells it is converted into organic substances such as proteins and lipids. The water that flows out of the reed bed is a lot cleaner than that flowing in. There is much less suspended organic material and a lower concentration of ions such as nitrates and phosphates.

Reeds are tall grasses which grow at the edge of lakes and rivers where they may form large swampy areas. Not only are the resulting 'wetlands' important habitats for wildlife, but they can also be used to help deal with certain types of pollution.

In chapter 5 we saw how energy was transferred from one trophic level in a food web to the next. Transfer of energy is inefficient and it is all eventually lost as heat. Energy is described as a renewable resource. Ions like the nitrates and phosphates in the waste from Trerice are, however, non-renewable. There is only a certain amount of them and, like all nutrients, they are recycled. In this chapter, we shall look at the passage of carbon and nitrogen through the various trophic levels in an ecosystem and at the role of microorganisms in converting organic molecules into inorganic substances which are made available to plants.

## Nutrient cycles

Look at Table 6.1 overleaf. You will see that living organisms such as animals and plants require different chemical elements. Plants take these elements up either as carbon dioxide in the case of carbon, or as ions from the soil. Inside the plant, they are involved in various chemical reactions and are eventually converted into the organic substances which form plant cells and tissues. Consumers obtain their supply of these elements from plants or from other animals which feed on plants.

An insect, for example, digests the complex organic molecules in its food and absorbs the products through its gut wall. In this way, elements are passed from organism to organism along the food chains which make up a food web (Figure 6.2).

Those organisms or parts of organisms which are not eaten as food will eventually die and decompose. Microorganisms such as fungi and bacteria break down the complex organic molecules which form the dead material. They absorb some of the products but the rest are released as inorganic substances which can be taken up by plants.

So, here we have the basic nutrient cycle. Elements are taken up by plants and passed from organism to organism in the various trophic levels. Death and decomposition result in microorganisms making these elements available to plants again.

| Chemical element | Plants | Animals |
|---|---|---|
| Carbon | The essential component of all organic substances. Proteins, carbohydrates, lipids and nucleic acids are all built up from carbon-containing molecules.<br><br>Many other substances such as ATP and chlorophyll also contain carbon. | |
| Nitrogen | An essential component of all amino acids, and of the nucleotides which make up DNA and RNA. | |
| Iron | Contained in the cytochromes which form part of the electron transport chain.<br><br>Needed for enzymes such as catalase to work.<br><br>Essential for the synthesis of chlorophyll. | Contained in the cytochromes which form part of the electron transport chain.<br><br>Needed for enzymes such as catalase to work.<br><br>Part of haemoglobin. |
| Iodine | | Contained in thyroxine. This is a hormone produced by the thyroid gland. |
| Molybdenum | Needed for the enzyme nitrate reductase to work. This enzyme reduces nitrates during the synthesis of amino acids. | |

Table 6.1
Some examples of chemical elements required by plants and animals

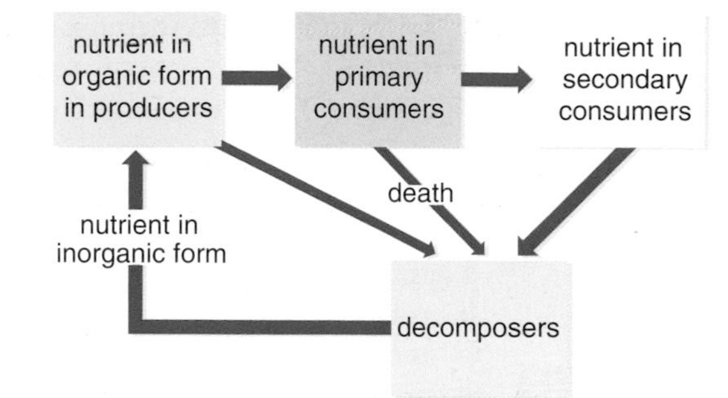


Figure 6.2
The basic nutrient cycle. All elements, whether they are carbon or nitrogen, iron or molybdenum, are cycled in this basic way. It is only the details of the process that differ in different cycles

**Q 1 Using the basic cycle shown in Figure 6.2 as a guide, draw a simple diagram representing an iron cycle.**

## The carbon cycle

Plants take up carbon dioxide from the atmosphere. It diffuses into the photosynthesising cells in the leaves where it is converted to sugars and other organic compounds. Primary consumers feed on plants, and carbon-containing compounds in their food are digested. Smaller, soluble molecules are produced and these are absorbed. They are built up into the carbohydrates and other large molecules which make up the tissues of the primary consumers. The same thing happens when a primary consumer is eaten by a secondary consumer. In this way, just as in other nutrient cycles, the element is passed from one trophic level to the next through the food web.

The next step involves the decomposers. Many soil bacteria and fungi are saprobionts. They secrete enzymes which break down the complex organic substances which make up dead plants and animals. The smaller molecules which result are absorbed by the microorganisms and some of them are used in respiration. Respiration releases carbon dioxide which can be used by plants for photosynthesis. The cycle (Figure 6.3) is complete.

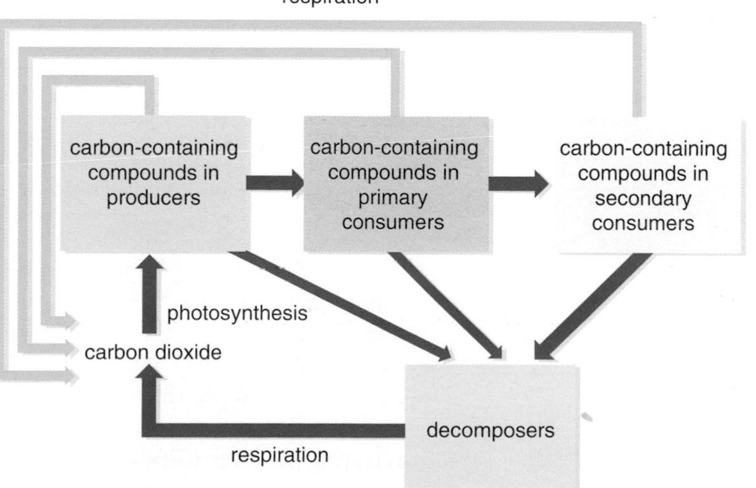


Figure 6.3
The carbon cycle. Note how similar this cycle is to the basic nutrient cycle shown in Figure 6.2

However, there is one complicating factor which we do not see in other cycles. We do not have to wait for an animal or plant to die before any of the carbon stored in its tissues is released. The producers and consumers all respire and release carbon dioxide into the atmosphere.

**Q** 2 **In which of the following ways does a maize plant take up carbon?**

**A as carbon dioxide in the air**
**B as hydrogencarbonate ions in the soil**
**C as carbonate ions in the soil**
**D as organic molecules from dead plants and animals**

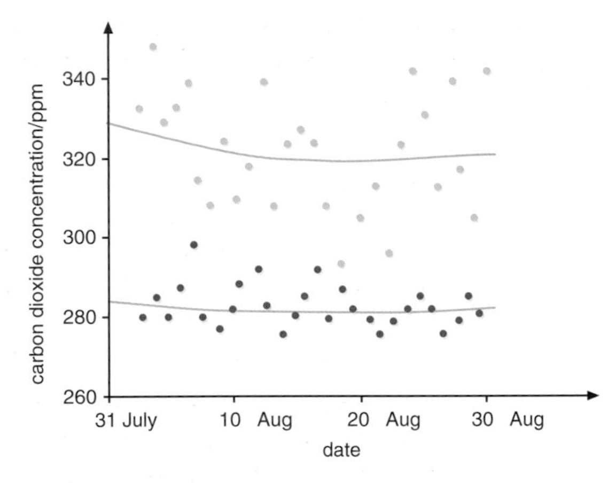


**Figure 6.4**
Graph showing how the concentration of carbon dioxide in a sugar beet crop varies. The blue circles (●) show the mean values during the daylight (08–16h) and the red circles (●) show mean values at night time (20–04h). The lines show the overall trends

## Respiration, photosynthesis and carbon dioxide

The concentration of carbon dioxide in the atmosphere is not constant. It changes according to the balance between photosynthesis and respiration. Look at the graph in Figure 6.4. It shows how the concentration of carbon dioxide varied in a sugar beet crop in which the plants had reached their full height. All the measurements were made at crop height, 50 cm above the ground.

This graph shows us that the concentration of carbon dioxide at crop height is greater at night time than during daylight. There is a simple explanation for this difference. At night, plants are unable to photosynthesise but they are still respiring. During the daytime, both photosynthesis and respiration take place. Since the rate of photosynthesis is usually greater than the rate of respiration, more carbon dioxide will be taken up than is released and its concentration falls.

If we look at the graph more carefully, there is another feature that we might notice. There is a lot of variation in the mean carbon dioxide concentration at night. The points are scattered quite widely. Much of this variation is due to the wind. On still nights, carbon dioxide accumulates round the plants and is not mixed with the surrounding air, so values are higher. On windy nights, it is mixed with air from higher up resulting in lower values.

## Human activities and the carbon cycle

So far we have only looked at the biological aspects of the carbon cycle. We have ignored the fact that large amounts of carbon are locked up in coal, oil and limestone formed millions of years ago. The ways in which human activities influence the carbon cycle are nearly all involved with releasing this 'stored' carbon as carbon dioxide. These include the burning of fossil fuels such as coal and oil (Figure 6.5) and the weathering of limestone (Figure 6.6).

Figure 6.5
These vapour trails were left by aircraft. It is not only our increasing reliance on motor vehicles which increases the rate at which we burn fossil fuels but also increased consumption by aircraft and power stations

Figure 6.6
The limestone from which these houses were built is a rock formed from the skeletons of tiny marine organisms. It consists mainly of calcium carbonate. Using limestone for building increases the surface area exposed to the weathering processes which produce carbon dioxide

We will look at one other human activity in a little more detail, the clearing of forests. In a rain forest such as that shown in Figure 6.7, there is a balance between the rate of photosynthesis and the overall rate of respiration. The carbon dioxide removed from the atmosphere in photosynthesis is roughly equal to the amount put back in by the respiration of all the plants, animals and microorganisms that make up the ecosystem.

Figure 6.7
An area of mature rainforest like this has a neutral effect on the concentration of carbon dioxide in the atmosphere. The amount of carbon dioxide removed in photosynthesis is more or less the same as the amount replaced by respiration

Forest may have been growing in a particular area for hundreds of years and, in that time, a lot of carbon has been locked up in the tissues of the trees and other plants as compounds like cellulose and lignin. If we want to clear the land for agriculture, we need to get rid of the forest and the easiest way to do this is by burning. Burning releases the carbon locked in the tissues into the atmosphere as carbon dioxide.

The result of all of these human activities is that there has been a steady rise in the atmospheric concentration of carbon dioxide. Figure 6.8 provides one piece of evidence for this. It shows a graph of the concentration of carbon dioxide in the atmosphere at the Mauna Loa observatory in Hawaii. Records have been kept here for over 50 years but, to see the detail, we will look at a ten-year period.

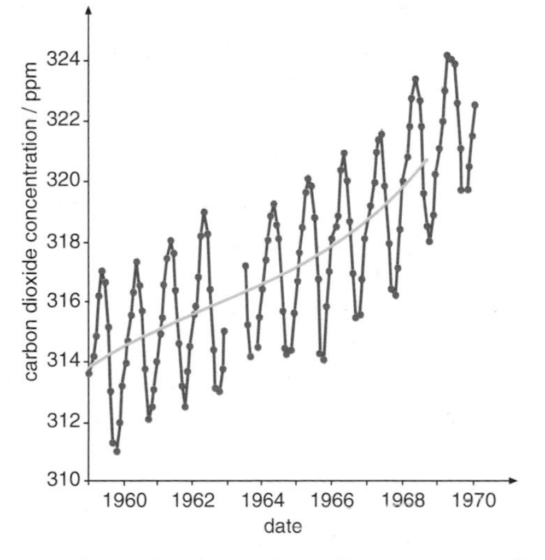


Figure 6.8
The individual points on this graph show the mean monthly carbon dioxide concentrations at the Mauna Loa observatory in Hawaii. The curve shows the general trend in the mean annual carbon dioxide concentration

In order to interpret these figures we need to be aware that the gases in the atmosphere are continually being mixed by wind and air currents. This means that the graph reflects the carbon dioxide concentration of the atmosphere over the whole of the northern hemisphere, not just over Hawaii. You will see that there is an annual fluctuation in carbon dioxide concentration. The peaks are in the winter months. There are two main reasons for this. First, oil consumption is much higher at this time of the year and, second, the overall rate of photosynthesis is lower during the winter.

The second aspect of this graph is that there is an overall trend towards an increase in the mean annual concentration of carbon dioxide in the atmosphere. This is a cause of considerable concern as increase in carbon dioxide has been linked to global warming and its consequences.

**Q 3 Would you expect the trend in atmospheric carbon dioxide concentration between 1990 and 2000 to have been the same, greater or less than in the period shown in the graph? Give a reason for your answer.**

## Extension box 1

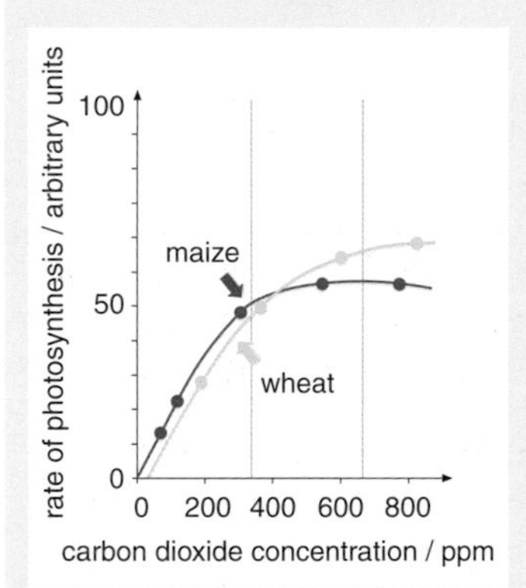


**Figure 6.9**
The effect of carbon dioxide concentration on the rate of photosynthesis in wheat and maize. Compare what happens to these two crops if we double the atmospheric carbon dioxide concentration from 330 ppm to 660 ppm

## Increasing carbon dioxide – a good thing?

Most people link an increase in the carbon dioxide concentration in the atmosphere with global warming, rising sea levels and climatic change. But could an increase in carbon dioxide concentration have a beneficial effect on the growth of crop plants? This is not a simple question to answer. An increase in carbon dioxide concentration is likely to be accompanied by a series of other changes and these separate factors will interact with each other in a way that is difficult to predict.

Plants depend on carbon dioxide for photosynthesis. If you followed the AS Biology course, you will probably remember that in bright conditions it is the concentration of carbon dioxide that usually limits the rate of photosynthesis. Increasing carbon dioxide concentration should therefore increase photosynthetic rate. Look at the graph in Figure 6.9. It shows the effect of an increase in carbon dioxide concentration on photosynthesis in two important cereal crops, wheat and maize.

Wheat is a $C_3$ plant. This means that it uses the biochemical pathway we described in Chapter 5 for photosynthesis. Many of the important food crops grown in temperate regions, crops such as rice and soya beans, are also $C_3$ plants. They respond to an increase in carbon dioxide concentration in a similar way to wheat. An increase in the concentration of carbon dioxide in the atmosphere ought then to result in a significant increase in food production.

Some plants, however, photosynthesise using a slightly different biochemical pathway, the $C_4$ pathway. These plants include important tropical plants such as maize and sorghum. Increasing carbon dioxide concentration only brings about a small increase in their rate of photosynthesis. In practice, this slight increase may not be enough to offset decreases brought about by other factors associated with increased carbon dioxide concentration and global warming.

So far the story is quite simple. Increased carbon dioxide concentration may result in a significant increase in yield in $C_3$ plants; it is likely to produce a much smaller one in $C_4$ plants. There are, however, complicating factors. Only a certain amount of the carbohydrates and other substances produced by photosynthesis go into the part of the plant that is eaten. Look at the figures in Table 6.2. All these data refer to $C_3$ plants and predict the effect of doubling the present concentrations of carbon dioxide.

| Crop | Percentage increase in biomass | Percentage increase in marketable yield |
|---|---|---|
| Cotton | 124 | 104 |
| Tomato | 40 | 21 |
| Rice | 20 | 36 |
| Cabbage | 37 | 19 |
| Weeds | 34 | not applicable |

**Table 6.2**
Effects of doubling the present concentrations of carbon dioxide on selected $C_3$ plants

In some of these crops, such as tomatoes, you can see that the increase in the yield of the part of the plant that we eat is nowhere near as high as the increase in total plant growth. It is worth noting as well that an increase in carbon dioxide concentration will increase weed growth.

What about crop quality? More does not necessarily mean better and there is evidence to suggest that plants grown under conditions where the carbon dioxide concentration is higher have a relatively lower proportion of nitrogen, and hence protein, in their tissues.

So, in terms of crop yield, is increased carbon dioxide concentration a good thing? We simply do not have the evidence to answer this complex question.

**Q** 4 **Give two reasons why an increase in carbon dioxide concentration may lead to a farmer gaining a smaller profit from a crop.**

## The nitrogen cycle

The general features of the nitrogen cycle differ very little from the basic nutrient cycle shown in Figure 6.2. Plants take up nitrate ions from the soil. They are absorbed into the roots by active transport and used to produce proteins and other nitrogen-containing compounds in plant cells. Primary consumers feed on plants and the proteins in their food are digested to form amino acids. The amino acids are absorbed from the gut and built up into the proteins which make up the tissues of the primary consumers. The same thing happens when a primary consumer is eaten by a secondary consumer. In this way nitrogen is passed from one trophic level to the next through the food web.

When organisms die, the nitrogen-containing organic substances which they contain are digested by saprobiotic bacteria and ammonia is released. Nitrogen from consumers is also made available to saprobiotic bacteria through excretory products such as urea. Another group of bacteria, the nitrifying bacteria, then convert ammonia to nitrites and nitrates. This complete nitrogen cycle is summarised in Figure 6.10.

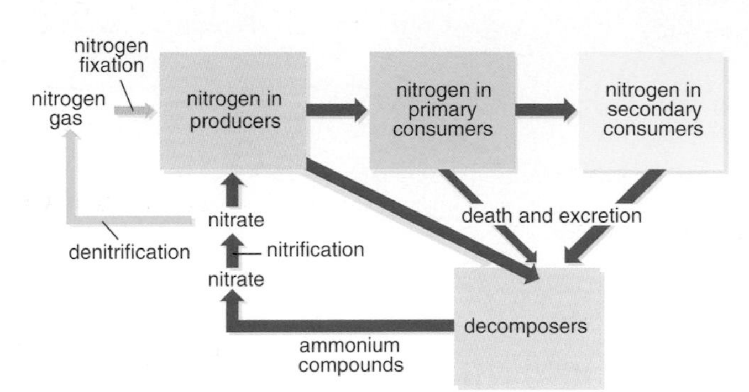


**Figure 6.10**
Note how similar this basic nitrogen cycle is to the basic nutrient cycle shown in Figure 6.2 and to the carbon cycle shown in Figure 6.3

Under anaerobic conditions, such as those that are found when soil becomes waterlogged (Figure 6.11), **denitrifying bacteria** are found in large numbers. These bacteria are able to use nitrate instead of oxygen as an electron acceptor in the respiratory pathway. The reaction involves reduction of nitrate to nitrogen gas. This nitrogen escapes into the atmosphere so it is no longer available to plants.

**Figure 6.11**
When fields are flooded, the spaces between soil particles are filled with water. No oxygen is available for soil microorganisms. Under these waterlogged conditions, denitrifying bacteria are common

**Q** 5 **Give two differences between nitrification and denitrification.**

Nitrogen gas can be made available to plants again by **nitrogen fixation**. This term refers to a number of processes which involve the conversion of nitrogen gas to nitrogen-containing substances. These processes all require a considerable amount of energy. A nitrogen molecule consists of two atoms, and the molecules have to be split into these atoms before nitrogen fixation can take place. Energy is required for this. Methods of fixing nitrogen include:

### Lightning

During a thunderstorm, the electrical energy in lightning allows nitrogen and oxygen to combine to form various oxides of nitrogen. Rain washes these compounds into the soil where they can be taken up by plants as nitrates. In the UK, the amount of nitrogen fixed by lightning is small. In some parts of the tropics, however, violent thunderstorms are common and this is an important way of fixing nitrogen.

### Industrial processes

The Haber process combines nitrogen and hydrogen to produce ammonia. The reactions involved take place at high temperatures and pressures and require the presence of a catalyst. A lot of the ammonia produced by the Haber process is used to make fertilisers which are added to the soil.

## Nitrogen-fixing microorganisms

Many microorganisms are able to fix nitrogen. Some of them live free in the soil. Others are associated with the roots of leguminous plants such as peas, beans and lupins (Figure 6.12).

The biochemistry of nitrogen fixation is very similar in all these organisms and is summarised in the equation:

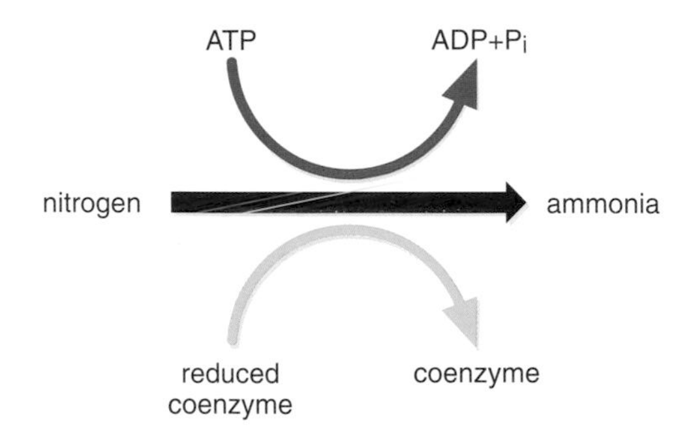


Nitrogen is reduced to ammonia. The reaction is catalysed by the enzyme nitrogenase. Nitrogenase, however, does not function in the presence of oxygen and many nitrogen-fixing microorganisms have adaptations which ensure that anaerobic conditions exist in the parts of the cells involved in nitrogen fixation.

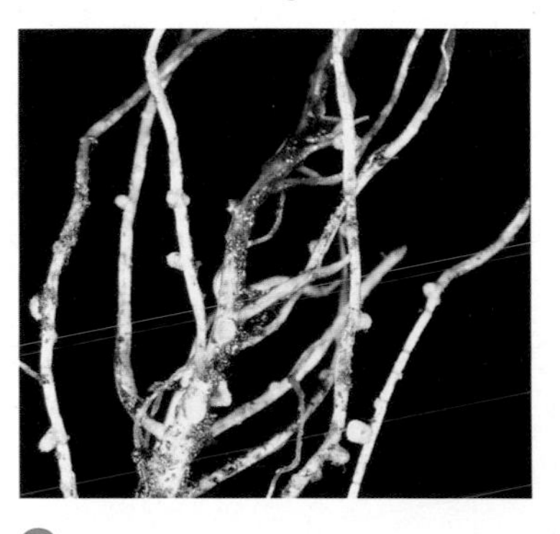

**Figure 6.12**
The swellings on these roots are nodules. Inside each nodule are large numbers of nitrogen-fixing bacteria. The bacteria and the plant have a mutualistic relationship. The bacteria get their carbon-containing compounds from the plant while the plant gains ammonium ions from the bacteria

**Q** 6 **What is the function of ATP in nitrogen fixation?**

## Human activities and the nitrogen cycle

When we grow a crop such as wheat, we do not allow the plants to complete their lives and then die and decompose. We harvest them. Instead of being recycled, nitrogen-containing substances in the grain and the straw are taken away for our use. To make up this loss we need to add fertiliser. Nitrogen-containing fertilisers contain nitrate and ammonium ions. These ions are very soluble so they can be readily leached from the soil into lakes and rivers. You may have studied the environmental effects of this as part of the AS course but you should look at this material again. It links closely with the nitrogen cycle described in this chapter. Extension box 2 is reprinted from the AS book *A New Introduction to Biology* and should remind you of some of the important details.

## Extension box 2

### Where do all the ions come from?

**Figure 6.13**
The vigorous growth of algae and water plants in this East Anglian stream is the result of pollution with nitrate and phosphate ions

The stream shown in Figure 6.13, which runs through an area of farmland in East Anglia, is very heavily polluted. The high concentrations of nitrate and phosphate ions in the water have led to increased growth of algae and water plants. As the stream is in a rural area, little of this pollution comes from sewage. It is mainly the result of agriculture. There are two main sources of these polluting nitrates and phosphates:

- fertilisers add nutrients to the soil and not all of these are taken up by crop plants. The rain leaches soluble ions from the soil, with the result that they drain into streams and ponds.
- cattle and other livestock produce large amounts of urine and faeces. This material is not always disposed of efficiently and can lead indirectly to an increase in the concentration of nitrate and phosphate ions in streams such as that shown in Figure 6.13.

Figure 6.14 shows the monthly concentration of nitrate ions in this stream. It also shows the total volume of water present. This is indicated by the curve showing the flow.

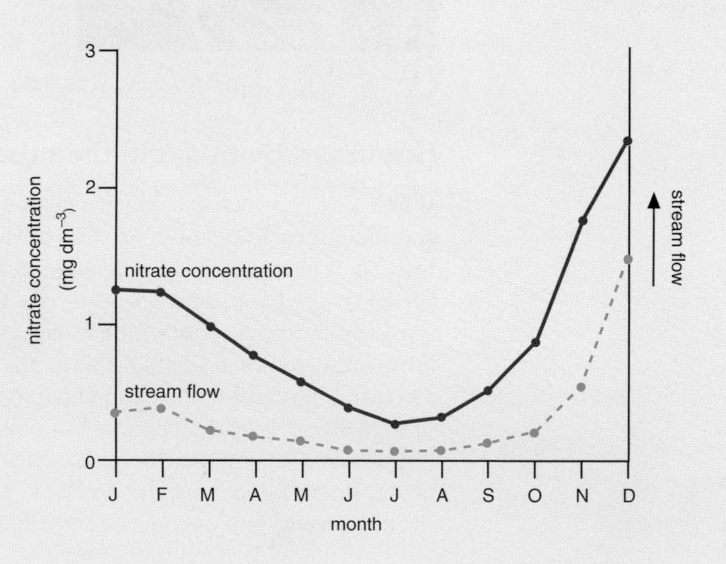


**Figure 6.14**
Graph showing how nitrate ion concentration and stream flow vary at different times of the year

This is a relatively small stream so the flow provides a good indication of the amount of rain falling on the land immediately around it. During the winter months stream flow and rainfall are high. More rainwater drains into the stream during these months so more nitrates are leached from the surrounding farmland. Other factors are also involved. Crops sown in the autumn are only just beginning to grow so there is a lot of bare soil. Heavy rain results in water running off the soil, taking soil nutrients with it. It is cold and plants only grow slowly. Consequently, at this time of the year they take up a relatively small amount of mineral ions from the soil.

In the summer months things are different. Crop plants are photosynthesising and growing rapidly in the warmer conditions. Consequently, they remove nitrates from the soil, leaving less to be leached into the stream. Those ions in the stream are also being absorbed by the water plants as they grow.

Now look at Figure 6.15. This graph shows that there is a positive correlation between the concentration of nitrate in the stream and the amount of nitrate added in fertiliser to the surrounding land. The more nitrate added as fertiliser in a particular year, the more nitrate there is in the stream. We have to be careful about the conclusions that we draw from sets of data, however. We must be aware that just because these two things are correlated, it does not mean that the fertiliser is the only cause of pollution in the stream.

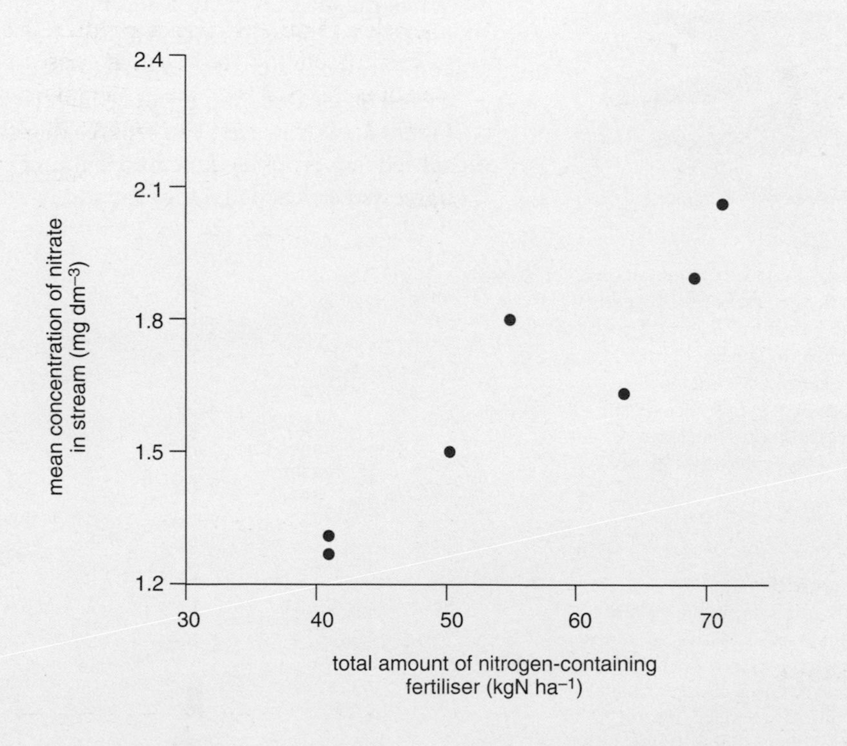


**Figure 6.15**
Graph showing the relationship between the mean concentration of nitrate in the stream and that added as fertiliser applied to the surrounding land

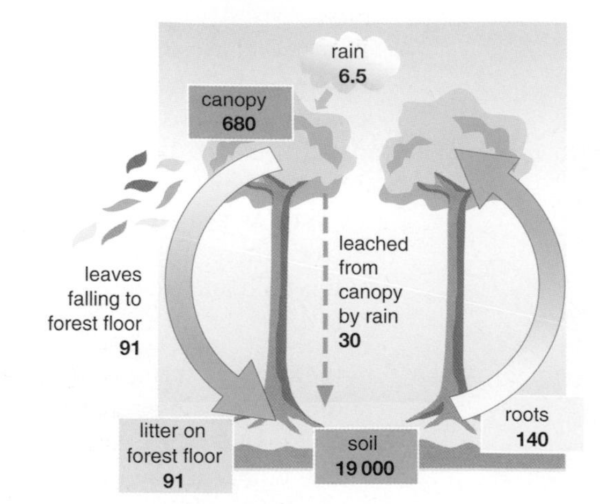


**Figure 6.16**
The cycling of nitrogen in a rain forest

The figures in boxes represent the amount of nitrogen contained in the soil and in the trees. The figures by the arrows show the flow of nitrogen over the course of a year. The units are kilograms per hectare

We will look now at another way in which human activities influence the cycling of elements such as nitrogen – the clearing of forests. Look at Figure 6.16. It shows the distribution of nitrogen in the soil and the trees which make up a rainforest. It used to be thought that the soil on which the forest grew was very poor in nutrients; most of the nitrogen was assumed to be locked up in the trees. We now know this is incorrect. There is a lot of nitrogen in the soil, but most of this is in forms which the trees are unable to use. What is true, however, is that nitrogen in dead animal and plant material is recycled very rapidly (Figure 6.17).

When the forest is cleared and burnt to make way for crops, a lot of the nitrogen in the soil is lost in smoke but the ash that remains is still comparatively rich in the nutrients required for crops. The first crops grown on cleared land produce high yields. If you look at the graph in Figure 6.18, however, you will see that the yield rapidly falls with subsequent crops as these nutrients are either removed as the crops are harvested or leached out of the soil by rain.

**Figure 6.17**
Fungi (a) are very common on the floor of the forest. Together with bacteria and animals such as giant millipedes (b), they break down the organic substances in fallen leaves and release nitrates which can be taken up by the forest plants

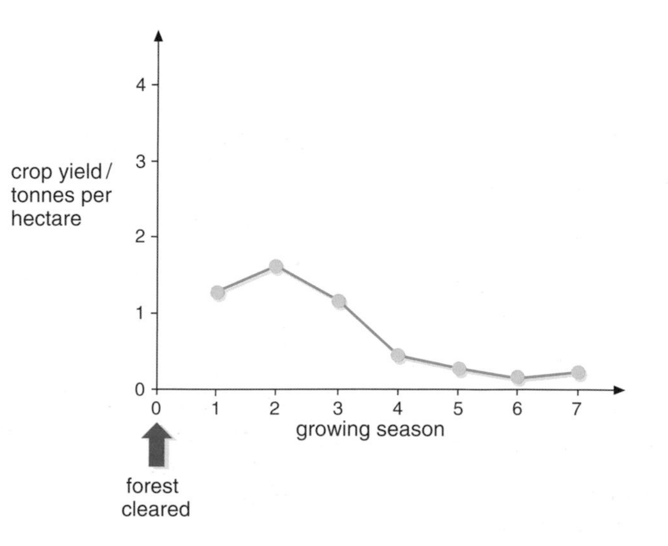


**Figure 6.18**
This graph shows the yield of grain from successive rice crops grown on an area where the forest has been cleared. Unless fertilisers are added, it soon becomes uneconomical to grow crops

## Summary

- Nutrients are examples of non-renewable resources and are constantly recycled. They are taken up by plants and passed from organism to organism in the various trophic levels. Death and decomposition result in microorganisms making them available to plants again.
- Carbon cycles follow the general pattern shown by other nutrient cycles. Carbon, however, is taken up by plants as carbon dioxide rather than as an ion from the soil, and carbon dioxide is returned to the air by respiration.
- Nitrogen cycles also follow the general pattern shown by other nutrient cycles. Nitrogen passes in organic compounds from organism to organism in the various tropic levels. Saprotrophic bacteria break down dead organic material and release ammonium ions. Nitrifying bacteria release nitrates from these ammonium ions. The nitrates are taken up by plants.
- Some nitrates are converted to nitrogen gas in the process of denitrification. Nitrogen gas can be made available to plants again by nitrogen fixation.
- Human activities affect the cycling of both carbon and nitrogen. The burning of fossil fuels increases the concentration of carbon dioxide in the atmosphere and is one of the main factors contributing to global warming. Nitrate ions are soluble and may leach into freshwater where their accumulation eventually leads to a fall in oxygen concentration and the death of many organisms.
- The clearing and burning of rain forest releases carbon dioxide from the carbon stored in forest trees. Nutrients released from the soil are rapidly leached out by rain or removed when crops are harvested.

**Figure 6.19**
Hedges can be thought of as long, narrow strips of woodland. They form a habitat for animals such as this wood mouse

## Assignment

### Hedges

Most of the United Kingdom was once covered in forest. Much of this natural woodland has now disappeared, cleared by an expanding human population for agriculture and industry. Many woodland plants and animals now find the only suitable habitat, the hedges which surround our fields (Figure 6.19). In this assignment we will look at some of the arguments for and against removing hedges. In order to answer the questions you will need to make use of some of the information from Chapter 4 as well that from this chapter.

The maps in Figure 6.20 are drawn from aerial photographs. They show the hedges around fields in one area in the Midlands in 1946 and in the same area in 1965.

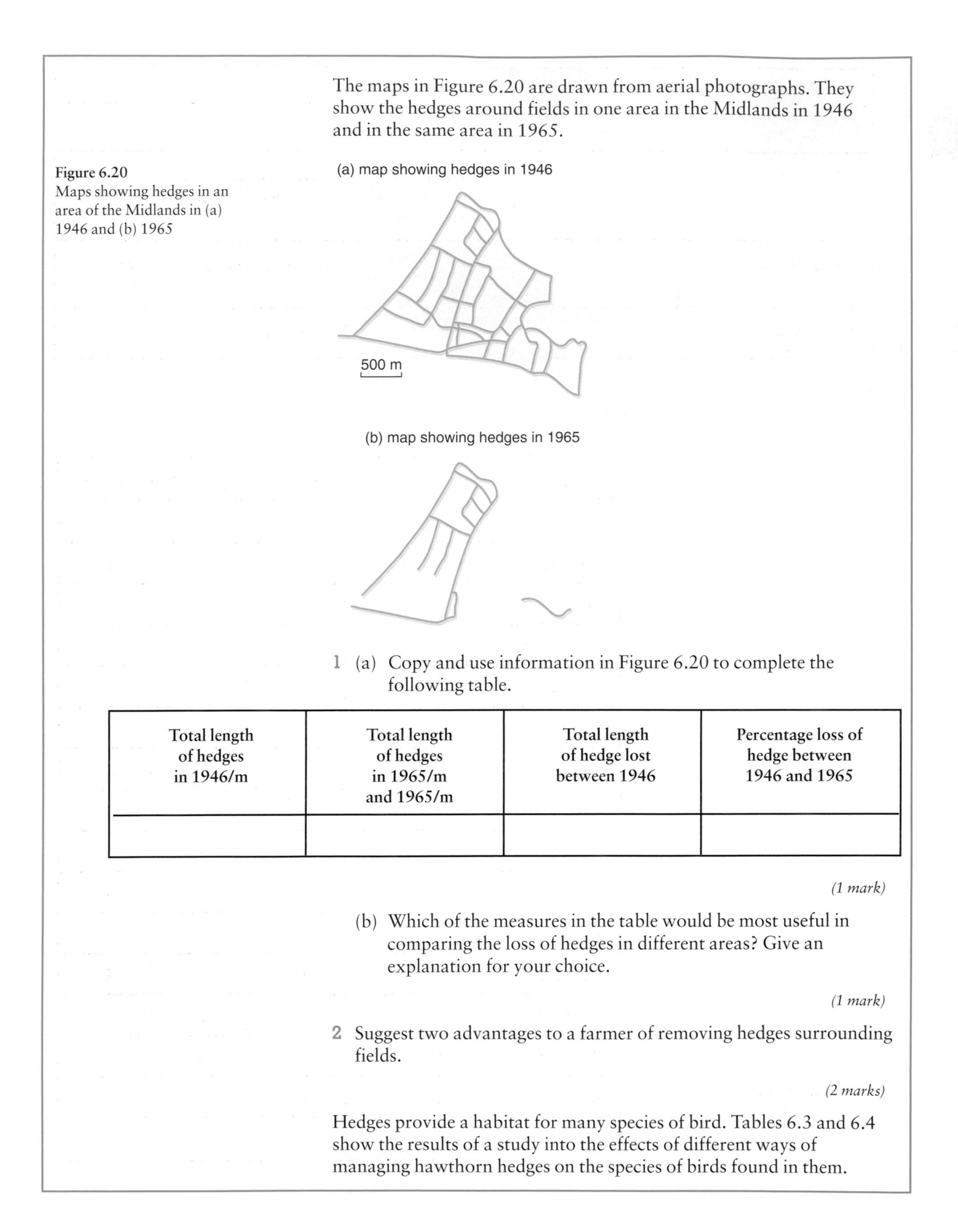


**Figure 6.20**
Maps showing hedges in an area of the Midlands in (a) 1946 and (b) 1965

1 (a) Copy and use information in Figure 6.20 to complete the following table.

| Total length of hedges in 1946/m | Total length of hedges in 1965/m and 1965/m | Total length of hedge lost between 1946 | Percentage loss of hedge between 1946 and 1965 |
|---|---|---|---|
| | | | |

*(1 mark)*

(b) Which of the measures in the table would be most useful in comparing the loss of hedges in different areas? Give an explanation for your choice.

*(1 mark)*

2 Suggest two advantages to a farmer of removing hedges surrounding fields.

*(2 marks)*

Hedges provide a habitat for many species of bird. Tables 6.3 and 6.4 show the results of a study into the effects of different ways of managing hawthorn hedges on the species of birds found in them.

| Type of field boundary | Total length/m | Notes on management |
|---|---|---|
| A No hedge present | 4130 | Not applicable |
| B Remnant | 2320 | Hedge neglected. A few isolated bushes and small trees with large gaps in between are all that remains of the original hedge. |
| C Mechanically cut | 2805 | Mechanically cut with tractor and hedge trimmer once a year |
| D Overgrown | 2645 | Hedge not managed in any way. |

Table 6.3
Hedges in the study area and their management

| Species | Pairs of breeding birds in type of field boundary | | | |
|---|---|---|---|---|
| | A | B | C | D |
| Red-legged partridge | 6 | 2 | 3 | 1 |
| Skylark | 4 | 0 | 0 | 0 |
| Corn bunting | 5 | 4 | 0 | 1 |
| Whitethroat | 2 | 2 | 5 | 6 |
| Yellow hammer | 0 | 0 | 6 | 6 |
| House sparrow | 0 | 1 | 0 | 2 |
| Song thrush | 0 | 1 | 0 | 2 |
| Blackbird | 0 | 0 | 4 | 11 |
| Linnet | 0 | 0 | 3 | 6 |
| Bullfinch | 0 | 0 | 0 | 2 |
| Dunnock | 0 | 0 | 4 | 5 |
| Chaffinch | 0 | 0 | 1 | 3 |
| Goldfinch | 0 | 0 | 0 | 1 |
| Blackcap | 0 | 0 | 0 | 1 |
| Wren | 0 | 0 | 0 | 3 |

Table 6.4
Breeding birds present in different types of field boundary

3 How do the different methods of management affect the number of species and the diversity of breeding birds?

*(3 marks)*

Hedges affect the abiotic environment of crops growing in the fields they surround. Figure 6.21 and Table 6.3 show some of the results of an investigation of these effects.

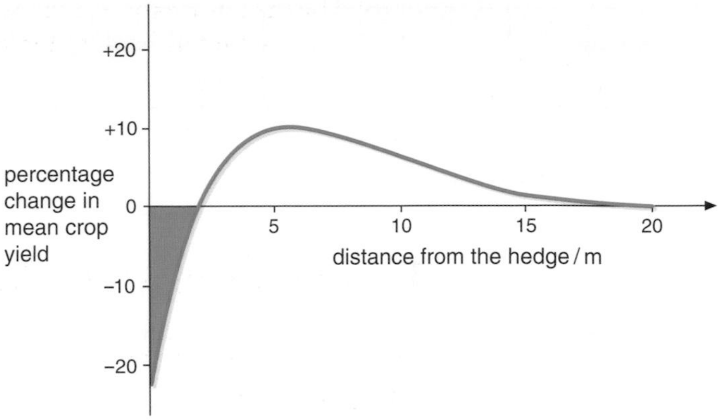

**Figure 6.21**
The effect of a hedge on crop yield. In this area water shortage normally limits crop yield



**Table 6.5**

| Factor | Effect of hedge as percentage of value in the open at different distances from the hedge | | | | |
|---|---|---|---|---|---|
| | 2 m | 4 m | 8 m | 12 m | 16 m |
| Soil moisture | 118 | 118 | 110 | 104 | 100 |
| Daytime air temperature | 114 | 112 | 109 | 102 | 100 |
| Evaporation | 74 | 65 | 78 | 90 | 97 |
| Wind speed | 61 | 39 | 52 | 65 | 80 |

The estimated effect of a hedge on various abiotic factors

4 The mean yield for the crop in this investigation was 5.3 tonnes hectare$^{-1}$. Use Figure 6.18 to calculate the mean yield of this crop:

(a) 2 metres from the hedge

(b) 7 metres from the hedge.

*(2 marks)*

5 Use the data in Table 6.4 to explain the difference between the figures in your answer to Question 4.

*(2 marks)*

6 Put yourself in the position of a conservation officer talking to a farmer who wants to remove a well-kept hawthorn hedge between two fields.

(a) Summarise the arguments the farmer may put forward for removing this hedge.

*(2 marks)*

(b) Use the information in this assignment to outline the argument you would put forward for leaving this hedge.

*(4 marks)*

## Examination questions

1 Read the following passage.

It is estimated that a thousand million tonnes of combined nitrogen are made available to the biosphere each year from the fixation of atmospheric dinitrogen by prokaryotic organisms.

The most effective nitrogen-fixing organisms are those which form
5 symbiotic (mutualistic) relationships with higher plants and the most important of these relationships is the symbiosis between plants of the legume family and several species of the bacterial genus *Rhizobium*. This symbiosis is obviously a true one and it is also highly specialised. Rhizobia live freely in the soil where they do not fix nitrogen, and are
10 attracted to roots of young legumes by secretions from these roots. The rhizobia synthesise compounds which deform the root-hair cells of the host allowing the bacteria to penetrate them. Following infection, the root-hair cell wall turns inwards and extends into a tube-like 'infection thread' which penetrates the walls of the cells of the cortex, allowing
15 the bacteria to migrate into the inner cortex of the root. They are finally released into the cytoplasm of these cells where they multiply and enlarge into swollen, irregularly shaped bacteroids. Groups of four to six of these become surrounded by membranes produced by the cell to form a number of isolated nitrogen-fixing colonies within the cell. Just
20 before the bacteria are released from the infection threads, the cortex cells of the host in the region of infection multiply rapidly to form the root nodule.

In its final structure, the root nodule consists of a central region containing the nitrogen-fixing bacteroids surrounded by a rhizobia-free
25 area into which the vascular tissue of the root has developed. The central region has a light pink colour due to the presence of a special sort of haemoglobin, leghaemoglobin, in the membranes surrounding each bacteroid group.

The mechanism of nitrogen-fixation is complex and not yet fully
30 understood. The overall equation for the reaction is

$$15ATP + 6H^+ + 6e^- + N_2 \rightarrow 2NH_3 + 15ADP + 15P_i \text{ (inorganic phosphate)}$$

This reaction in the bacteroids is catalysed by the enzyme nitrogenase and makes heavy demands on the photosynthetic product of the host. The enzyme is sensitive to the presence of oxygen and will only fix 35
35 nitrogen in an anaerobic environment. Although ammonia is the first product, it is rapidly converted into nitrogenous compounds and exported by the nodule through the xylem connected to the host.

(Adapted from *Plants and Nitrogen*, Lewis)

Using information in the passage and your own knowledge, answer the following questions.

(a) What is the meaning of the following terms as used in the passage:
(i) atmospheric dinitrogen (line 3);

*(1 mark)*

(ii) combined nitrogen (line 1)?

*(1 mark)*

(b) Draw a fully labelled diagram to show the structure of one of the cortex cells containing bacteroids.

*(3 marks)*

(c) Explain why this symbiosis is described as 'obviously a true one' (line 8).

*(2 marks)*

(d) (i) What is the precise source of the nitrogen fixed by the bacteroids?

*(1 mark)*

(ii) By what process does the nitrogen reach the bacteroids?

*(1 mark)*

(e) Suggest an explanation for the presence and distribution of leghaemoglobin.

*(4 marks)*

(f) (i) Explain the connection between the nitrogen-fixing reaction and the 'photosynthetic product of the host' (line 33).

*(3 marks)*

(ii) Why are the demands on the photosynthetic product described as 'heavy' (line 33)?

*(1 mark)*

(g) Describe **one** pathway by which the nitrogen fixed in the nodule might become available for uptake by plants of other species.

*(4 marks)*

(h) Naturally occurring vegetation on a soil with high nitrate content was compared with similar vegetation on a soil with low nitrogen

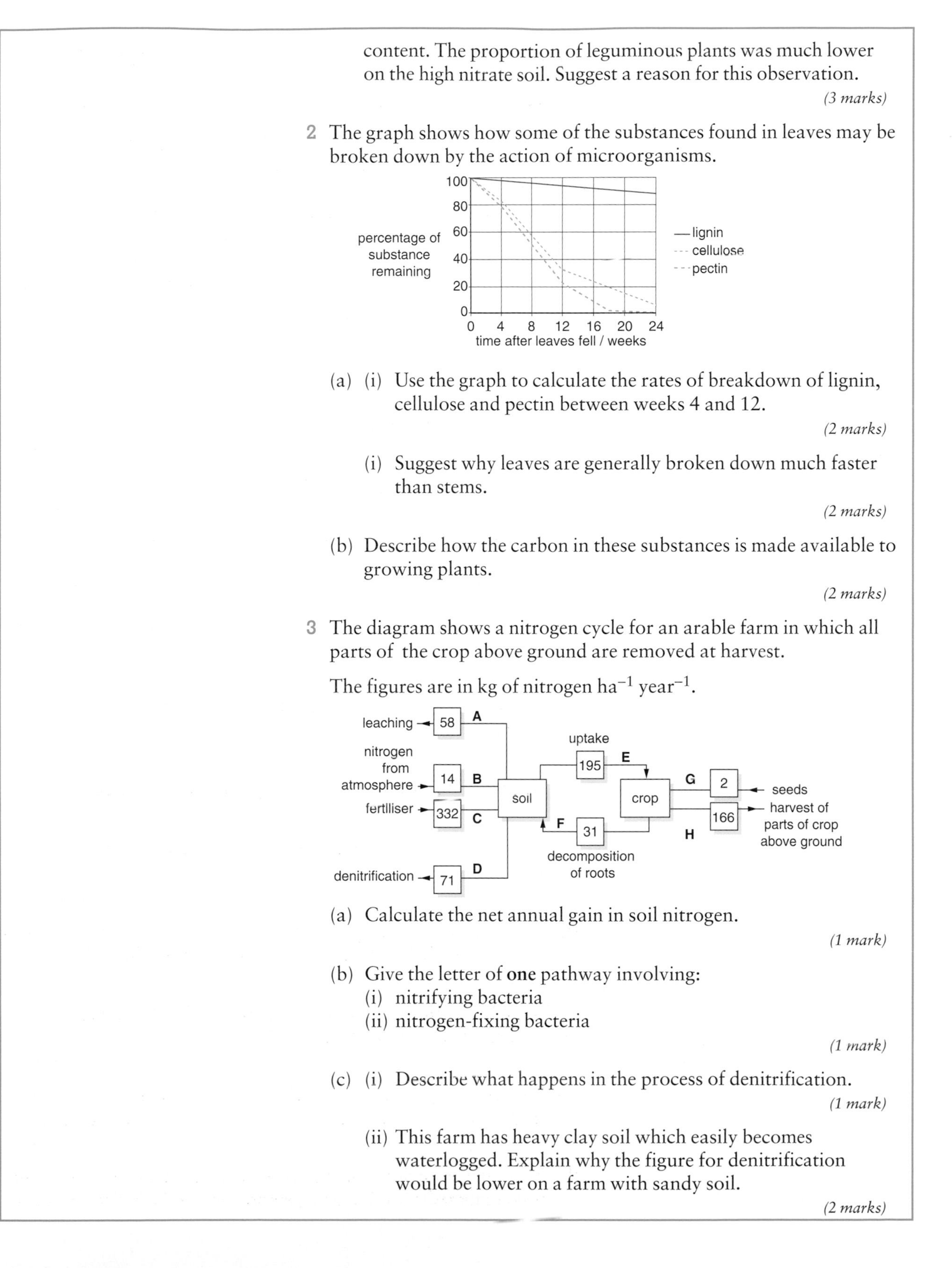

content. The proportion of leguminous plants was much lower on the high nitrate soil. Suggest a reason for this observation.

*(3 marks)*

2 The graph shows how some of the substances found in leaves may be broken down by the action of microorganisms.



(a) (i) Use the graph to calculate the rates of breakdown of lignin, cellulose and pectin between weeks 4 and 12.

*(2 marks)*

(i) Suggest why leaves are generally broken down much faster than stems.

*(2 marks)*

(b) Describe how the carbon in these substances is made available to growing plants.

*(2 marks)*

3 The diagram shows a nitrogen cycle for an arable farm in which all parts of the crop above ground are removed at harvest.

The figures are in kg of nitrogen $ha^{-1}$ $year^{-1}$.



(a) Calculate the net annual gain in soil nitrogen.

*(1 mark)*

(b) Give the letter of **one** pathway involving:

(i) nitrifying bacteria

(ii) nitrogen-fixing bacteria

*(1 mark)*

(c) (i) Describe what happens in the process of denitrification.

*(1 mark)*

(ii) This farm has heavy clay soil which easily becomes waterlogged. Explain why the figure for denitrification would be lower on a farm with sandy soil.

*(2 marks)*

# 7 Uptake and Loss of Water in Plants

Everybody knows that plants need water. Gardens have to be watered regularly during long hot summers and a bouquet of flowers will last much longer if put in a vase of water. Although water is important to plants, most of the water that a plant draws from the soil is lost by transpiration. Water evaporates from the wet surfaces of the spongy mesophyll cells in the leaf and diffuses into the atmosphere through tiny pores called stomata.

In 1984 astronauts took into space some sunflower seeds, a variety known as 'Teddy Bear'. They wanted to find out how seedling plants would develop in conditions of near weightlessness. They wondered whether the plants would even start to grow since the water in the soil can go in all directions in weightless conditions. There was therefore a possibility that the roots of the young seedlings might either drown or fail to absorb any water at all.

Instead of roots growing downwards and shoots growing upwards, they randomly changed direction as they grew, creating a haphazard growth pattern. So it was discovered that water could still enter the plant and move through roots and stems even in weightless conditions.

## Transport in plants

Plants are relatively inactive and their metabolic requirements are such that they do not have the type of transport system characteristic of most animals. However, the distances over which substances are moved can be far greater than in animals.

**Figure 7.1**
Water is transported over 65 metres from root to leaves in this redwood tree

Flowering plants have two distinct transport systems, both of which consist of tubes. One system, **xylem**, is concerned with the movement of water and mineral ions, obtained from the soil, from the roots through the stems to the leaves. The other system, composed of **phloem** tissue, is concerned with the transport of sugars and other soluble organic products of photosynthesis from where they are formed in the leaves to where they are needed in the developing shoots, flowers, fruits and roots.

### Life depends on water

Plants and animals require water for essentially the same reasons.

- It is the medium in which all metabolic reactions take place.
- It is needed for hydrolysis; the breaking down of a large molecule into smaller molecules by the addition of water.
- It is needed for the transport of solutes around the organism.
- It cools the organism when it evaporates from its surface.

In plants it performs the additional function of creating pressure in the cells, which helps to support the plant.

## Structure of a root

As well as anchoring the plant, the **roots** provide the surface through which water is taken up. The surface area is greatly increased by the presence, just behind the tip of each root, of thousands of **root hairs** (Figure 7.2).

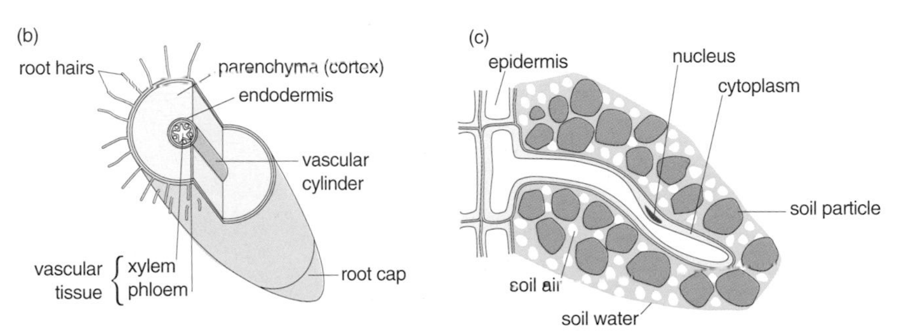


**Figure 7.2**
(a) Photomicrograph of root and root hairs (b)Transverse section of root showing the position of various tissues (c) Root hair in detail, showing its close association with the soil from which it absorbs water and mineral ions

To understand the pathways involved in water transport you need to know about the structure of a root. A transverse section is illustrated in Figure 7.3, which shows the arrangement of the tissues in a young root of a plant such as a buttercup.

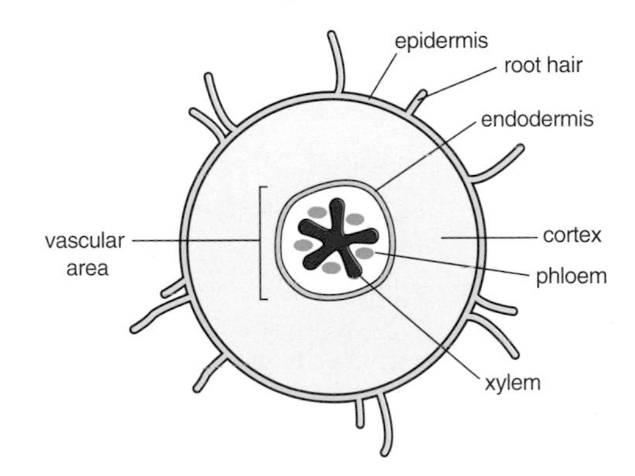


**Figure 7.3**
The arrangement of tissues in a transverse section of a young buttercup root

The single layer of cells around the outside of the root is called the epidermis. A root hair is a slender extension of a single epidermal cell and can be up to 4 mm long. The root hairs are immediately behind the tip of the root. They penetrate between the soil particles and are in close contact with the soil water. There are thousands of root hairs near each root tip and they greatly increase the surface area for absorption.

The central part of the root contains the xylem. Xylem provides a continuous system for the transport of water and dissolved mineral ions. The xylem and phloem are enclosed in a cylinder made of a single layer of cells called the **endodermis**. The spaces between the endodermis and the xylem and phloem are packed with thin walled unspecialised cells.

Between the endodermis and the epidermis, there are several layers of relatively large cells, called parenchyma, forming the **cortex.** The walls of these cells are permeable to water and dissolved solutes. There are also many air spaces in the cortex, which allow the diffusion of oxygen across the root for respiration (Figure 7.4(a)).

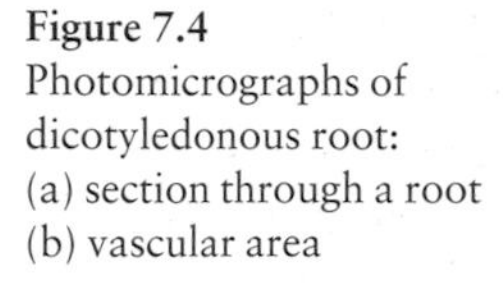

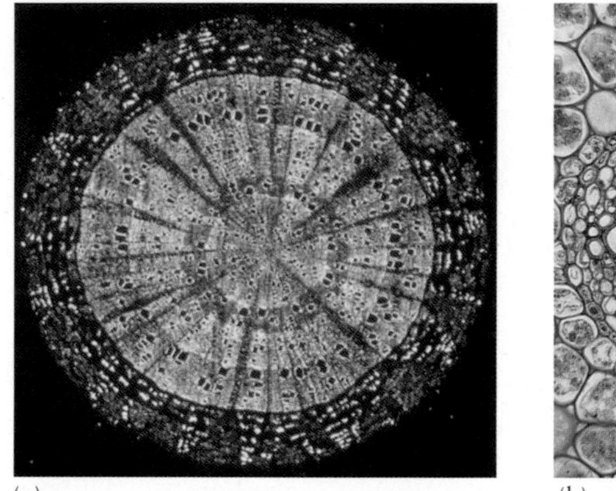

(a)

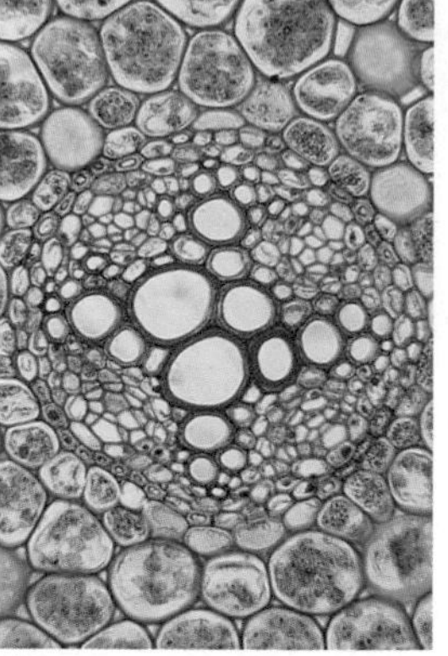

(b)

**Figure 7.4**
Photomicrographs of dicotyledonous root:
(a) section through a root
(b) vascular area

## Extension box 1

## Water and plant cells

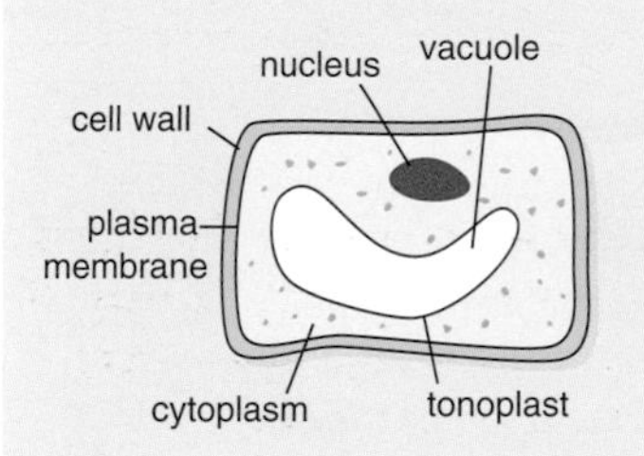


**Figure 7.5**
Principal structural features of a typical plant cell

The outer cell wall consists of cellulose fibres embedded in a matrix. It presents little resistance to the movement of water and water-soluble substances and so provides a relatively fast route for water transport. Touching the inside of the cell wall there is the plasma membrane, a protein phospholipid bilayer that permits the passage of water but restricts the movement of other molecules on the basis of their size, charge and solubility. It is said to be partially permeable. The cytoplasm consists of a solution in which substances can move from one part of the cell to another. In this solution substances can come into contact and reactions may occur.

Water molecules are continually in motion and are colliding with each other in a completely random way. Water potential is a measure of the free energy of water. The more molecules of water present compared to solute; that is the weaker the concentration of a solution, the higher the water potential. Pure water is given a water potential value of 0 and so a solution always has a value less then 0. Water moves from a region where it is in high concentration, a less negative water potential, to an area where it is in low concentration, a more negative water potential. In biological systems, osmosis is a special case of diffusion, in which water molecules move down their water potential gradient across a partially permeable membrane. If water and a solution of sugar and water are separated by a partially permeable membrane, then there will be a net movement of water into the sugar solution.

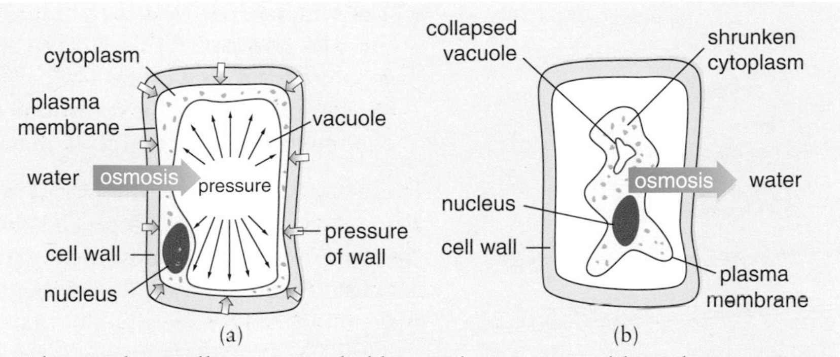


**Figure 7.6**
Effects of external water concentration on plant cells.
(a) When a plant cell is placed in more dilute surroundings, water enters the cell by osmosis. The cell swells and becomes turgid
(b) When the cell is placed in more concentrated surroundings water leaves the cell by osmosis. First the cell shrinks slightly and becomes flaccid, and then the membrane pulls away from the cell wall resulting in plasmolysis

When a plant cell is surrounded by a solution more dilute than its own cell contents, water moves into the cytoplasm and the vacuole. This causes the volume of the cell to increase and the contents to press against the cell wall. How much outward movement occurs, and thus how much increase in cell volume takes place, depends on the elasticity of the cell wall. The pressure from the cell contents resists any further net uptake of water into the cell. A cell in this state is described as fully **turgid** (Figure 7.6(a)). Turgor is important for maintaining mechanical support in young plants and leaves.

When a plant is surrounded by a solution more concentrated than its own contents, water will move out of the cell by osmosis and the volume of the cell's cytoplasm will shrink. If the water loss continues the plasma membrane will be pulled away from the cell wall. The cell in this state is **plasmolysed** (Figure 7.6(b)). If water becomes available plasmolysis can be reversed but if the cell wall dries out, the cell will die. In this way, prolonged drought will result in the death of the plant.

## Uptake of water

Water is taken up mainly by the younger parts of the roots, in the region of the root hairs. The water potential of the water in most soils is close to zero because the concentration of solutes, such as mineral ions, is very low. Most mineral ions are in a lower concentration in the soil surrounding the roots than they are in the cells, and are therefore taken into the roots by active transport. The higher concentration of mineral ions inside the root cells means that the cells have a more negative water potential than the surrounding soil water and therefore a water potential gradient exists. Water moves by osmosis from a higher (less negative) water potential in the soil to a lower (more negative) water potential in the cells.

Once inside the root, water and mineral ions can move from cell to cell across the cortex of the root from the epidermis to the central tissues by two main pathways.

- The **apoplast pathway** is probably the most important of the routes by which water moves through plant tissues. Water passes through the continuous system of adjacent cell walls. There are no barriers to movement so it is able to diffuse freely.

- The **symplast pathway** is the route by which water moves through the cytoplasm from cell to cell. The cells of the cortex are not completely separate from each other. The cytoplasm of one cell is continuous with the cytoplasm of the next through fine channels in the cell walls. These channels are called **plasmodesmata** (singular, plasmodesma).

These two pathways are illustrated in Figure 7.7. The symplastic route of water flow between cells is much slower than the apoplastic route since the resistance to water movement in the cytoplasm is four times greater than that of the cell wall. Remember that water, together with dissolved mineral ions may move along both of these pathways.

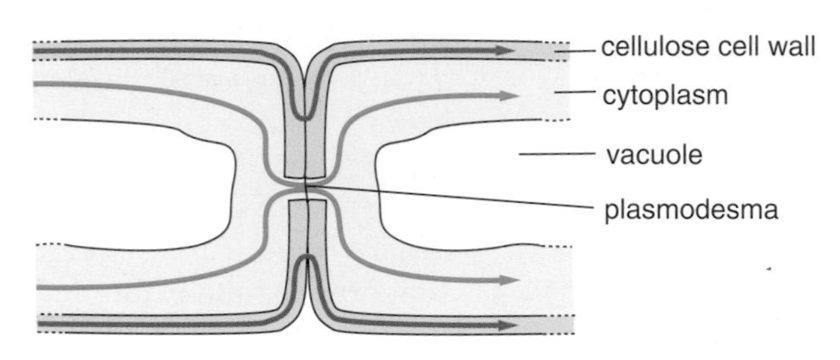


Figure 7.7
Apoplast and symplast pathways

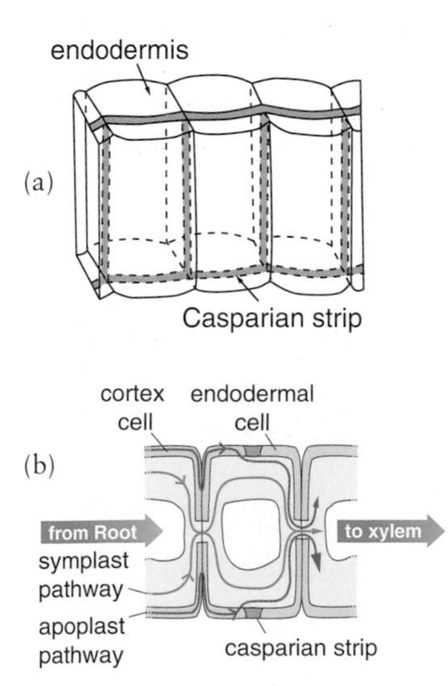


Figure 7.8
(a) Endodermal cells of the root showing the Casparian strip in the walls (b) The Casparian strip diverts the water from the cellulose cell wall (apoplast pathway) to the cytoplasm (symplast pathway)

Movement toward the centre of the root by the apoplastic pathway is stopped when water reaches the endodermis by a waterproof layer in the cell walls called the **Casparian strip** (Figure 7.8(a)). This strip is impregnated with **suberin,** a waxy compound, which is impermeable to water. Water is therefore prevented from passing around the endodermal cells through the cell walls, but instead it must pass through the plasma membrane and into the cytoplasm (Figure 7.8(b)). So the Casparian strip forces water to take the symplast pathway through the endodermal cells. This is important when we come to consider root pressure in a later section.

## From root to leaf

Water is transported from the roots to the leaves via the **stem.** The xylem and phloem are called **vascular tissue.** In a stem such as that in a buttercup, the xylem and phloem are grouped together into **vascular bundles.** These are arranged in a ring as shown in Figures 7.9 and 7.10.

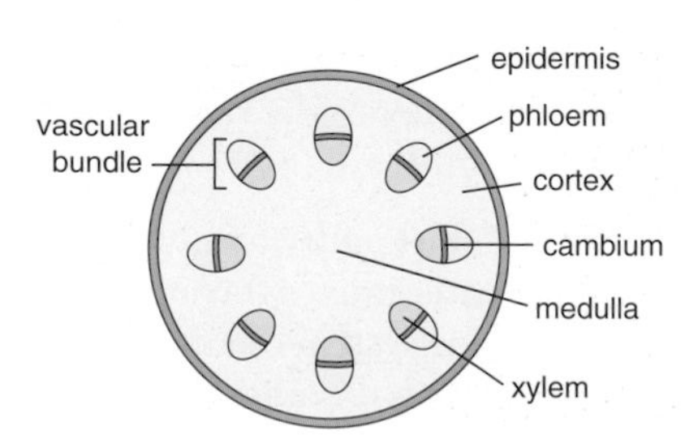


Figure 7.9
A transverse section through a dicotyledonous stem showing the arrangement of vascular bundles. This arrangement is typical of broad-leaved plants

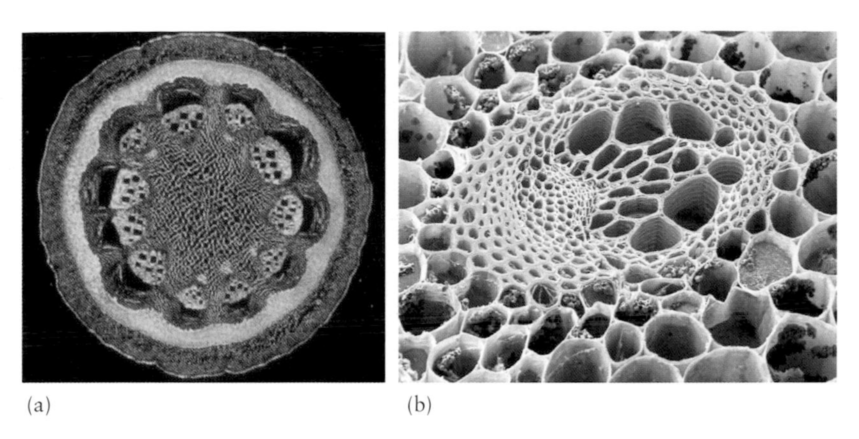


Figure 7.10
Photomicrograph of a transverse section of a dicotyledonous stem showing (a) the stem structure and (b) a vascular bundle

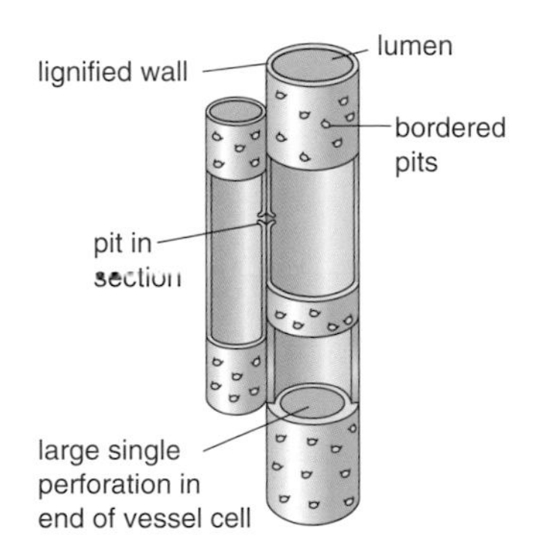


Figure 7.11
The important features of a xylem vessel

Xylem and phloem run the entire length of the plant, from the roots to the midrib and veins of the leaves.

## Xylem

Xylem contains tubes called **vessels**. These are made up of dead cells called **vessel elements** arranged end-to-end. They begin life as elongated, living cells. Once they have reached their full size, their cellulose cell walls become impregnated with **lignin**, which is impermeable to water. The living contents of the cells die, leaving the cell walls surrounding a water-filled cavity (**lumen**). The end walls break down so that the cells are in open communication with each other. Water moving up the plant encounters little resistance as it moves through them. The structure of vessels is shown in Figure 7.11.

As xylem cells develop lignin is laid down on the immediate inside of the cellulose walls often in the form of rings or spirals (Figure 7.12). Older xylem vessels have a continuous layer of lignin in their walls except for perforations called **pits**. Where a pit occurs, lignin fails to be deposited and only the cellulose cell wall remains. Pits of neighbouring cells match up, so the cell cavities of adjacent cells are connected. This permits the passage of water sideways as well as upwards.

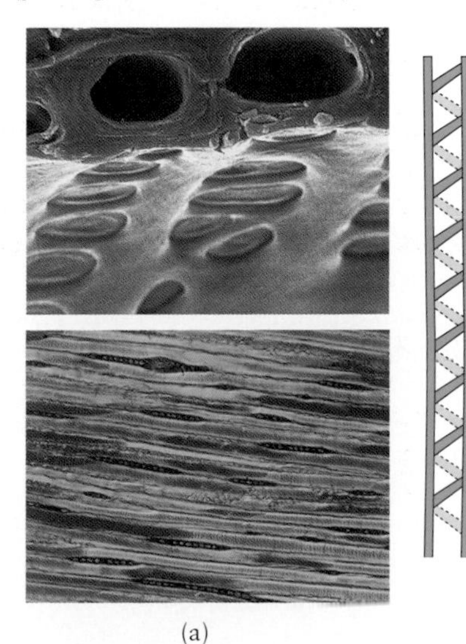
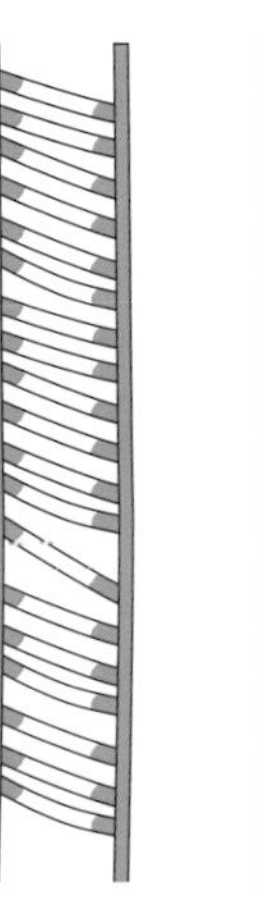
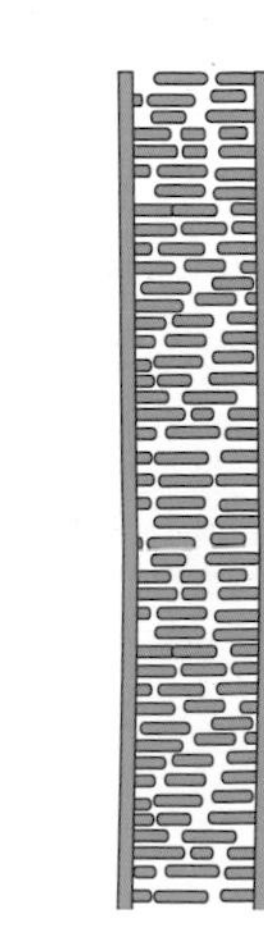


Figure 7.12
(a) Photomicrographs of different types of xylem (b) Drawings of different types of xylem

**Q** 1 **Explain how each of the following features of xylem vessels adapts them for their function of transporting water from roots to leaves.**

(a) **lack of cell contents**

(b) **no end walls in individual xylem elements**

(c) **lignified walls**

(d) **pits**

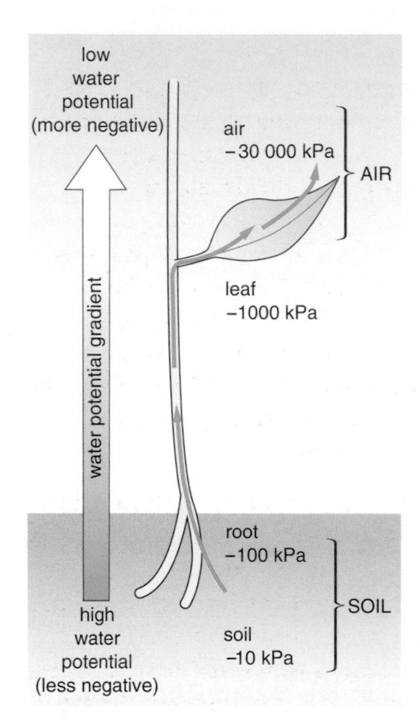


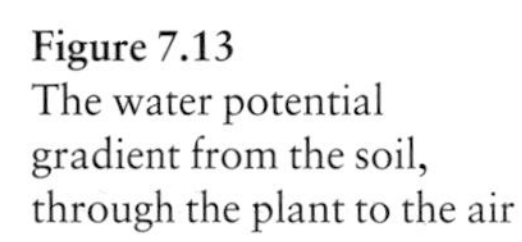

**Figure 7.13**
The water potential gradient from the soil, through the plant to the air

## Transpiration

The root hairs are in contact with soil water, which has a high (less negative) water potential, whilst the leaves and stems are exposed to the atmosphere, which has a much lower (more negative) water potential. Consequently, there is a water potential gradient through the whole plant, so water is drawn through and lost from the plant (Figure 7.13). Thus movement of water is a passive process; no energy is required from the plant for it to occur.

Water vapour may be lost from three sites on the aerial parts of the plant.

- leaves (through the cuticle and the stomata)
- flowers
- stems in herbaceous (non-woody) plants and lenticels on woody stems

Most of the water vapour loss from a plant occurs through the open stoma. The stomata of a leaf are open during the day allowing the uptake of carbon dioxide for photosynthesis. Evaporation occurs from the cellulose cell walls of the mesophyll cells into the intercellular spaces, from where it diffuses out through the stomata, from a higher (less negative) water potential inside the leaf to a lower (more negative) water potential outside. In dicotyledonous (broad-leaved) plants, there are usually more stomata on the lower surfaces of leaves than the upper.

Evaporation through the cuticle will vary with its thickness but may account for 10% of the total water loss.

The loss of water from a plant by evaporation through its surface is called **transpiration**. The water lost this way causes water to be drawn up through the xylem in what is called the **transpiration stream.**

Water reaches the mesophyll cells in the leaves from the xylem of the vascular bundles in the leaf veins. At their extreme ends, these veins consist of little more than one or two vessels with little lignification, so water can easily pass to the adjacent mesophyll cells via the apoplast or symplast pathways.

### Mechanisms responsible for the movement of water in xylem

How does water get to the tops of tall trees? This question puzzled plant biologists for many years. They came up with three possible mechanisms, all of which may contribute to some extent to the movement of water up the xylem vessels.

### Capillarity

If you dip a narrow glass tube into water, you will see that the water rises up the tube. This is called capillarity and is due to the adhesion of water to the sides of the tube. Xylem vessels are very fine tubes, as small as 20 μm in diameter. Even in tubes as narrow as this, water rises less than 50 mm. So capillarity alone cannot account for the transport of water through the xylem.

### Root pressure

If the shoot of a plant is cut off close to the ground, sap may exude from the xylem of the stump, suggesting that there is a force pushing water up the stem from the roots. This force is known as **root pressure.** The endodermis in the root appears to be important in the development of root pressure. Water with dissolved mineral ions can pass along the apoplast pathway as far as the Casparian strip. The only route past this barrier is via the symplast. Once inside a living cell, the ions can be actively secreted. The endodermal cells secrete the ions into the cells around the xylem which in turn secrete them into xylem vessels. This creates a water potential gradient and by osmosis, water diffuses from the cortex, through the endodermis and into the xylem. Root pressure is caused by this accumulation of water in the xylem.

Anything that inhibits active transport, such as metabolic inhibitors, low temperature or shortage of oxygen also reduces root pressure. It can be demonstrated that a positive hydrostatic pressure of about 150 kPa may be generated by root pressure, so although it could not account for all the water movement in the xylem, it may have a contributory role. Root pressure provides a force, which in effect, pushes water up the stem, but it is not enough to account for the movement of water to leaves at the top of trees.

### The cohesion–tension theory

The **cohesion–tension theory** is the one which most adequately accounts for the movement of water through the plant. The theory can be divided into four main stages.

1 Leaves **transpire.** Water evaporates from the moist conditions inside a leaf, through the stomata to the drier surrounding air.

2 Water molecules demonstrate a property known as **cohesion.** Hydrogen bonds form between neighbouring water molecules, causing them to stick to each other. This means that as water is lost by transpiration, more is pulled up the xylem to replace it.

3 The pulling action of transpiration stretches the water column in the xylem so that it is under **tension.**

4 Water molecules also cling to the walls of the xylem. This is **adhesion** and also helps to pull the water column upwards.

Removal of water from the xylem in the veins of the leaf creates a pulling force which draws water through the vascular tissue from the roots through the stem in continuous columns. This works as long as the columns of water do not break. The maintenance of these continuous columns is due to the structure of the xylem and the properties of water.

Xylem vessels are small in diameter and have strong, rigid walls which are able to withstand tension. As there is an attraction between water molecules and lignin the water molecules stick (adhere) to the walls. Adhesion is defined as the force of attraction between **unlike** molecules. The narrower the vessel, the greater will be the proportion of water molecules in contact with its walls. Narrower columns of water are thus less likely to break.

There are also strong forces between the water molecules. Cohesion is defined as the force of attraction between **like** molecules. Cohesive forces between water molecules hold them together in continuous columns and when water transpires from the leaf, the whole of the water column moves up the xylem. However, there is a problem. If you suck water up a straw, the walls collapse if you suck too hard and the column of water breaks. In the xylem, the thickened lignified walls normally prevent this happening. The cohesion–tension theory offers an explanation for the rise of water in the xylem of all plants including tall trees. The tension in the xylem of a very tall tree must be colossal, but the tensile strength of the water column must be sufficient or such trees could not exist.

Further evidence that the water in the xylem is under tension comes from taking measurements of tree trunks over 24 hours. The measurements show that the diameter of a tree trunk decreases during the day. The diameter reaches the minimum size in the afternoon, after which it increases again to reach a maximum in the early morning (Figure 7.14).

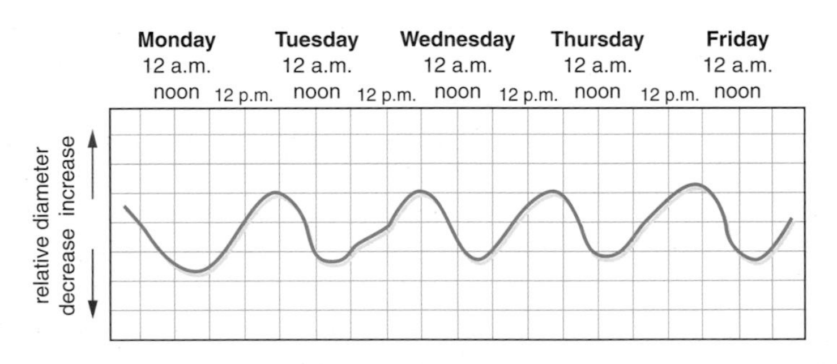


**Figure 7.14**
Daily variations in the diameter of the trunk of a pine tree

The decrease in diameter of the trunk during the day occurs because water loss (transpiration) puts the xylem contents under tension and the vessels contract. At night, transpiration is greatly reduced, the tension is released and the cells increase in diameter.

Sometimes water columns do break. In stems bending the wind, xylem vessels can be suddenly stretched, causing air bubbles to form. Air bubbles also form when water freezes in the xylem. These block individual vessel elements. But the pits in the xylem vessel walls allow water to pass sideways into an adjacent vessel and so bypass the blockage (Figure 7.15).

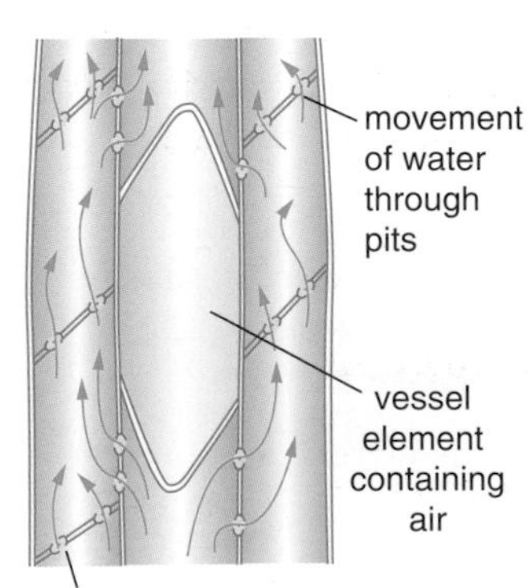


**Figure 7.15**
The upward and sideways movement of water in xylem

**Q** 2 **Why would water not pass up a tree if water molecules had very little cohesion?**

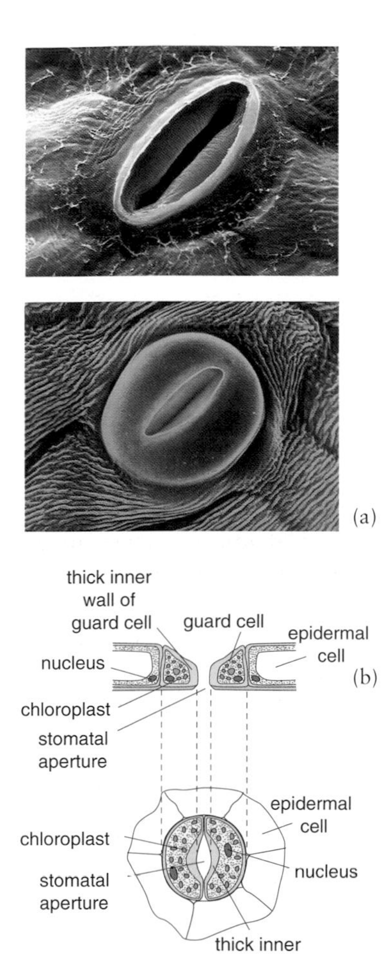


**Figure 7.16**
(a) Scanning electron micrograph showing the daylight conditions for most stomata – open (top) and the night-time or water-stressed condition of stomata – closed (bottom) (b) Vertical section of a stoma (c) Surface view of a stoma

## Stomata and guard cells

Stomata are pores in the epidermis of leaves, flowers and herbaceous stems through which exchange of gases occurs. They are found most abundantly on leaves and may occur on both surfaces but are more common on the lower surface. Surrounding each pore are two kidney-shaped guard cells (Figure 7.16). These are able to control the size of the stomatal pore by changes in their turgidity.

The guard cells are different from the rest of the epidermal cells in that they contain chloroplasts, they are not linked to adjacent cells by plasmodesmata and their walls are unevenly thickened. The part of the cellulose cell wall which borders the pore is thicker and less elastic than the outer wall. Most of the cellulose microfibrils of the cell wall run across the width of the cell. This allows the two cells to bend and so open and close the pore. As water is taken into the guard cells, increasing their turgidity, they become more curved and the pore between them opens wider. When guard cells lose turgidity, they become less curved and the pore closes.

Heat from the Sun causes water to evaporate from the surface of mesophyll cells and the water vapour formed diffuses into the air spaces. The water lost from the mesophyll cells is replaced by water from the xylem. This water moves to the mesophyll cells in the same ways as we described earlier for water movement in the root, along the apoplast or symplast pathways.

The water vapour in the air spaces diffuses out of the leaf through the stomata. The water vapour is diffusing from a high water potential inside the leaf to a lower water potential outside. A layer of stationary air called the boundary layer lies immediately next to the leaf. The thickness of this layer depends on a number of factors including the size, shape and hairiness of the leaf and also wind speed. Water vapour diffuses through the boundary layer (Figure 7.17) before being carried away in moving air by mass flow. Any factor that reduces the thickness of the boundary layer or increases the water potential gradient tends to increase the rate of diffusion of water from a leaf.

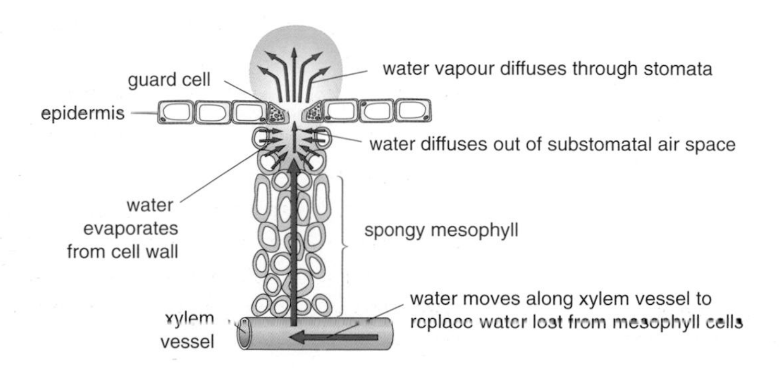


**Figure 7.17**
Diffusion of water from mesophyll cells out of the leaf

Successful exchange of gases by the leaf depends on stomata being open. However, there is a problem as these pores also allow water to escape by diffusion into the air so when a plant is exchanging gases, it will lose water. Water lost has to be replaced and if it is in short supply, the plant can suffer stress. Water loss can be reduced by closing the stomata, but this also, inevitably, reduces gas exchange. Stomata are involved in a continual balancing act; they open and close, according to the external conditions and the needs of the plant.

A number of factors influence the opening and closing of the stomata. These include the prevailing light conditions, the supply of water to the plant and the supply of respiratory substrates. In plants that are well supplied with water, the stomata are open at dawn and closed at dusk and this pattern appears to be controlled by light.

**Q** 3 **Leaves sometimes drip water at night. Suggest a cause.**

## Extension box 2 Mechanism of stomatal opening and closing

Even when all stomata are fully open, the total area of their pores rarely exceeds 2% of the leaf surface but the ability plants have to alter this area allows delicate control over transpiration rates.

We know that stomata open and close, and that small movements of the guard cells are responsible. The cellulose cell walls of the guard cells that surround the stoma are thicker and less elastic than those in contact with the epidermis. It is this, which makes the inner wall less able to stretch, and results in the typical kidney shape of these cells. Moreover, an increase in the volume of a guard cell, because of uptake of water, causes these cells to expand unevenly. The walls farthest from the stomatal opening are more flexible and expand more than the walls that line the stomatal opening. This causes the guard cells to bend and the stomatal aperture enlarges. Conversely, when the guard cells lose water the stomata close. This gain and loss of water is an osmotic effect brought about by altering the concentration of ions in the guard cells.

An early theory proposed that, when the guard cells photosynthesise, they make enough sugar to cause water to enter them by osmosis from the surrounding cells. This theory could not explain how some plants closed their stomata at midday, when maximum photosynthesis was possible. It is now thought that stomatal opening and closing depends on the generation of a potassium ion gradient with blue light activating the system. During the day, the blue light stimulates ATPase found in the guard cells. The ATPase then hydrolyses ATP into ADP and Pi and the energy released drives a proton ($H^+$) pump. This pumps $H^+$ out of the guard cells.

The protons return on a carrier that brings chloride ions ($Cl^-$) with them. Potassium ions ($K^+$) also enter the guard cells. Due to increased ion concentration in guard cells, water enters by osmosis. The guard cells swell and bend, as described earlier, causing the pore to open.

At night or in the absence of light the proton gradient is no longer maintained so that the flow of ions and water is reversed and the pore closes.

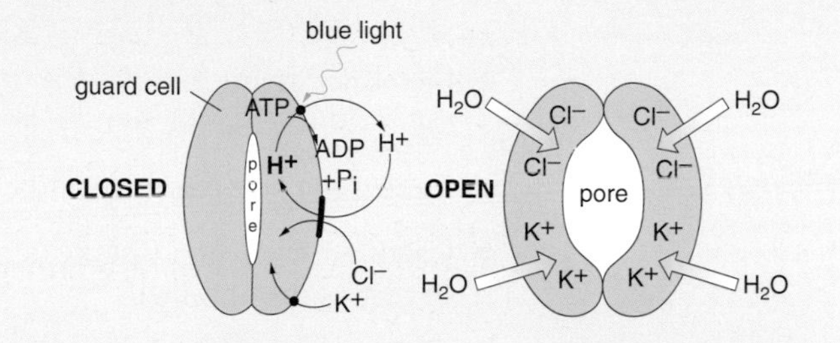


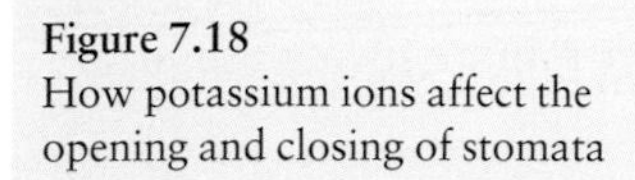
**Figure 7.18**
How potassium ions affect the opening and closing of stomata

It is quite common for stomatal closure to occur around midday in leafy trees. When air temperatures are high and the humidity is low, transpiration increases and water loss may exceed water uptake. Under these circumstances, the loss of water results in a decrease in turgor of the leaf cells. This triggers the synthesis of a plant growth substance, **abscisic acid (ABA)** in the chloroplasts. When the level of ABA is sufficiently high, it affects the plasma membrane of the guard cells, preventing the proton pump from operating and bringing about closure of the stomata. As soon as more water is available to the plant, the ABA is broken down. This pattern of closure will avoid air locks in the xylem vessels that could impede the transpiration stream.

**Q** 4 **Why should midday closure of stomata affect the plant's temperature?**

## Factors affecting the rate of transpiration

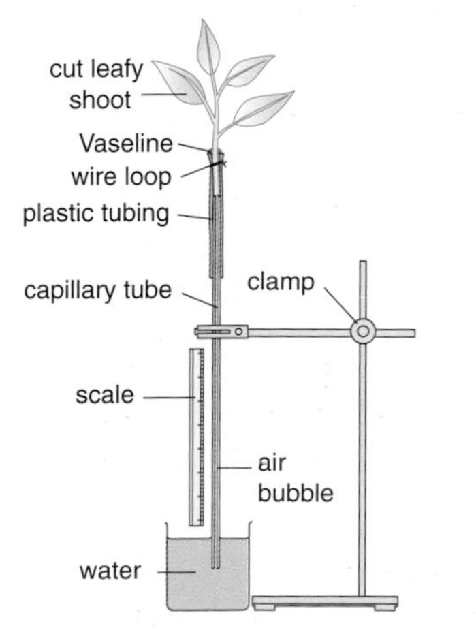


**Figure 7.19**
Simplified potometer

To find out which factors affect the rate of transpiration we must be able to measure its rate. This can be done in various ways. A commonly used method is to measure the rate at which the plant takes up water, the assumption being that this is the same rate as that at which water evaporates from the leaves. A little of the water taken up may be used in photosynthesis and other metabolic processes, but most is transpired. The apparatus for measuring water uptake is a **potometer** (Figure 7.19).

There are various types of potometer but they all work on the same principle and are used in the following way.

- Cut the end of the stem of a leafy shoot. As water within the stem is under tension, cutting the shoot will cause air to enter the xylem, so if possible cut the stem under water.
- Submerge the potometer and fill it with water, using a syringe if necessary to help pump out any air bubbles.

- Attach the leafy shoot by a short length of rubber tubing to a capillary tube, ensuring a tight fit. There must be no air locks; the water in the stem and the capillary tube should form a continuous system.
- Seal the joints around the rubber tube with Vaseline to keep the apparatus watertight.
- Introduce an air bubble into the capillary tube and time how long it takes for the bubble to travel a certain distance along the tube.
- Repeat the procedure a number of times.
- Calculate the water uptake in millimetres per minute, using the mean of the results obtained.
- The experiment can be repeated under differing conditions, e.g. at different air temperatures, in still or moving air.

The rate of transpiration can be affected by environmental factors, and also by a number of structural features of the plant.

The environmental factors can be divided into two categories:

**Factors affecting the supply of water**

- Dry soil conditions or increased soil salinity
  Both these conditions lead to a small water potential gradient between the soil and the root cells. The movement of water into the root will be reduced. In some cases water may even move out of the plant into the soil.
- Damage to roots
  If roots are damaged by pests the surface for the uptake of water is reduced and less water will enter the plants.

**Factors affecting loss of water from the plant**

- Light
  Light affects the rate of transpiration because it has an effect on the opening and closing of stomata. Even when stomata are closed some transpiration may occur through the cuticle. As stomata open in the morning, the rate of transpiration increases, decreasing at dusk when the stomata close. Since 90% of water lost in transpiration passes through the stomatal pores, light intensity will have an important effect on the rate of transpiration.
- Temperature
  An increase in temperature increases the kinetic energy and the random movement of molecules, leading to a faster rate of diffusion. The rate of transpiration increases with an increase in temperature, because water evaporates more rapidly from the cell walls of the mesophyll tissue. This increases the concentration of water vapour molecules in the air spaces in the leaf producing a greater water potential. Higher temperatures are associated with a lower humidity of the air outside the leaf. This leads to an increase in the water potential gradient between the leaf and the atmosphere, so water will diffuse out more rapidly.

- Wind speed
  Wind speed affects the water potential gradient between the inside of the leaf and the outside. In still air, water vapour molecules build up around the leaves, with the effect of reducing the rate of transpiration, as the water potential gradients are less steep. Any air movement can move the water vapour molecules away from the surface of the leaves thus creating steeper water potential gradients and therefore increasing the rate of transpiration.

- Relative humidity
  The humidity of the atmosphere has an effect on the rate of transpiration. A low humidity means the air is relatively dry and there is a steeper water potential gradient between the external atmosphere and the atmosphere inside the leaf, so the rate of transpiration is higher. A high humidity means the air is more saturated with water vapour molecules so transpiration is reduced.

**Structural features which affect the rate of transpiration include:**

- the surface area of the leaf: the greater the surface area the higher the total rate of transpiration.
- the thickness of the cuticle: the thicker the cuticle the less water lost through it.

**Q 5 What environmental factors cause the highest rate of transpiration?**

### Extension box 3

### Diffusion shells

Water vapour tends to build up close to the surface of the leaf as it diffuses out of the stomata. Obviously the atmosphere will be most saturated immediately outside each stoma and become progressively less saturated as water vapour diffuses away. Water vapour molecules are deflected by the perimeter of a stoma, and the closer they are to the edge the greater is the deflection. The diffusion paths of the water vapour molecules therefore describe a hemisphere around the stoma, called a **diffusion shell** (Figure 7.20). Points of equal water pressure (joined by dotted lines) form a series of hemispheres over the pore. Arrowed lines show the paths of molecules down the water potential gradient. This means that the molecules near the edge of the pore escape more readily than those in the centre. If the air is still, diffusion shells build up around the stomata and the rate of evaporation from the mesophyll cells inevitably decreases. Air movements blow away these diffusion shells, thereby increasing the rate of evaporation from the leaf.

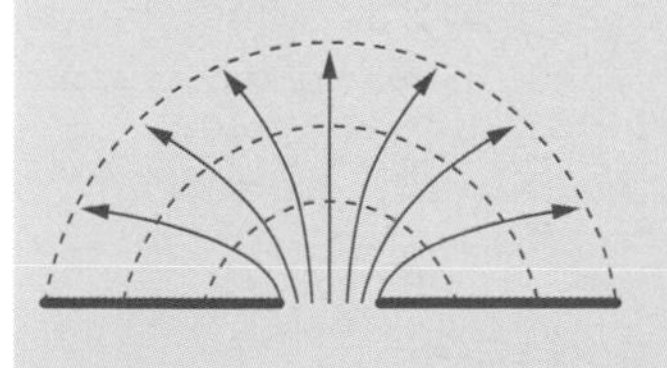

Figure 7.20
A diffusion shell

## Structural adaptations to dry conditions

Several physical factors affect the distribution of organisms in their habitats. These are often referred to as **abiotic** factors. In terrestrial habitats, light, temperature, soil type and the availability of water are important abiotic factors.

All organisms are adapted to survive in particular environmental conditions. Many adaptations are structural.

In land plants the rate of transpiration may exceed the rate of water uptake from the soil. Plants have little control over the volume of water they absorb, and therefore their water balance depends on their ability to limit water loss. Any feature that significantly reduces the evaporation of water from the aerial parts of a flowering plant can be considered to be a xeromorphic adaptation. Plants with these adaptations are called **xerophytes**.

Adaptations present in flowering plants include:

- thicker cuticles on leaves and herbaceous stems: for example on the upper epidermis of leaves of heather and holly and also on the lower epidermis in marram grass.
- reduction in the surface area of leaves: there often is a reduction in the surface area: volume ratio, resulting in a decrease in the area of the leaf blade where most of the stomata are situated. There may be a corresponding increase in the thickness of the leaf blade. For example, heather has small leaves whilst in other plants, such as cacti (Figure 7.21a), leaves are reduced to spines and photosynthesis takes place in the green stems. The needle-shaped leaf of the pine tree is another example.
- curling or rolling of the leaves into a cylindrical shape, reducing the surface area of the leaves: leaf rolling, for example in marram grass, encloses the upper epidermis, where the stomata are situated (Figures 7.21b and c). The lower epidermis has a thicker, waxy cuticle and no stomata. Humid air is trapped inside the rolled-up leaf, reducing the diffusion gradient and hence the rate of transpiration. Specialised cells known as hinge cells are present on the upper epidermis and lose water rapidly when transpiration rate is high causing the leaf to roll up.
- the number and distribution of the stomata: stomata may be confined to pits or grooves on the underside of the leaves, so that humid air is trapped and transpiration decreases (Figure 7.21d).
- the presence of epidermal hairs: these trap humid air so reducing the water potential gradient and therefore the rate of transpiration (Figure 7.21d).

(a)

(b)

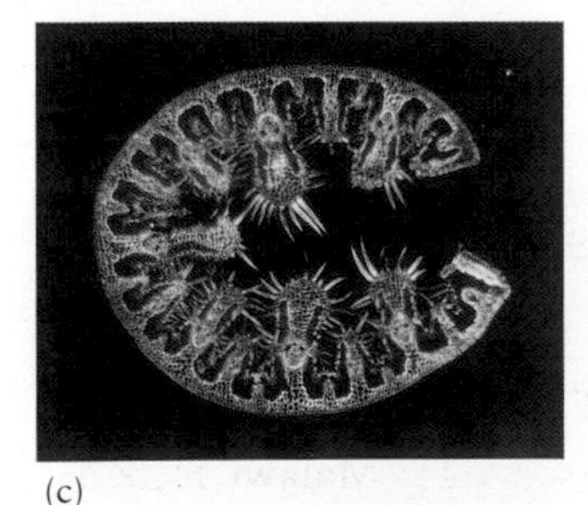
(c)

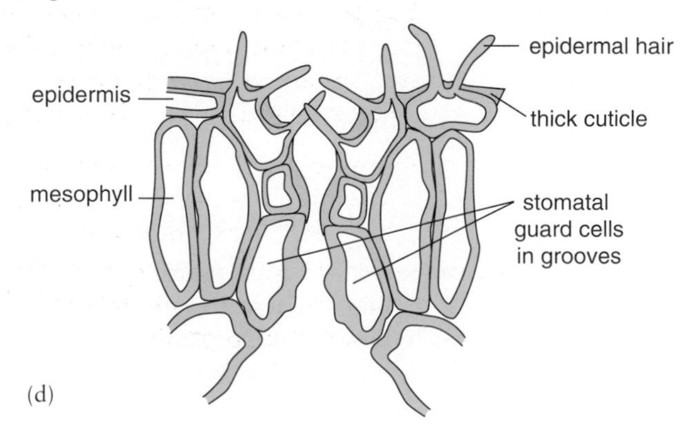

(d)

**Figure 7.21**
(a) A cactus, leaves are replaced by spines; (b) Marram grass leaves curl or roll up; (c) Photomicrograph of marram grass leaves curled up; (d) Sunken stoma often found in desert plants

## Summary

- A dicotyledonous root has a number of structures adapting it for the uptake of water.
- Water moves across the root by apoplastic and symplastic pathways
- Capillarity, root pressure and cohesion–tension are all involved in the movement of water up the stem.
- The rate of water lost by transpiration can be calculated using a potometer and is affected by air temperature, humidity and wind speed.
- Xerophytes have many structural adaptations which promote the uptake of water and control its loss.

## Assignment

### Growing at the limits

As human populations grow, more and more food is required. This often involves cultivating areas which are only just suitable for growing crops. Many of these areas are semi-arid regions in the tropics. These semi-arid regions receive some rain, but the main feature of their pattern of rainfall is that it is irregular and very variable.

In this assignment we will look at some of features of plant growth in semi-arid regions and at how we can improve crop yield by irrigation. You will need to use material from various parts of the specification to answer the questions. Before you start, it would be worth looking up the following key topics either in your notes or in your textbook:

- photosynthesis
- water potential.
- competition

The questions in this assignment also require the analysis and interpretation of data. Some of them will require you to select appropriate data to support your arguments. Because of this, the data for these questions have been put separately in Extension box 4.

1 Figure 7.22 (overleaf) shows groundnuts growing in Malawi. Here the rainfall is sufficient for a good crop. Some time ago, it was suggested that groundnuts would be a good crop to grow in central Tanzania. This 'groundnut scheme' failed because it was based on figures for mean annual rainfall. Suggest why this measure of rainfall was not appropriate.

*(2 marks)*

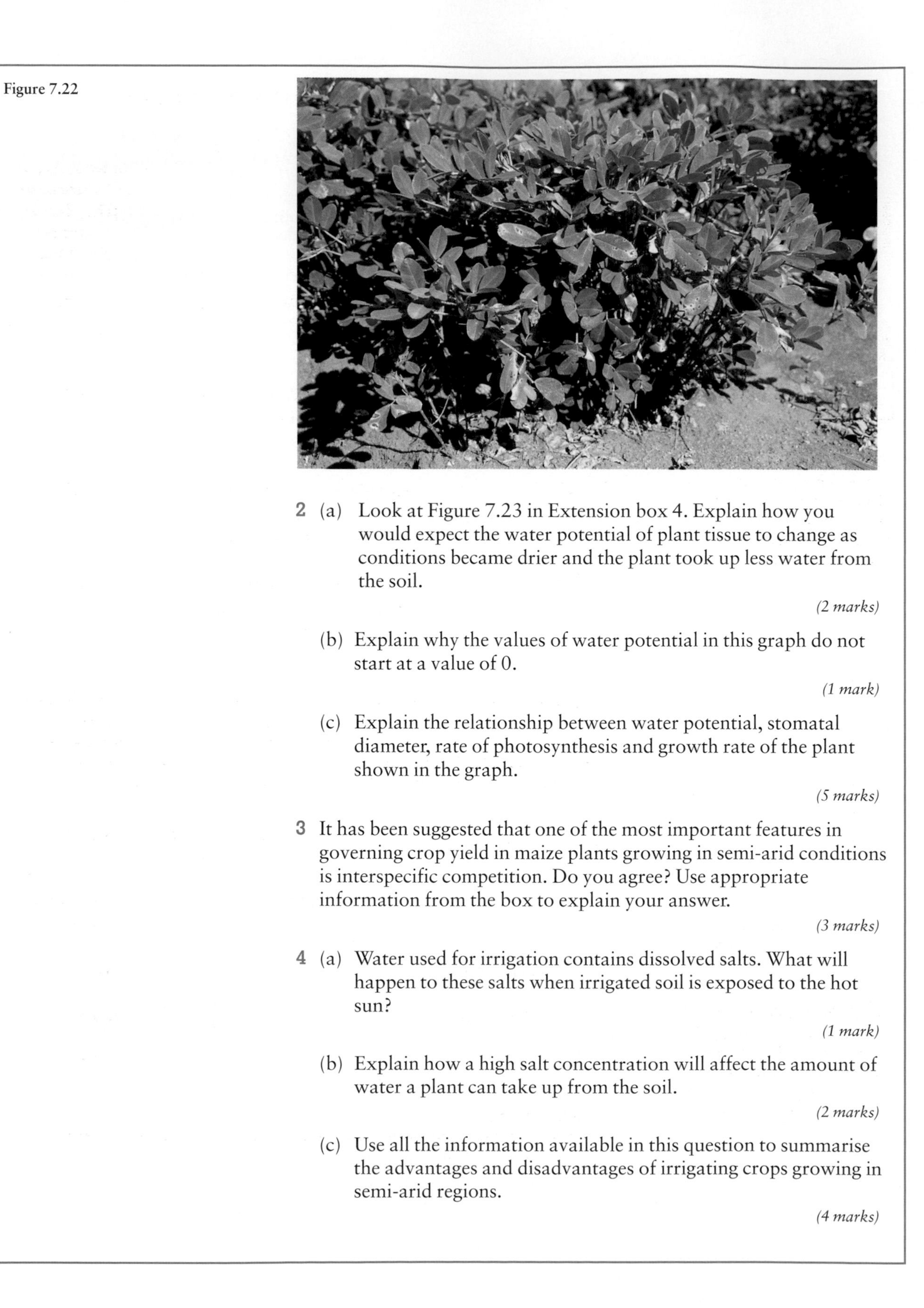

Figure 7.22

2 (a) Look at Figure 7.23 in Extension box 4. Explain how you would expect the water potential of plant tissue to change as conditions became drier and the plant took up less water from the soil.

*(2 marks)*

(b) Explain why the values of water potential in this graph do not start at a value of 0.

*(1 mark)*

(c) Explain the relationship between water potential, stomatal diameter, rate of photosynthesis and growth rate of the plant shown in the graph.

*(5 marks)*

3 It has been suggested that one of the most important features in governing crop yield in maize plants growing in semi-arid conditions is interspecific competition. Do you agree? Use appropriate information from the box to explain your answer.

*(3 marks)*

4 (a) Water used for irrigation contains dissolved salts. What will happen to these salts when irrigated soil is exposed to the hot sun?

*(1 mark)*

(b) Explain how a high salt concentration will affect the amount of water a plant can take up from the soil.

*(2 marks)*

(c) Use all the information available in this question to summarise the advantages and disadvantages of irrigating crops growing in semi-arid regions.

*(4 marks)*

## Extension box 4

## Some data concerning the growth of crops in semi-arid regions

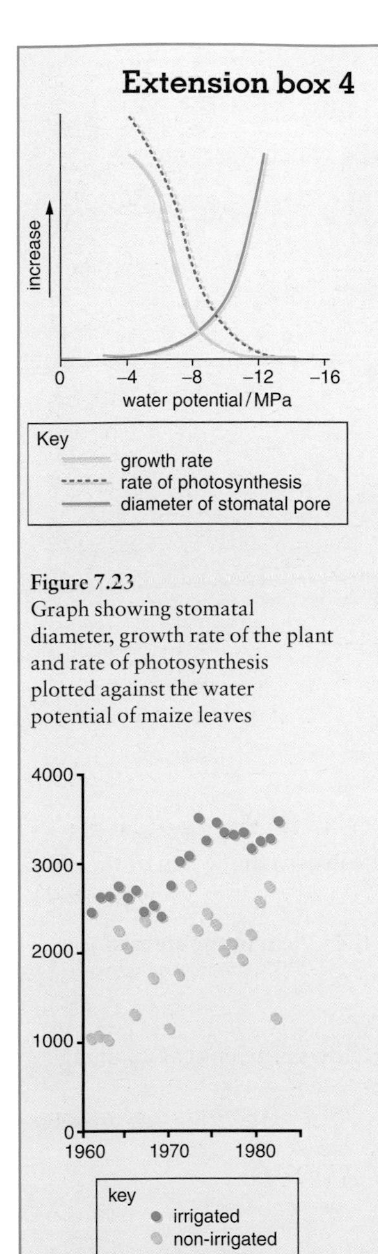


**Figure 7.23**
Graph showing stomatal diameter, growth rate of the plant and rate of photosynthesis plotted against the water potential of maize leaves

**Figure 7.24**
Graph showing the wheat yield in irrigated and non-irrigated plots of land over the period 1960–1985. Both plots of land are in the same semi-arid region in the Near East

| Plot | Treatment | Water potential of leaf tissue of maize plant in centre of plot/Mpa |
|---|---|---|
| A | All maize and grass growing around the central maize plant were left | −3.91 |
| B | All maize and grass growing around the central maize plant were removed | −3.37 |
| C | All maize growing around the central maize plant was removed. All grass was left | −3.93 |
| D | All grass growing around the central maize plant was removed. All maize was left | −3.46 |

**Table 7.10**
Maize was grown in experimental plots together with a species of grass. This grass is a common weed. The single maize plant growing in the centre of each of these plots was left. The plants growing round these maize plants were treated in the various ways described in the table. After two weeks, the water potential of the leaf tissue of the maize plant in the centre of the plot was measured

| Time of year | Population density of plant-sucking insects/ number per $m^2$ | |
|---|---|---|
| | Irrigated grassland | Non-irrigated grassland |
| Beginning of wet season | 208.0 | 13.7 |
| End of wet season | 398.2 | 136.5 |
| Dry season | 430.0 | 37.1 |

**Table 7.11**
Density of plant-sucking insects feeding on irrigated and non-irrigated grassland at different times of the year

# Examination questions

1 The diagram shows the main pathways by which water moves through a plant.

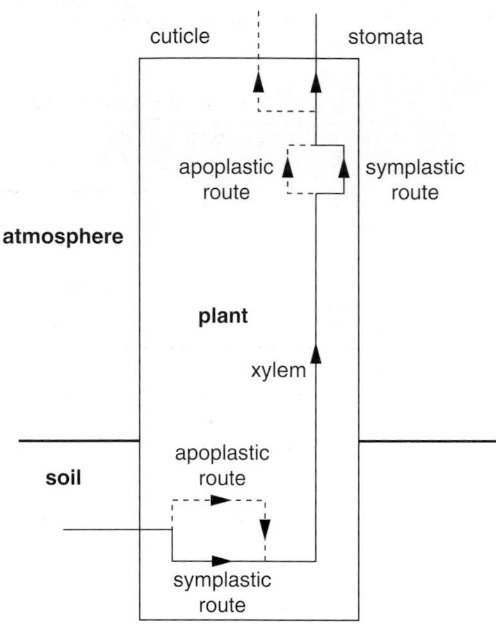


(a) Explain how water moves from the soil into the xylem of the root.

*(6 marks)*

(b) Explain how water moves through the xylem in the stem of the plant.

*(5 marks)*

(c) Explain how the structure of a leaf allows efficient gas exchange but also limits water loss.

*(6 marks)*

2 The diagram shows some cells from a plant root.

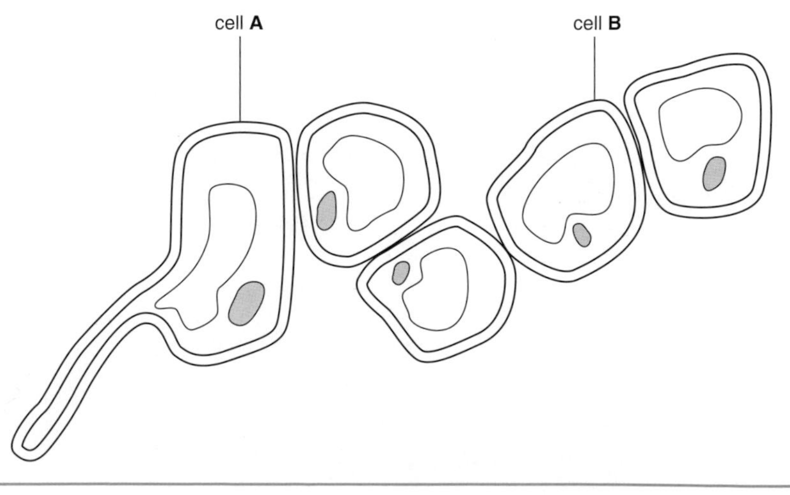

(a) In terms of water potential, explain how water moves from the soil into cell A.

*(2 marks)*

(b) Draw a line on the diagram to show the apoplastic pathway through these cells.

*(1 mark)*

(c) The table shows the concentration of certain mineral ions in the soil and in cell A.

| | Concentration (mmol $dm^{-3}$) | | |
|---|---|---|---|
| | **Potassium** | **Sodium** | **Chloride** |
| Soil | 0.1 | 1.1 | 1.3 |
| Cytoplasm of cell A | 93 | 51 | 58 |

What is the evidence from this table that uptake of mineral ions is:

(i) by active transport

*(1 mark)*

(ii) selective?

*(1 mark)*

# 8 Liver and Kidney

If one of your kidneys were injured or diseased and stopped working, you could probably still lead a normal life. However, if both your kidneys fail you would be faced with a crisis; water, urea and potassium build up rapidly in your body. You may continue to pass some urine, but you could not get rid of all the waste produced by normal cell processes.

Most people suffering from kidney failure hope for a transplant but there is a shortage of donors. Until a suitable organ becomes available, patients must rely on dialysis (Figure 8.1) to filter their blood and balance their fluid intake.

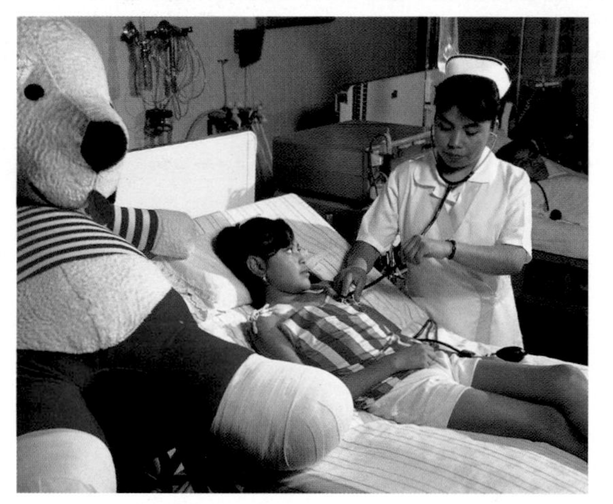

Figure 8.1
This girl is undergoing dialysis. This treatment takes a few hours and is usually required three times a week

The principle is straightforward. The person's blood passes on one side of a special membrane and a solution called a dialysate passes on the other side. The membrane allows some substances to pass through but not others. This is similar to the filtering action of the kidney. Substances that are small enough cross the membrane by diffusion, down a concentration gradient. Substances that are too large to pass through the membrane remain in the blood. This process is called dialysis. The blood and dialysate flow in opposite directions (a countercurrent system), maximising diffusion gradients. Clearly the composition of the dialysate is very important; it must match the components of the blood plasma that need to be retained in the blood. These substances include glucose and ions such as sodium. Substances that are not required by the body, such as urea, are absent from the dialysate.

Dialysis is uncomfortable and inconvenient. To reduce the time they need to spend in dialysis, patients with non-functional kidneys must stick to a strict diet. They must limit their fluid intake to only half a litre a day, about a quarter of the normal daily intake of a human adult. They must also keep their protein intake down to about 30 to 40 grams per day, the amount of protein in a small egg.

**Perhaps the biggest problem is the need to regulate potassium. This ion is a normal constituent of the body, but large amounts cause serious problems, including heart failure. Potassium-rich foods include citrus fruits, bananas, instant coffee, peanuts and chocolate and must be avoided.**

## Structure of the liver

The liver is the largest internal organ of the body. It lies in the abdomen just below the diaphragm. It is involved in regulating many metabolic activities, particularly those which help maintain the blood composition in a steady state. The liver is richly supplied with blood from two vessels, the **hepatic artery** and the **hepatic portal vein**. The hepatic artery delivers oxygenated blood, so the liver cells, can undergo aerobic respiration. The hepatic portal vein takes all blood from the small intestine to the liver (Figure 8.2)

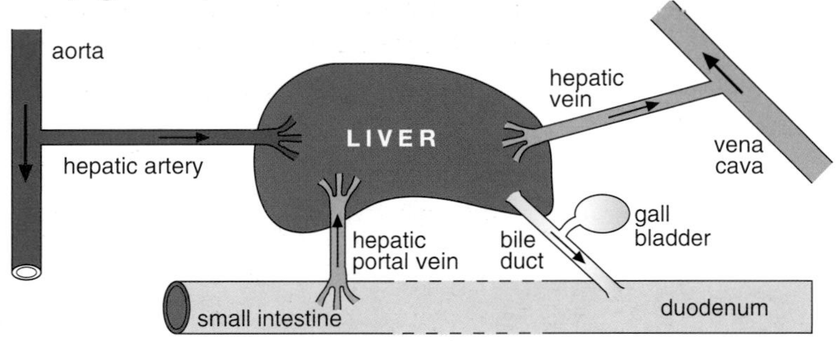


**Figure 8.2**
Blood enters the liver in the hepatic artery and the hepatic portal vein. It leaves the liver through the hepatic vein. Bile made in the liver passes into the duodenum through the bile duct

### Liver cells and lobules

Most of the liver is made of cells called **hepatocytes**. These are packed together into columns called lobules (Figure 8.3).

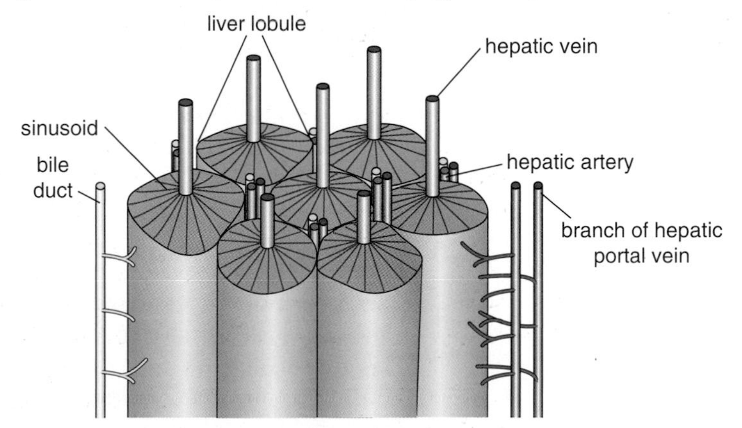


**Figure 8.3**
Liver cells are arranged into lobules and these lobules fit together. Each lobule is roughly cylindrical in shape and approximately 1 mm in diameter

In each lobule, blood from the hepatic artery and from the hepatic portal vein mixes and passes through **sinusoids** that run past each hepatocyte. Hepatocytes have many mitochondria to provide sufficient ATP for all the energy-requiring functions of these cells. On the plasma membrane bordering the sinusoids, the hepatocytes have microvilli which increase the surface area for uptake of substances from the blood.

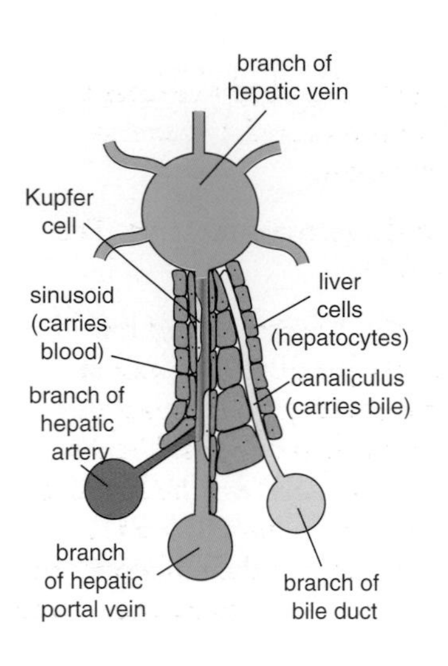


**Figure 8.4**
The arrangement of cells, sinusoids and canaliculi in the liver

There are specialised cells called **Kupfer cells** in the lining of the sinusoids. These are phagocytic cells, which ingest debris and bacteria and help defend against disease. They also break down red blood cells that are too old to perform their function.

Lying between the liver cells are channels called **canaliculi.** These collect the bile as it is formed in the liver cells and carry it towards the bile duct.

The flow of blood in the sinusoids is in the opposite direction to the flow of bile in the canaliculi. The blood flows inwards towards the centre of a lobule and the bile flows toward the outside of the lobule (Figure 8.4).

## Functions of the liver

The functions of the liver include:

- the production of bile and the synthesis of proteins such as those involved in the clotting of blood
- the production of excreting products such as ammonia, uric acid and urea
- the breakdown of haemoglobin from old red blood cells, hormones and toxic substances such as ethanol
- storage of vitamins and minerals such as iron
- metabolism of carbohydrates, lipids and amino acids. Many of these reactions are involved in keeping conditions in the body constant.

### Carbohydrate metabolism

The liver helps to regulate blood glucose concentration. The glucose concentration of blood leaving the liver is not necessarily the same as that entering the liver – it may be less or it may be more. Glucose can be added to the blood by converting glycogen stored in the liver to glucose (**glycogenolysis**) or by converting non-carbohydrate substances such as amino acids and glycerol into glucose (**gluconeogenesis**). Glucose can be removed from the blood by storing it as glycogen (**glycogenesis**), converting the glucose to fat or using the glucose as a fuel for respiration.

### Lipid metabolism

Liver cells process fatty acids so they can be transported and deposited in the body. The liver also regulates the amount of phospholipid and cholesterol, making them or removing them as required. Excess cholesterol and phospholipid are removed in the bile. The liver produces large amounts of bile, which helps to keep fats emulsified and then make them easier to digest (see page 232). Humans produce 0.5 $dm^3$ of bile per day, which is passed to a storage organ, the **gall bladder**, before it is secreted into the duodenum.

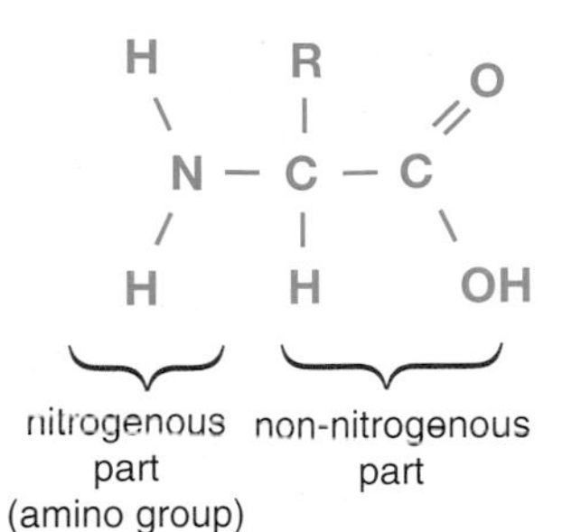


Figure 8.5
An amino acid showing the nitrogenous part and the non-nitrogenous part

## Amino acid metabolism

The liver is involved in dealing with amino acids absorbed in excess of the body's needs. If more protein is eaten than is required, the excess amino acids cannot be stored and are broken down.

An amino acid has a **non-nitrogenous** part and a **nitrogenous** part. The nitrogenous part is the **amino group** (Figure 8.5).

The first stage in breaking down an amino acid is the removal of the amino group. This is called **deamination** and the product is **ammonia**. Mammals convert ammonia to urea in a cycle of biochemical reactions called the **ornithine** cycle. This cycle occurs in the hepatocytes. Urea is transported in the blood from the liver to the kidneys where the blood is filtered and the urea passes out in the urine. The formation of urea is summarised in Figure 8.6. The non-nitrogenous part or the remainder of the amino acid can then either be respired or converted into carbohydrate or fats.

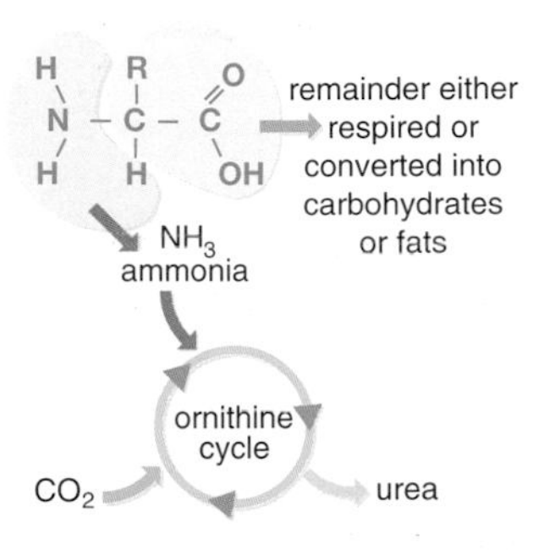


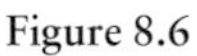

Figure 8.6
The outline of deamination and the ornithine cycle, which converts ammonia into urea

## Excretion

Excretion is the removal of the waste products resulting from metabolic processes in an organism. Waste products include urea, ammonia and uric acid, all of which contain nitrogen. Excretion also involves the removal of other waste products of metabolism such as carbon dioxide from respiration.

### Nitrogenous excretory products

The immediate nitrogenous waste product of the deamination of amino acids is ammonia. Ammonia may be excreted immediately or converted into one of the major nitrogenous excretory compounds, urea or uric acid, which differ in their solubility and toxicity. The exact nature of the excretory product is determined by which enzymes are present, the availability of water to the organism and the extent to which water loss can be controlled by the organism.

#### Ammonia

Ammonia is an extremely soluble molecule. It is very toxic to animals and cannot be stored in the body. Mammals are very sensitive to ammonia and cannot tolerate concentrations in excess of 0.02 mg per 100 $cm^{-3}$ of blood. As it is soluble, ammonia can be removed from the body rapidly and safely if diluted in a sufficient volume of water. Removal presents no real problem to organisms like fish, which have ready access to fresh water. The large volume of water lost in the urine is easily replaced from the environment. Organisms living in drier environments convert ammonia to other substances.

**Q** 1 **Why would terrestrial organisms have difficulty in removing nitrogenous waste as ammonia?**

### Urea

Urea is formed in the liver by the ornithine cycle. It is less soluble and less toxic than ammonia thus less water is needed to dilute urea to safe levels. It is the main chemical form in which nitrogen-containing compounds are excreted in mammals.

### Uric acid

When compared with urea, uric acid is extremely insoluble and much less toxic. This means that little water is required for its excretion and it is excreted more as a sort of paste than as a solution. This makes it an ideal excretory product in those animals where maximum conservation of water is important. It is the main excretory product in many insects, reptiles and birds.

## Mammalian kidney

The functions of the kidney include:

- removal of urea from the blood
- regulation of the water content of the body fluids

### Mammalian urinary system

The human urinary system is shown in Figure 8.7.

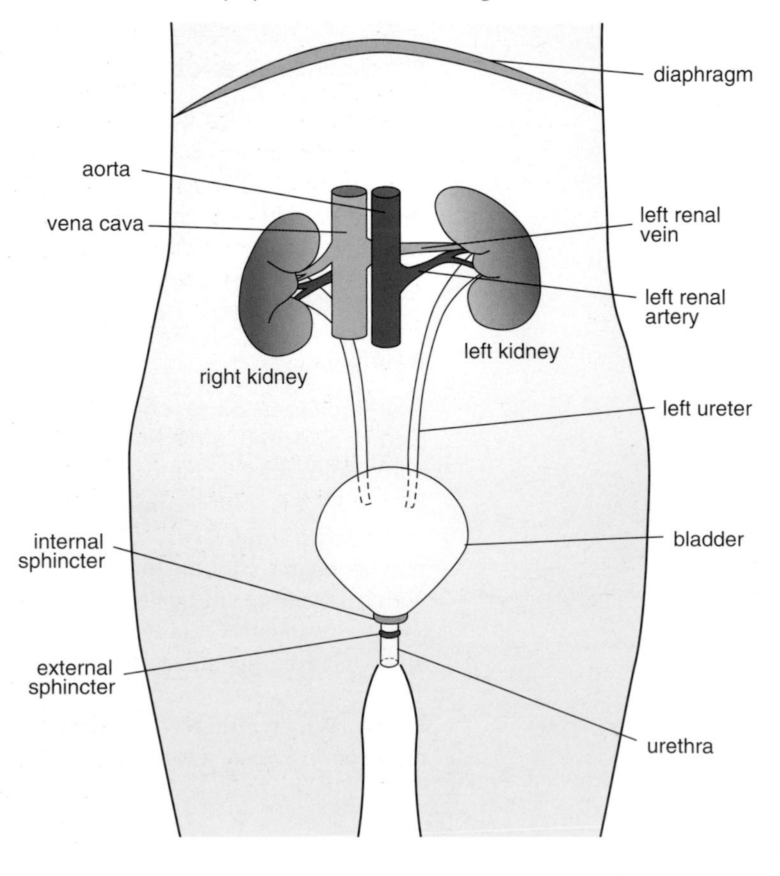


**Figure 8.7**
The human urinary system

The two kidneys lie at the back of the body, at about waist level.

The kidneys:

- receive blood through the **renal arteries** and blood is taken away by the **renal veins**
- have the largest supply of blood of any organ, about 1200 $cm^3$ of blood each minute
- require a blood supply at high pressure if they are to function properly; they fail quickly if the pressure drops.

**Urine** made by the kidneys flows into the **ureter** and is pushed by peristalsis into the **bladder** where it is stored until it is convenient to release it. When full, sphincter muscles are relaxed allowing the urine to pass into the **urethra** and then out of the body.

**Q** 2 **A person has 4 litres of blood and the kidneys receive 1 litre per minute. How many times on average does the total volume of blood in the body pass through the kidneys every hour?**

## Extension box 1

## Controlling the loss of urine from the body – micturition (the act of emptying the bladder)

Urine passes from the kidneys along tubes called ureters to the bladder, partly because of gravity and partly because of the squeezing action of the muscular ureter walls called peristalsis. The bladder wall is mostly made of muscle, with an epithelial lining. When it is almost empty the inner surface of the wall is folded. This allows stretching as the bladder fills. At the neck of the bladder are two rings of muscle or **sphincters**. The internal sphincter is controlled by a branch of the nervous system called the autonomic nervous system and cannot be controlled voluntarily. The external sphincter can be controlled voluntarily.

As the bladder fills to around 200 $cm^3$ the pressure inside it rises. Further filling to around 400 $cm^3$ does not cause further increase in pressure because the muscle layer relaxes and is stretched. As urine continues to enter the bladder there is an increase in pressure and the sensation of a 'full bladder' becomes more intense. Once the bladder contains around 600–700 $cm^3$ of urine, the emptying response cannot be voluntarily suppressed any longer.

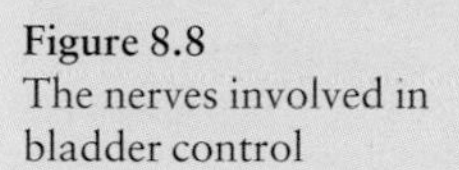

Figure 8.8
The nerves involved in bladder control

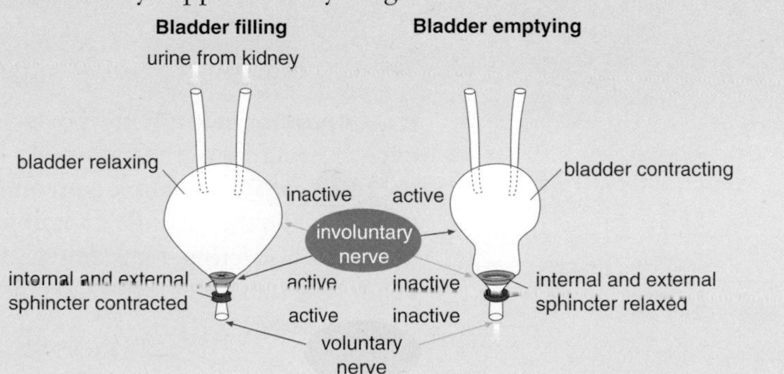

Stretch receptors in the bladder wall transmit impulses increasingly as the bladder fills. This stimulates the parasympathetic motor nerve fibres to the bladder. The bladder wall then contracts and at the same time the internal sphincter relaxes and opens. As the external sphincter relaxes at this stage urine passes out of the body.

Normally, we learn to control the external sphincter at about the age of two years.

## The structure of the kidney

A section through the kidney shows three distinct regions, an outer **cortex**, a middle **medulla** and an inner **pelvis** (Figure 8.9).

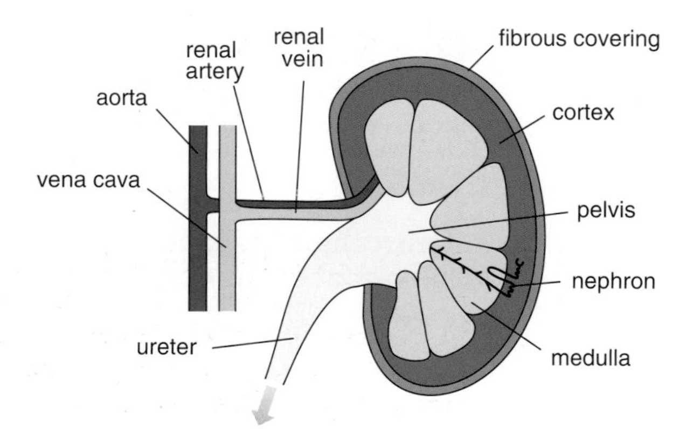


**Figure 8.9**
Section through a kidney showing its three distinct regions and the position of the nephrons

Each of our kidneys contains between 1 and 2 million microscopic multicellular structures called **nephrons**. Each nephron is a long tube about 3 cm in length.

Each nephron is divided into a number of distinct regions:

- renal capsule
- first convoluted tubule
- loop of Henle
- second convoluted tubule
- collecting duct.

The structure of a nephron is shown in Figure 8.10.

It is important to know the position of the nephrons in relation to the overall structure of the kidney. As Figure 8.10 shows, the cortex contains the **renal capsule,** the **first convoluted tubule** and **second convoluted tubule,** while the medulla contains the **loop of Henle** and the **collecting duct.** The collecting ducts deliver urine into a space called the **pelvis.** From here the urine drains into the ureter. The ureter connects the kidney to the bladder.

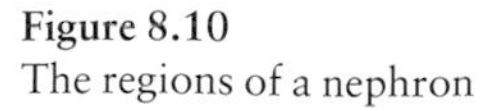

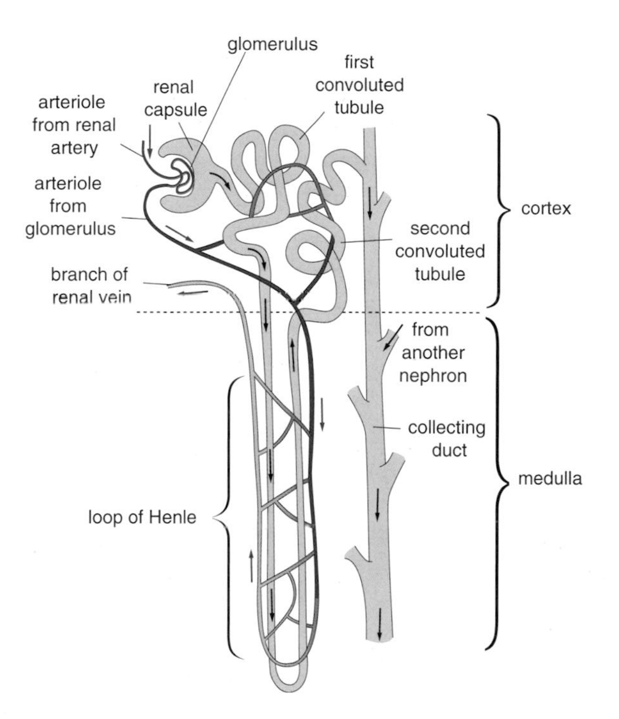


**Figure 8.10**
The regions of a nephron

Blood enters each kidney by the renal artery, which progressively branches, finally into **arterioles**. Each arteriole divides into a knot of **capillaries** called the **glomerulus**. The glomerulus is found in the cup of each renal capsule. Unusually, instead of blood leaving the capillary in a venule, it leaves the glomerulus by another arteriole and divides again into a second capillary network. This network surrounds the first and second convoluted tubules and the loop of Henle. The blood in these capillaries now drains into **venules** and finally into the **renal vein**.

## Kidney function

Kidney function involves three main processes.

- **Ultrafiltration:** this takes place in the renal capsule. Water and soluble substances pass from the blood into the lumen of the renal capsule. The **filtrate**, which is now in the nephron, contains excretory products but also substances useful to the body.
- **Selective reabsorption:** this takes place as the filtrate passes down the first convoluted tubule. The composition of the filtrate is changed as useful substances such as glucose and amino acids are reabsorbed into the blood.
- **Concentration:** this takes place in the second convoluted tubule and collecting duct. The loop of Henle create the right conditions in the medulla allowing concentration to occur.

### Ultrafiltration

A **renal capsule** is a double-walled cup-shaped structure surrounding a knot of capillaries called a **glomerulus**. The space inside the renal capsule is called the **lumen**. The capillary walls of the glomerulus are composed

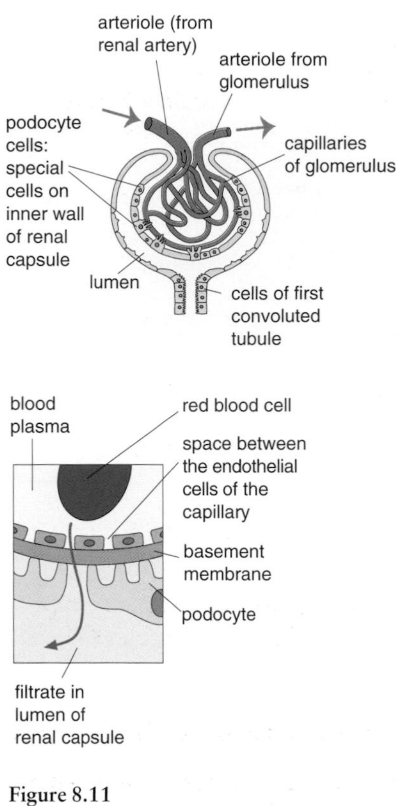


**Figure 8.11**
Ultrafiltration takes place through the basement membrane. Only small molecules can pass from the blood into the renal capsule

of a single layer of endothelial cells with spaces between the cells of between 50–100 nm. These cells sit on a **basement membrane.** This is the filtration membrane. It is the only continuous structure separating the blood in the capillary from the lumen of the renal capsule. The cells of the inner layer of the renal capsule do not fit tightly together. These cells, called **podocytes,** have many foot-like projections in contact with the basement membrane. Between the projections are gaps through which the filtrate can flow. The only barrier between the lumen of the capillary and the lumen of the renal capsule is the basement membrane. This is the filter, preventing the passage of cells and large molecules, mainly proteins, from the blood (Figure 8.11).

For ultrafiltration to occur there must be sufficient pressure (**hydrostatic pressure**) in the capillaries of the glomerulus to force solution through the basement membrane and into the lumen of the capsule. The blood pressure in the capillary of the glomerulus is much higher than in capillaries of other organs. One reason for this is that the arteriole which leaves the glomerulus is much narrower than the arteriole that enters. This leads to a high hydrostatic pressure which forces the blood plasma, minus the large proteins, into the renal capsule.

The hydrostatic pressure of the blood in the capillaries of the glomerulus is opposed by the hydrostatic pressure of the fluid already in the capsule and by the tendency for water to flow from the capsule space into the capillaries by osmosis. This osmotic effect is caused by the presence in the blood of the large plasma proteins; the blood has a high solute concentration and therefore a low water potential.

However, the hydrostatic pressure of the blood in the capillaries of the glomerulus is greater than these opposing forces. So, there is a **net** pressure from the capillaries to the capsule lumen, forming the filtrate.

The fluid in the capsule is called the **filtrate.** All the blood in the circulatory system passes through the kidneys every 4–5 minutes. This means that the rate of filtrate production is high, about 125 $cm^3$ per minute. Obviously, we do not produce anything like this volume of urine or we would dehydrate in minutes. On average we produce 1 $cm^3$ per minute. The rest, over 99%, of the filtrate is reabsorbed.

Ultrafiltration results in all kinds of small molecules passing into the nephron, whether the molecules are a waste product, like urea, or useful, like glucose. As the fluid passes through the rest of the nephron, useful substances are reabsorbed into the blood while urea stays in the nephron.

The rest of the nephron is now concerned with the reabsorption of small but valuable molecules like glucose. Necessary substances are reabsorbed, while toxic compounds, some water and any solutes in excess are excreted.

| Substance | Amount in blood plasma entering the glomerulus/arbitrary units | Amount present in filtrate in renal capsule/arbitrary units |
|---|---|---|
| water | 90–93 | 97–99 |
| large proteins | 7–9 | none |
| glucose | 0.10 | 0.10 |
| urea | 0.03 | 0.03 |
| sodium ions | 0.32 | 0.32 |

Table 8.1
The amounts of different substances in the blood plasma entering the glomerulus and in filtrate in the renal capsule

**Q 3 Using the table compare the amounts of substances in the blood plasma with that of substances in the filtrate in the renal capsule. Answer the following questions.**

**(a) Which substance is not filtered out of the blood plasma? Give reasons.**

**(b) For those substances which are filtered out of the blood plasma what can you say about their concentrations in the plasma and in the filtrate in the renal capsule? How can you account for this?**

**Q 4 A person lost a large amount of blood in an accident. What effect could this have on kidney function?**

## Selective reabsorption

The first convoluted tubule receives filtrate from the renal capsule. The cells of this part of the tubule, shown in Figure 8.12, show a number of adaptations that can be linked with their function.

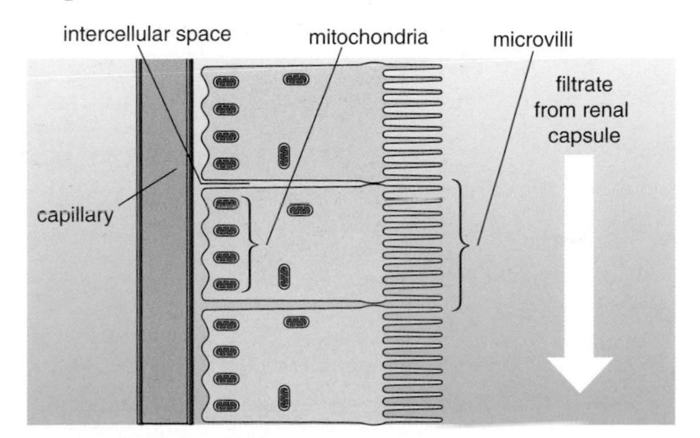


Figure 8.12
Cells of the first convoluted tubule showing their special features for selective reabsorption

The cells of the first convoluted tubule:

- have **microvilli** which increase their surface area
- contain many mitochondria which provide the ATP necessary for active transport
- lie next to the capillaries, with a continuous layer of fluid in the intercellular spaces between adjacent cells and between them and the capillaries.

All these are adaptations for the reabsorption of substances from the filtrate.

Over 80% of the filtrate is reabsorbed here, including **all** of the glucose, amino acids, vitamins, most hormones, 85% of salts (sodium chloride) and water.

The active uptake of sodium, glucose and other solutes allows a water potential gradient to be established. Water will therefore be reabsorbed from the nephron by osmosis.

Urea is not reabsorbed. However, as the fluid flows along the tubule, water has been removed and the concentration of urea in the first convoluted tubule increases.

**Reabsorption of glucose**

1 Glucose which has been filtered out of the blood is present in the filtrate in high concentrations. By the process of facilitated diffusion, the glucose is removed from the filtrate and enters the cells of the first convoluted tubule down a concentration gradient.

2 The glucose, now in a cell of the first convoluted tubule, is actively transported out of the cell into the intercellular spaces around the cell. This has two effects:

- it maintains a concentration gradient between the filtrate and the cells of the first convoluted tubule so that glucose will continue to diffuse out of the filtrate
- the glucose can enter the extremely permeable capillaries by diffusion. It is then carried away in the blood (Figure 8.13).

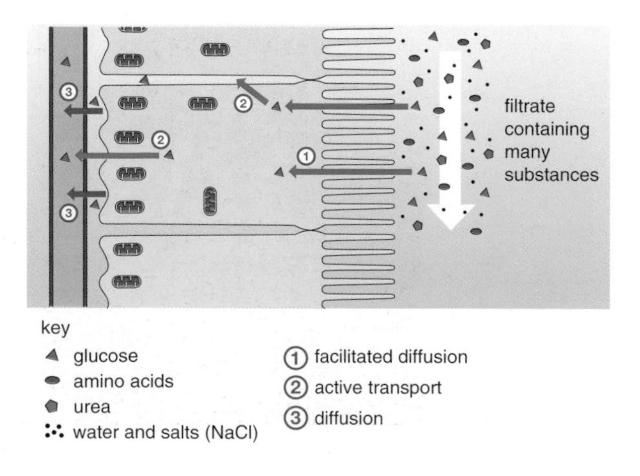


Figure 8.13
The processes involved in the reabsorption of glucose

Urine from a healthy person should not contain glucose. Glucose does appear in the urine when blood glucose becomes very high, as in diabetes for example (Figure 8.14).

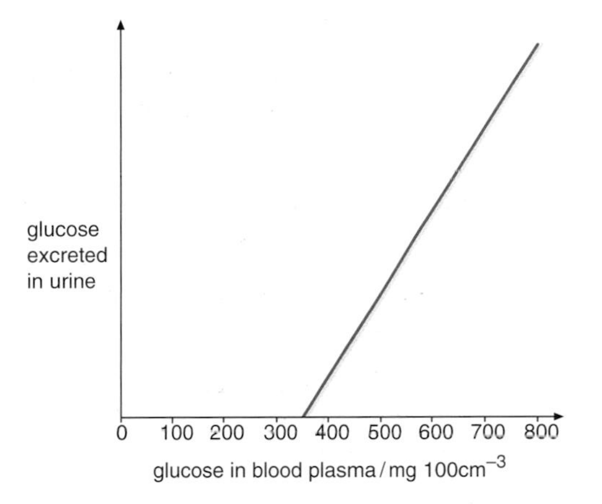


**Figure 8.14**
Only when glucose concentrations are very high does glucose appear in the urine

**Q 5 (a) From the graph in Figure 8.14, what is the threshold of sugar in the blood plasma before it appears in the urine?**

**(b) Explain the reason why all the glucose is not reabsorbed at high concentrations.**

### Producing concentrated urine

The function of the loop of Henle is to create a gradient of salt concentration in the medulla. This creates the conditions which enable a mammal to excrete urine which is more concentrated than blood.

The loop of Henle is a hairpin-shaped region that descends into the medulla and then ascends back into the cortex (Figure 8.15). The loop of Henle is surrounded by capillaries. By the time the filtrate reaches the beginning of the loop of Henle its volume is relatively low. The filtrate at this point is **isotonic** with blood in the capillaries (Box 2).

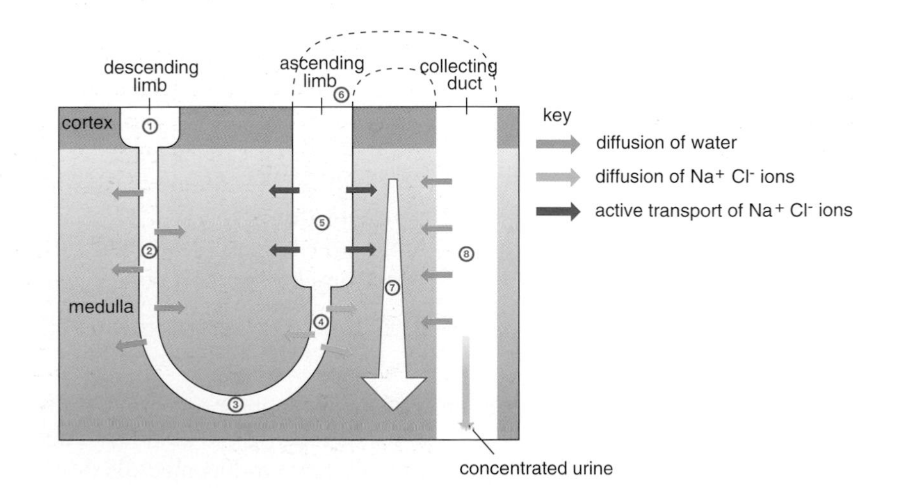


**Figure 8.15**
Loop of Henle and collecting duct

To help you understand how the loop of Henle and the collecting duct produce urine that is more concentrated than blood plasma, you will need to use Figure 8.15 as well as the text.

**Extension box 2**

**You will need to understand the following terms.**

- **Isotonic**
  Isotonic solutions have the same solute concentration as each other. If two solutions separated by a partially permeable membrane are isotonic, it means that the water potential is the same on either side. There will be no net osmotic movement of water between the two solutions.
- **Hypertonic**
  One solution may be described as hypertonic to another. A hypertonic solution has a greater solute concentration. Thus if a cell is surrounded by a hypertonic solution, the concentration of the surrounding solution is greater than the concentration of dissolved substances in the cytoplasm of the cell. The water potential of the cytoplasm will therefore be higher (less negative) than the water potential of the surrounding solution. As a result, water will move out of the cell by osmosis.
- **Hypotonic**
  One solution may be described as hypotonic to another. A hypotonic solution has a lesser solute concentration. Thus if a cell is surrounded by a hypotonic solution the concentration of the surrounding solution is less than the concentration of dissolved substances in the cytoplasm of the cell. The water potential of the cytoplasm will therefore be lower (less negative) than the water potential of the surrounding solution. As a result, water will move into the cell by osmosis.

- The first part of the descending loop of Henle ① is short and impermeable to both water and salts (sodium and chloride). This region brings the filtrate from the first convoluted tubule in the cortex into the medulla.
- The thin descending limb ② is freely permeable to water, but impermeable to sodium and chloride ions. Since there is a higher concentration of dissolved substances in the fluid in the medulla, water moves out of the nephron by osmosis, down the water potential gradient, and is carried away by the capillaries. The loss of water not only reduces the volume of the filtrate but it also results in the ion concentration increasing as the filtrate passes down the descending limb of the loop of Henle. The filtrate becomes increasingly **hypertonic**, being most concentrated at the apex ③.
- The filtrate now returns towards the cortex up the ascending limb of the loop of Henle. The ascending limb is impermeable to water.
- The first part of the ascending limb ④ is permeable to sodium and chloride ions. As the fluid flows up the ascending limb, the ions diffuse into the surrounding medulla down a concentration gradient. Water cannot leave the tubule. So, compared with the concentration of ions at

the apex of the loop of Henle, the fluid becomes increasingly hypotonic as it passes up the ascending limb.

- The cells in the second part of the ascending limb ⑤ have many mitochondria and actively transport sodium and chloride ions out of the filtrate into the surrounding medulla. The result of this is that the filtrate continues to become less concentrated. By the time it passes into the second convoluted tubule ⑥, it is hypotonic to blood in the surrounding capillaries.
- Within the medulla, from the cortex to the pelvis, the concentration of ions increases ⑦.
- The collecting duct ⑧ passes through the region of increasing salt concentration in the medulla. Water leaves the collecting duct by osmosis and is reabsorbed into the blood.
- The concentration gradient in the medulla ensures that there is always a water potential gradient between the collecting duct and the surrounding fluid. Water will therefore continue to be removed and returned to the blood. This will result in concentrated (hypertonic) urine.

### Countercurrent multiplier

The mechanism that occurs in the loop of Henle is referred to as a countercurrent multiplier. This is because the loop of Henle has two limbs, descending and ascending, side by side and the fluid in them flows in opposite directions (countercurrent). The descending limb is permeable to water but impermeable to salts. Water is drawn out of the descending limb by osmosis so that as the filtrate goes down the limb it becomes more and more concentrated and the water is taken away in the blood. As the filtrate passes up the ascending limb it becomes less and less concentrated. This is because sodium and chloride ions are able to diffuse out. At any point in the medulla there is only a little difference in the salt concentration between the inside and outside of the tubule. But there is a big difference between the top and bottom of the medulla. The effect is multiplied along the hairpin. The longer the loop of Henle, the greater the concentration that can be achieved at the bottom of the medulla. Remember that the saltiness of the medulla affects how much water can be removed from the collecting ducts.

Animals that live in dry conditions have very long loops of Henle. To accommodate this extra length, the medulla of their kidneys is very thick.

| Species | Habitat | Thickness of medulla/arbitrary units |
|---|---|---|
| Beaver | Freshwater | 1.0 |
| Human | Land | 2.6 |
| Kangaroo rat | Desert | 7.8 |

**Table 8.2**
The thickness of the medulla of the kidney in mammals that live in different habitats. Many mammals that live in desert conditions have very long loops of Henle and can produce extremely concentrated urine.

### Adjusting the concentration of the urine

The permeability of the walls of the second convoluted tubule and the collecting duct is affected by hormones. This hormonal effect, together with the concentration gradient in the medulla, determines whether more concentrated or less concentrated urine is released from the kidney. If the walls of the collecting duct and second convoluted tubule are permeable, water leaves and more concentrated urine is produced. If they are impermeable to water the final urine will be less concentrated.

We will consider the hormonal control of the permeability of the walls of the collecting duct and second convoluted tubule in the next section.

### The kidney and water balance

Table 8.3 shows a typical water balance sheet for an average person, assuming a normal level of activity and a comfortable external temperature.

| Water gain | Volume/$cm^3$ | Water loss | Volume/$cm^3$ |
|---|---|---|---|
| Food and drink | 2100 | Skin | 350 |
| Metabolic waste | 200 | Sweat | 100 |
| | | Breath | 350 |
| | | Urine | 1400 |
| | | Faeces | 100 |
| **Total** | **2300** | **Total** | **2300** |

Table 8.3
How water is gained and lost in an average person

**Q 6 How would the water balance sheet change if an individual was suffering from a serious bowel infection causing diarrhoea?**

Almost two thirds of our water comes from drink and the remaining third from food. We obtain a small but important proportion from metabolic reactions, such as respiration. Some of the water loss is unavoidable and depends on factors such as the environmental temperature and activity. Nitrogenous waste must be removed in solution, and so some water loss as urine is inevitable. The amount of water lost in removing the waste can vary. Like most homeostatic mechanisms, maintenance of water balance involves a negative feedback loop that has a detector and a correction mechanism (see page 177).

A part of the brain, the hypothalamus, contains osmoreceptor cells that are sensitive to the solute concentration of the blood.

A rise in solute concentration indicates that water loss is greater than intake. This may be due to:

- little water being taken in
- much sweating due to very hot conditions
- large amounts of salt being taken in.

The hypothalamus responds in two ways:

- it stimulates the thirst centre in the brain
- it stimulates the pituitary gland to secrete **antidiuretic hormone** (**ADH**)

ADH increases the permeability of cells of the collecting duct and second convoluted tubule to water. When ADH makes the cells of the collecting duct more permeable to water, the region of high solute concentration created by the loops of Henle has a dramatic effect. More water leaves the tubule and re-enters the blood. A much more concentrated urine is produced and vital water is conserved. When the concentration of the blood returns to normal less ADH will be produced.

Conversely, when fluid intake exceeds loss, the blood becomes more dilute.

This could be due to:

- large volumes of water being taken in
- little sweating
- low salt intake.

When dilute blood is detected by the hypothalamus, ADH secretion is reduced. The walls of the collecting duct and second convoluted tubule remain impermeable to water. As a result, less water is reabsorbed and larger volumes of dilute urine are produced. Figure 8.16 summarises this mechanism.

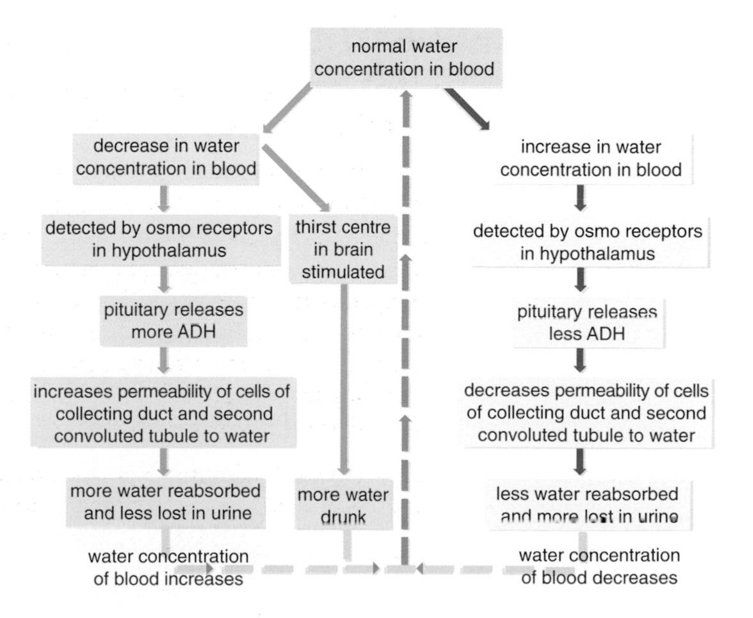


**Figure 8.16**
ADH and the control of water balance

**Q 7 In desperation, a castaway drinks some seawater. Why would this not be a good idea?**

### Water regulation in desert mammals

A kangaroo rat is a small desert-living mammal (Figure 8.17). Table 8.4 shows how water is gained and lost in a kangaroo rat.

**Figure 8.17**
The kangaroo rat lives in the very harsh environment of a desert. It spends the hottest part of the day in a burrow deep underground. This means that it can keep its body temperature from rising without having to lose valuable water as sweat

| Water gain | Volume/$cm^3$ | Water loss | Volume/$cm^3$ |
|---|---|---|---|
| Metabolic water | 54.0 | Urine | 13.5 |
| Dry food | 6.0 | Faeces | 2.6 |
| | | Evaporation (breath) | 43.9 |
| **Total** | **60.0** | **Total** | **60.0** |

**Table 8.4**
How water is gained and lost in a kangaroo rat

These animals can survive on dry food and no drinking water for long periods. Their ability to control water loss enables them to survive in such a harsh environment.

This animal is well suited to life in the desert. Not only does it avoid dehydration, but it also avoids overheating even though it does not sweat.

The kangaroo rat shows the following adaptations to its hot dry environment.

- Its kidneys have very long loops of Henle. This allows the kangaroo rat to produce very concentrated urine. Urea can be removed without losing significant amounts of water.
- It spends long periods underground where the air is cooler and more humid. This reduces its water loss by evaporation.

**Q 8 How is metabolic water formed?**

## Summary

- Different organisms excrete different forms of nitrogenous waste depending on their environment: ammonia in fish, uric acid in birds and insects and urea in mammals.
- The deamination of urea occurs in the liver and involves the removal of an amino group from an amino acid forming ammonia, which is converted to urea.
- The kidney is made up of many multicellular structures called nephrons that are associated with blood vessels.
- The ultrastructure of the glomerulus and renal capsule filters blood under pressure (ultrafiltration) to produce the filtrate.
- The structures of the cells lining the first convoluted tubule (epithelial cells) are adapted for reabsorption with microvilli and many mitochondria.
- The countercurrent multiplier mechanism of the loops of Henle creates a concentration gradient through the medulla, allowing water to be reabsorbed from the nephron.
- ADH affects the permeability of the second convoluted tubule and the collecting duct, allowing the kidney to regulate water levels in the body.
- Small desert mammals face particular problems in regulation of water. They minimise respiratory, cutaneous and excretory loss of water as well as adapting their behaviour.

## Assignment

### Writing synoptic essays

When you have studied the material in the next chapter as well as this one, you will appreciate that homeostasis means keeping conditions in the body constant. So separating a chapter about the kidney and the liver from one on homeostasis doesn't make a lot of biological sense. These two organs play a very important part in maintaining a constant internal environment. For example, the liver is involved in regulating the glucose concentration in the blood and the kidney helps to control the water balance of the body.

The idea of homeostasis is one that links many areas of an organism's biology and, as biologists, it is important that we understand links like this. One of the ways in which AQA Specification A assesses your ability to bring together different areas of the subject is with an essay. In this essay, marks are awarded not only for your scientific knowledge but also for selecting appropriate material from different parts of the specification. In order to write a good essay you need to practise this skill.

In this assignment we will look at writing an essay. You will probably find it better to attempt it after you have completed this chapter and the next one. You will then have more knowledge of homeostasis.

Writing any essay, involves three steps. These are

- analysing the question so that you know exactly what you have to do
- planning your answer
- writing the essay

In this assignment we will concentrate on the first two of these steps – analysing and planning.

Here is your task.

**Write an essay on the following topic:**

**The ways in which a mammal maintains constant conditions inside its body.**

## Analysing the question

1 Start by looking at the instruction. In this case it requires you to "Write an essay" so you need to write out your answer in continuous prose. You shouldn't be using bullet points or writing in note form. If it helps to make a particular point, however, you can use a diagram, providing that it is relevant and adequately explained.

2 Look at the topic you are required to write about. The wording of the essay title tells you that you should confine your answer to mammals – so there is no place here for all that interesting stuff about how a crocodile maintains a constant body temperature. A crocodile is not a mammal. It also tells you that you have to look at maintaining constant internal conditions, so you should not, for example, include a lot of irrelevant detail about how the liver and the kidney function.

3 Finally, what about the mark allocation? The essay is worth 25 marks but, perhaps more important than this, you have about 40 minutes and that includes time for planning as well as writing.

## Planning your answer

Right. You know exactly what you have to do. The next step is to plan your answer. Planning is one of those awkward pieces of advice which everyone is given but few act upon. In an examination there never seems enough time for this. What we will try to do in this assignment is to show you how to produce an effective plan in a very short time.

You need a framework to help you to explore your knowledge of biology. Otherwise it is all too easy to concentrate on a few aspects of a particular topic. We can divide biology up in a number of different ways. Look at Table 8.5. It shows some of the ways in which we can do this:

| Prokaryotes | Protoctists | Fungi | Plants | Animals |
|---|---|---|---|---|
| Cell biology | Biochemistry | Physiology | Genes and genetics | Ecology |
| Gas exchange and transport | Nutrition | Homeostasis and excretion | Co-ordination and movement | Reproduction and growth |

Table 8.5

Each row in the table shows a different way of dividing up the subject. The first row is based on classification and shows the five kingdoms of living organisms; the second row shows the main areas of biology that make up this specification; the third concentrates on physiology. Now let's go back to our essay. We need to look at this table and choose the row that will provide us with the best framework for a plan. We can't really select the first row as the essay only concerns mammals. The second row is similarly of little use as we have to write about the ways in which a mammal maintains constant conditions inside its body. The third row seems the best option.

The next step is to use this as the basis of a brain storming exercise. Draw a table with the items in this third row forming your column headings. Now, try to write something under each heading - one way in which this topic is related to maintaining constant conditions. If you are stuck, don't worry. Move on to the next column. The object is to work as quickly as possible. Set yourself a limit of five minutes - you won't have more time than this in an examination. You might end up with something like Table 8.6.

| Gas exchange and transport | Nutrition | Homeostasis and excretion | Co-ordination and movement | Reproduction and growth |
|---|---|---|---|---|
| pH of the blood<br><br>Exercise and blood flow to the brain | | Temperature<br><br>Blood glucose concentration<br><br>Water balance | | Concentration of reproductive hormones such as FSH |

Table 8.6

We certainly haven't managed to write something in every column but if you look carefully at this table you will see that we have covered a good range of topics and we should be able to gain most of the marks available for demonstrating a range of knowledge.

## Writing the essay

You could try writing this essay but, before you do, look at the hints below:

- Structure your essay. Start with an introduction. In this case you could relate the title of the essay to homeostasis and, possibly, explain why maintaining constant internal conditions is so important. Follow this with the main content, devoting a separate paragraph to each of the different aspects you plan to write about.
- Make sure that you include enough detail. In practice this means that you can't afford to write more than about four lines without bringing in something that you have learnt as part of your A-level Biology course. Don't forget, you have only got about 40 minutes altogether.
- Keep it relevant. Don't get side-tracked and include things "just in case".
- Use the right language. To get your marks, you must use proper scientific terms.

# Examination questions

1 The diagram shows some important features of homeostatic mechanisms in the body.

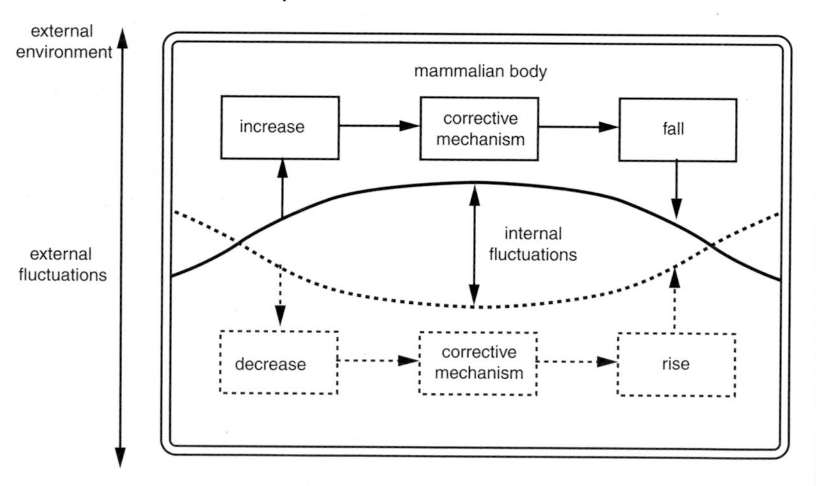


(a) Use the information in the diagram to help explain the importance of a mammal maintaining a constant internal temperature.

*(4 marks)*

(b) Explain the role of the hypothalamus and nervous system in the regulation of body temperature.

*(5 marks)*

(c) Explain why, in a normal healthy individual, the blood glucose level fluctuates very little.

*(6 marks)*

2 A person fasted overnight and then swallowed 75 g of glucose. The graph shoes the resulting changes in the concentration of insulin and glucose in the blood.

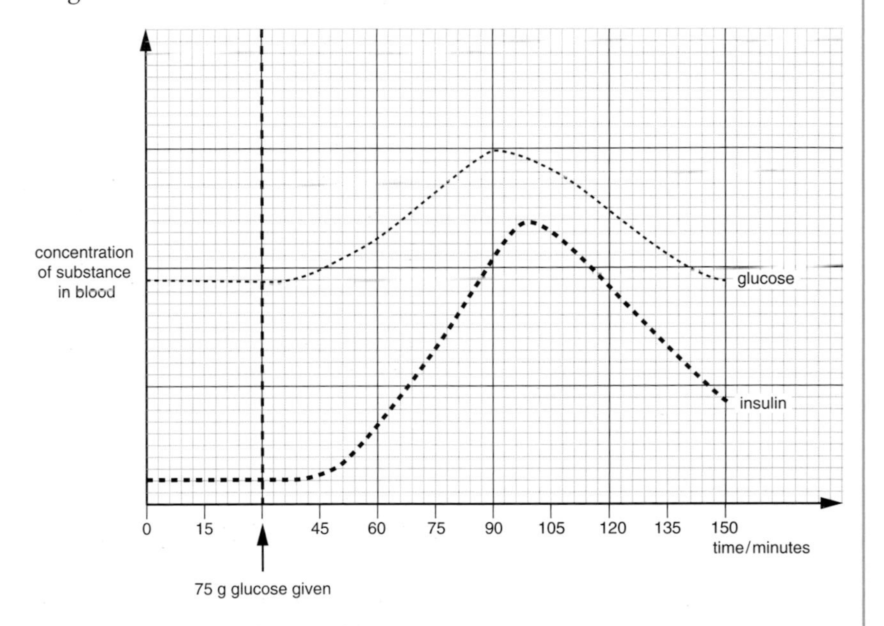


(a) (i) Explain the relationship between the concentration of glucose and insulin in the blood in the first 30 minutes after the glucose was swallowed.

*(2 marks)*

(ii) Use information from the graph to explain what is meant by the term negative feedback.

*(1 mark)*

(b) Explain why the concentration of glucagen in the blood rises during exercise.

*(2 marks)*

# Homeostasis

Figure 9.1
Meerkats standing to expose their bodies to the warming rays of the Sun

After a cold night in Africa's Kalahari Desert, animals small enough to fit in a coat pocket emerge stiffly from their burrows. These 'meerkats' stand on their hind legs and face east, exposing their chilled bodies to the warm rays of the morning sun (Figure 9.1). This simple behaviour helps to maintain internal body temperature.

Meerkats do not know it, but sunning behaviour helps the functioning of their enzymes. If their body temperature were to fall below or to exceed a tolerable range, enzyme activity and thus metabolism would suffer.

Once meerkats warm up, they fan out from the burrows, looking for food. Into the meerkats' guts go insects and an occasional lizard. These are digested and the products absorbed into their bloodstreams and then to cells throughout their bodies. These nutrients are used in respiration and so release vital energy, not only for all functions of their bodies, but also in the form of heat.

The emperor penguin lives in the Antarctic. It is a warm-blooded animal which breeds in the middle of winter when temperatures are at their lowest: minus 40 °C.

Once the female has laid her single egg she returns to the sea to feed, leaving the male to incubate the egg for two months. Though the male stands on the ice, the egg is held on his feet and covered by a fold of fat. Despite the very cold conditions, the core body temperature of the male penguin remains at 38 °C. The penguins are able to maintain their body heat by respiring stored fat and by huddling together (Figure 9.2).

Figure 9.2
Snuggling up to a neighbour greatly reduces the surface area that is exposed to the freezing air. They take it in turns to stand on the outside of the huddle where the air temperature is at its lowest

## Why is homeostasis important?

The cells in any living organism will only function properly in the correct conditions. The rates of chemical reactions change according to environmental conditions and most organisms live in changeable environments. Temperature, for example, on a cold day in Britain might vary from 0 °C outside to 20 °C indoors. This variation could be a major problem if cells were at the mercy of changes in the external environment. The solution is to resist these changes, to keep conditions inside the cell constant, whatever may be going on outside.

Homeostasis involves mechanisms which keep conditions inside an organism within narrow limits and thus allow independence from fluctuating external conditions.

- Biochemical reactions are controlled by enzymes. Changes in pH and temperature affect the rate of enzyme-controlled reactions. Extreme temperatures and pHs can lead to denaturing of enzymes and other proteins.
- Water moves in and out of cells by osmosis. By maintaining a constant water potential in the fluid surrounding the cells, osmotic problems, which could lead to cellular disruption, are avoided.

## Keeping the internal environment constant

The blood cells are suspended in the blood plasma and other cells are surrounded by tissue fluid, which is derived from the plasma. A range of conditions in the blood (and therefore in the tissue fluid) is maintained close to the value at which cells function best. Keeping constant conditions in the tissue fluid around the cells is an example of homeostasis. The word homeostasis means steady state. Features of an organism's internal environment such as temperature, pH and the concentration of many dissolved substances have a set level, often referred to as the **norm**. Negative feedback is the process in which a fluctuation from this set level sets in motion changes, which return it to its original value. Although we talk about maintaining a constant internal environment, it is rarely constant, but fluctuates around the norm. Blood temperature in mammals is allowed to fluctuate only within very narrow limits, while blood glucose concentration can rise considerably above the norm without any harmful effects. Several different mechanisms are responsible for homeostasis but the basic features are always the same (Figure 9.3).

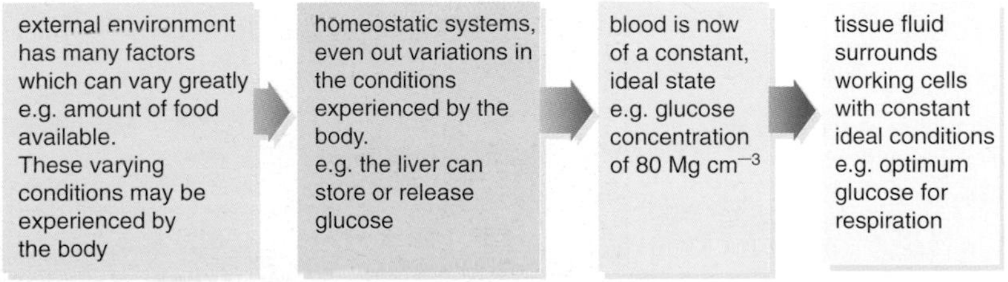

**Figure 9.3**
The basic features of homeostasis

external environment has many factors which can vary greatly e.g. amount of food available. These varying conditions may be experienced by the body
→ homeostatic systems, even out variations in the conditions experienced by the body. e.g. the liver can store or release glucose
→ blood is now of a constant, ideal state e.g. glucose concentration of 80 Mg $cm^{-3}$
→ tissue fluid surrounds working cells with constant ideal conditions e.g. optimum glucose for respiration

## Extension box 1 Regulation of blood cholesterol

### Cholesterol in the body

Most cholesterol found in the human body is made by liver cells. Some also enters the blood in foods such as butter and cheese. Despite concerns about health and high cholesterol concentrations, all cells need it as a component of membranes. The liver uses a lot to make bile salts, vital for fat digestion. Significant amounts are deposited in the skin, helping to make it waterproof, and some is used to make steroid hormones like oestrogen and testosterone. Acetylcholine, a neurotransmitter, is also made from it.

### Cholesterol in the blood

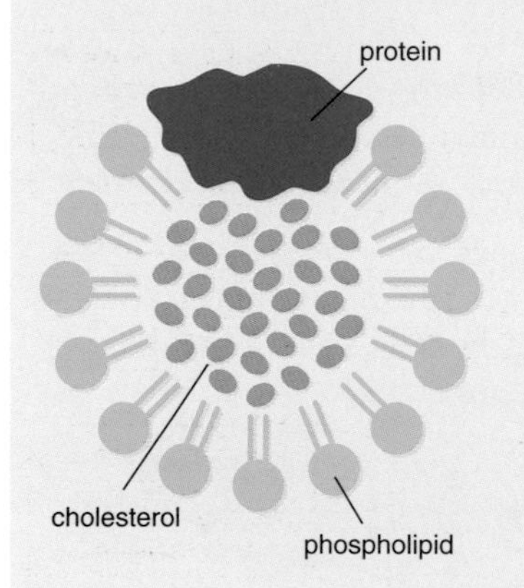


**Figure 9.4**
Lipoprotein

Cholesterol is insoluble in water. As it needs to pass to all cells it is carried in the blood in complexes called **lipoproteins**, mostly low density lipoprotein (LDL). A protein and phospholipid coating separates a ball of cholesterol from the aqueous environment (Figure 9.4).

### Why regulate cholesterol?

Cholesterol is needed by all cells, so must circulate in reasonable concentrations in the blood. However, at higher concentrations it tends to be deposited in the linings of artery walls. This narrows the artery and could make a heart attack or stroke more likely. Patients with an inherited inability to regulate blood cholesterol would not survive beyond childhood without treatment.

### Regulation

Liver cells are able to regulate cholesterol concentrations in their own cytoplasm as cholesterol inhibits one of the enzymes involved in its own synthesis. This is a typical example of negative feedback within the cell. However, by doing this, liver cells are also regulating the concentration in the blood. They export cholesterol to the blood, so when levels rise in the cells, levels in the blood rise as well. Just like other cells, liver cells also take in LDL, containing cholesterol, by endocytosis. This means that when blood cholesterol rises, cytoplasmic cholesterol increases too. In terms of cholesterol concentration, the blood can be seen as an enormous extension of the liver-cell cytoplasm. By regulating levels inside the cell, they also regulate the amount in the blood. This is an unusually simple example of whole body homeostasis. One cell type both detects the change and brings about the response.

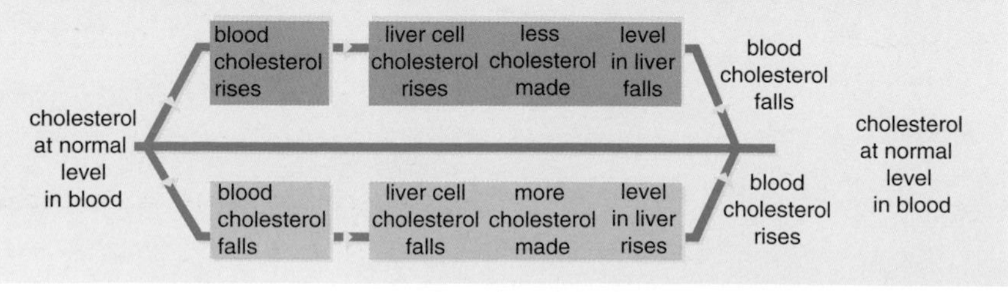


**Figure 9.5**
Summary of cholesterol regulation by the liver

## How is homeostasis brought about?

The work of a number of organs must be coordinated to achieve homeostasis. Information about the conditions in the body is continuously fed to the brain from sensory receptors around the body. For example, if the external temperature rises, temperature receptors in the skin send nerve impulses to the brain. In response, the brain initiates mechanisms that will lower the body temperature again. The same temperature receptors send nerve impulses to the brain when the temperature is back to normal. Homeostasis depends on the continual feedback of information (Figure 9.6).

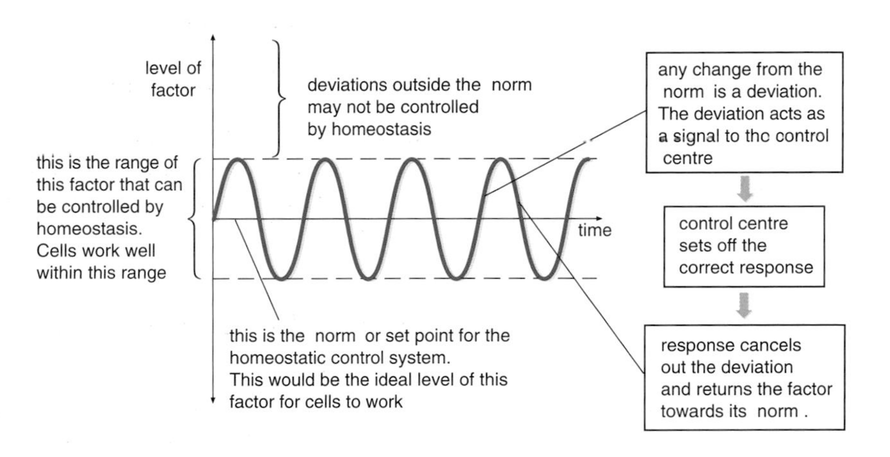


**Figure 9.6**
In homeostasis, deviation from the norm acts as the signal that sets off the correction mechanism. This negative feedback keeps variable factors within the narrow ranges suitable for life

### Negative feedback

In biological systems, homeostasis is usually achieved by a process called negative feedback. (Figure 9.7).

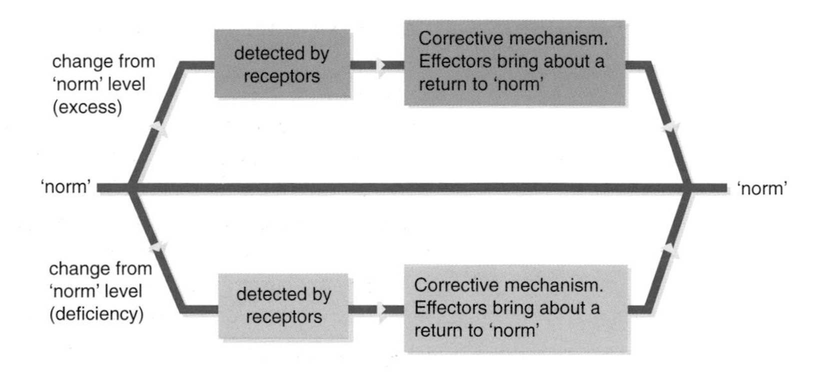


**Figure 9.7**
How negative feedback keeps a system in a stable condition

Negative feedback involves:

- a change in the level of an internal factor
- detected by receptors
- these activate effectors
- restoring the factor to its set point.

### What must be kept constant?

Some of the features of the blood and tissue fluid that must be kept within narrow limits are:

- the concentration of glucose
- temperature
- pH
- water potential
- the concentration of ions, such as sodium, potassium and calcium.

## The control of blood glucose concentration

### Blood glucose, supply and demand

In 1859 Claude Bernard, a French physiologist, was the first to recognise the importance of homeostasis in mammals. He studied variations in glucose concentrations in the blood of dogs and he found that the concentration remained remarkably stable despite dramatic variations in diet.

Humans eat a variety of foods, which are sources of energy and raw materials. In normal circumstances, we obtain most of our energy by respiring glucose. Some vital organs, notably the brain, are unable to store carbohydrates and so cannot function for even a short period of time without glucose provided by the blood. Thus a lack of glucose in the blood can cause the brain to malfunction.

There are three sources of blood glucose:

- Digestion of carbohydrates in the diet

In the UK, on average about 45% of our diet consists of carbohydrates. The carbohydrates we eat include sugars, starch and cellulose. Carbohydrates such as maltose, sucrose and starch are converted to glucose during digestion. Therefore the fluctuation in input of glucose to the blood is influenced by both the amount and type of carbohydrate we eat.

- Breakdown of **glycogen**

This storage polysaccharide is made from excess glucose in a process called **glycogenesis**. Glycogen is particularly abundant in the cells of the liver and muscles. When needed, glycogen can be quickly broken down to release glucose.

- Conversion of non-carbohydrate compounds

Glucose can also be produced from substances which are not carbohydrates. This process is called **gluconeogenesis**.

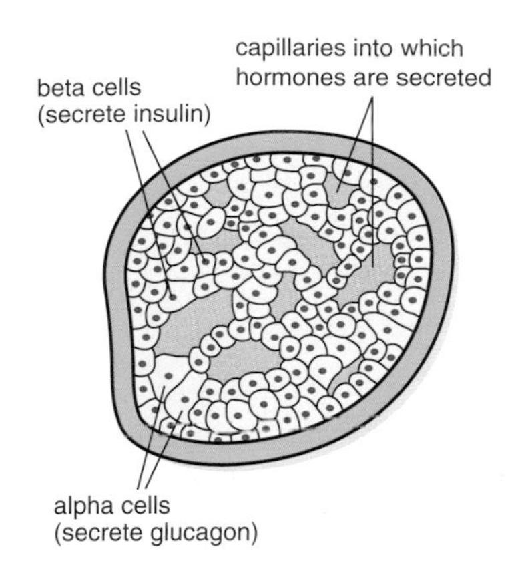


**Figure 9.8**
Detail of an islet of Langerhans

Glucose is oxidised during respiration and is broken down into carbon dioxide and water. During this complex series of chemical reactions, chemical potential energy in glucose is used to produce ATP. This in turn can be broken down to release energy for activities in the body. Some activities, for example vigorous exercise, demand a greater energy supply than others. This means that the rate of respiration varies, affecting how fast cells take up glucose from the blood.

The normal concentration of glucose in human blood is approximately 90 mg per 100 $cm^3$, and even after the largest carbohydrate meal it rarely exceeds 150 mg per 100 $cm^3$.

**Q** 1 **One person eats a bar of chocolate and another person eats the same mass of toast. Explain why there is a more rapid rise in blood glucose concentration in the person who has eaten the toast.**

## The mechanism of blood glucose control

### The pancreas

The **pancreas** plays a central role in the control of blood glucose concentration. The digestive functions of this organ are covered in Chapter 12, but here we are concerned with its endocrine role, the production of hormones.

The pancreas contains cells which are sensitive to blood glucose concentration and produces the hormones insulin or glucagon accordingly. These groups of cells are called **islets of Langerhans** (Figure 9.8). There are two types of islet cells. Glucagon is produced by $\alpha$-cells and insulin is produced by **$\beta$-cells**. The islet cells are surrounded by other cells, which produce digestive enzymes.

### High blood glucose concentration

When blood glucose concentration becomes too high, $\beta$-cells in the islets of Langerhans detect this rise and respond by releasing insulin. This hormone travels to all parts of the body in the blood but mainly affects cells in muscles, liver and adipose tissue. Insulin has a number of effects on the body, all of which tend to lead to a reduction in the concentration of glucose in the blood.

- Insulin speeds up the rate at which glucose is taken into cells from the blood. Glucose normally enters cells by facilitated diffusion through protein carrier molecules in the plasma membrane. Cells have extra carrier molecules present in their cytoplasm. Insulin causes these carrier molecules to move to the membrane where they increase the rate of glucose uptake by the cell (Figure 9.9).
- It activates enzymes, which are responsible for the conversion of glucose to glycogen.
- It activates enzymes which promote fat synthesis.

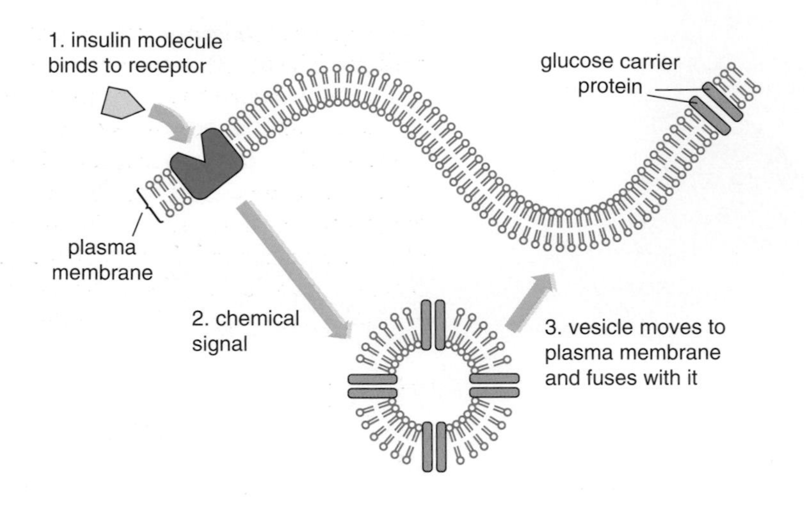


Figure 9.9
Insulin increases glucose uptake by increasing the number of carrier molecules in the plasma membrane

### Low blood glucose concentration

If the concentration of blood glucose falls too low, this is detected by α-cells in the islets of Langerhans which then secrete glucagon. The main effect of this hormone on the body is to activate enzymes in the liver which are responsible for the conversion of glycogen to glucose. It also stimulates the formation of glucose from other substances such as amino acids. The glucose then passes out of the cells and into the blood, raising the blood glucose concentration. The control of blood glucose concentration is summarised in Figure 9.10.

**Q 2 In the control of blood sugar concentration:**

**(a) what are the receptors**

**(b) how do the receptors activate the effectors**

**(c) what are the effectors?**

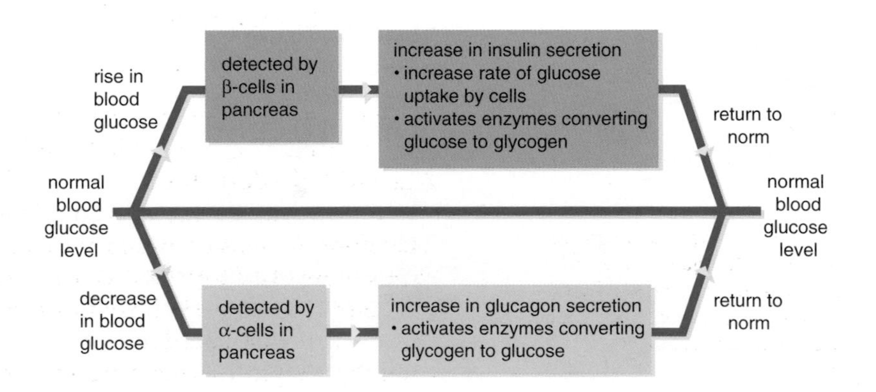


Figure 9.10
Summary of the homeostatic control of blood glucose concentration

## When homeostasis goes wrong – diabetes mellitus

Should the pancreas become diseased and fail to secrete insulin, or if target cells lose their responsiveness to insulin, the blood glucose concentration can reach dangerously high levels. The inability to control blood glucose concentration results in a condition called **diabetes mellitus.** The blood concentration becomes so high that the kidney is unable to reabsorb back into the blood all the glucose filtered into its tubules. Diabetes mellitus is thus characterised by the excretion of large amounts of glucose in the urine. Other diagnostic features of this condition include a craving for sweet food and persistent thirst. The main diagnostic test is a glucose tolerance test in which the patient swallows a glucose solution and then the blood glucose concentration is measured at regular intervals. Graphs of the results from a diabetic and a person with normal glucose metabolism are distinctly different (Figure 9.11).

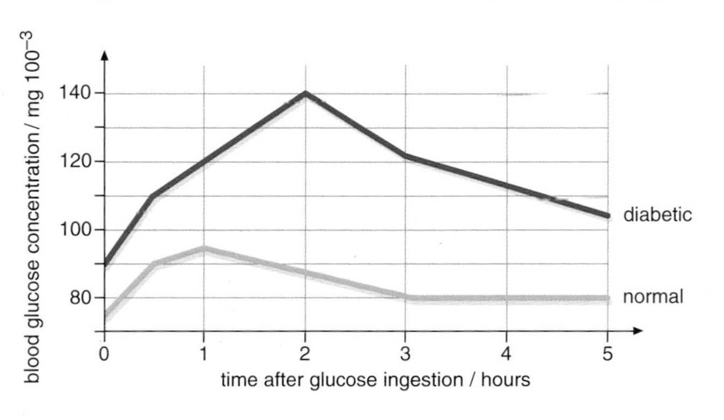


**Figure 9.11**
The results of a glucose tolerance test in a person with normal glucose metabolism and a diabetic

There are two main types of diabetes mellitus:

- **Type I** – also known as insulin-dependent or juvenile-onset. This usually occurs suddenly in childhood and results from the body being unable to make insulin. This is often caused by an **autoimmune reaction** (one in which the body's immune system attacks and destroys its own cells) which destroys the β-cells in the islets of Langerhans.
- **Type II** – also known as insulin-independent or late-onset. This usually occurs later in life and is more common than type I, accounting for more than 70% of the cases in the UK. It is often caused by a gradual loss in the responsiveness of cells to insulin.

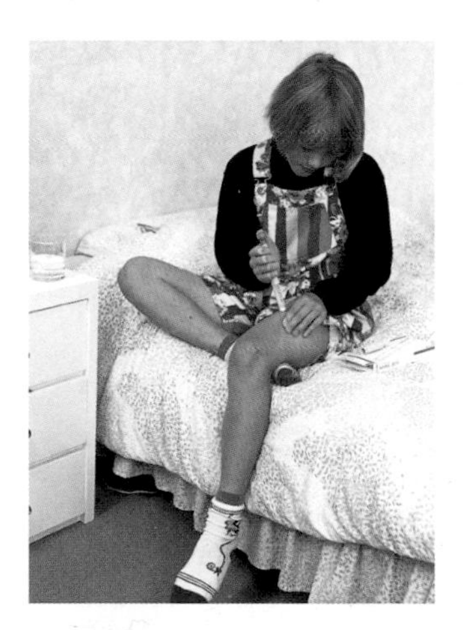

**Figure 9.12**
A person with type I diabetes has to inject with insulin regularly. Insulin cannot be taken orally because, being a protein, it would be digested in the alimentary canal

### Treating type I diabetes

Before insulin treatment became available, about 75 years ago, type I diabetes would have been fatal within a year of diagnosis. Care has to be taken with the doses because an overdose causes too much glucose to be withdrawn from the blood, a condition known as **hypoglycaemia.** Therefore amounts of insulin given must match glucose intake and expenditure. All diabetics need to monitor their blood glucose concentration regularly. Easy-to-use biosensors are now available for this. Diabetics also need to manage their diet and levels of exercise very carefully. Despite this most diabetics lead normal lives.

### Treating type II diabetes

Most type II diabetics control their blood glucose concentration by carefully regulating their diet, especially sugar intake and balancing this with the amount of exercise taken.

**Extension box 2**

## Insulin patches

The chemical structure of insulin (Figure 9.13) shows two small polypeptide chains containing a total of 51 amino acids.

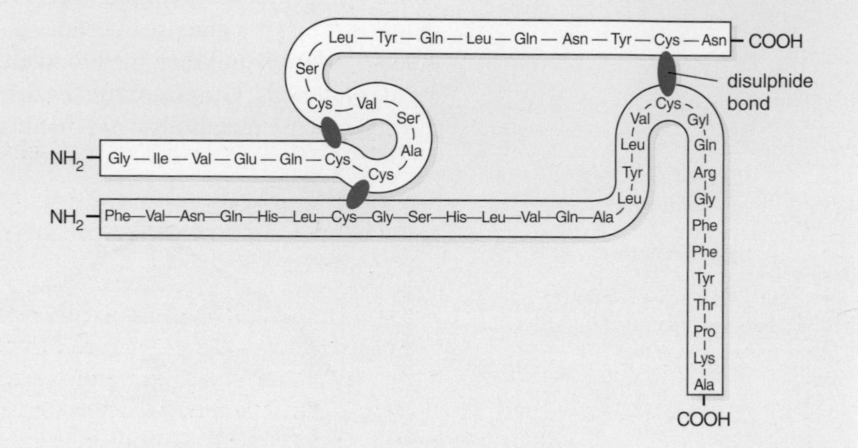


**Figure 9.13**
Chemical structure of insulin

If insulin is swallowed, it is digested in the same way as other proteins in the diet. This means that the only way of using insulin at present is to inject it directly into the body. This is a straightforward procedure, but is not very pleasant.

Now there may be a better alternative for getting insulin into the body. Scientists have been experimenting with the use of ultrasound and skin patches. A skin area is first treated with ultrasound to disrupt the underlying fat tissue. This is necessary because insulin is not soluble in fat, so disrupting the fat tissue allows it to move in through the skin more easily. A patch containing insulin is then applied to the area, and can supply insulin for several days .

## Temperature control

Animal life can exist at almost all the temperatures encountered on the Earth's surface, from the extreme cold of the poles to the intense heat of the deserts. This is possible because it is the internal, not the external temperature that is important to an organism. If an animal's temperature falls too low biochemical reactions will be too slow for it to remain active. If it goes too high there is a risk of enzymes and other proteins being denatured. To a certain extent, all animals exert some control over their internal temperature.

### How heat is lost or gained

An organism's temperature changes as it gains or loses heat (Figure 9.14). If the heat gain is greater than the heat loss, the temperature of the organism will rise, and vice versa. To maintain a steady temperature, an organism must balance its heat gains and losses.

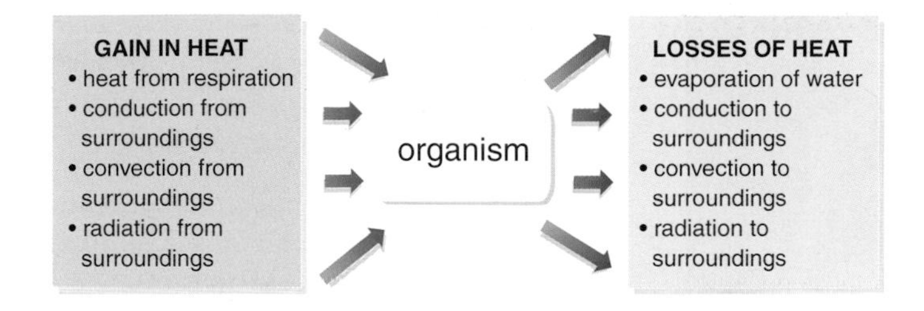


Figure 9.14
How an organism gains and loses heat

## Temperature regulation in ectotherms

Ectothermic organisms can control their body temperature only by changing their behaviour. All animals except birds and mammals are ectotherms.

Crocodiles are ectotherms and rely largely on moving between land and water to remain in the environment with the most suitable temperature. Many lizards move between sun and shade. Ectothermic animals therefore rely on altering their external environment to maintain a reasonably constant internal environment (Figure 9.15).

Other methods of controlling their body temperature include:

- colour; darker colours absorbing heat more readily
- orientation to the sun; exposing larger areas to benefit from the sun's radiation
- metabolic heat generation; cells respire producing heat as a waste product

Sometimes ectotherms find it easier to gain heat than lose it. Moving to a cooler environment, which may be below their present body temperature may be their only option. Sometimes this may not be possible.

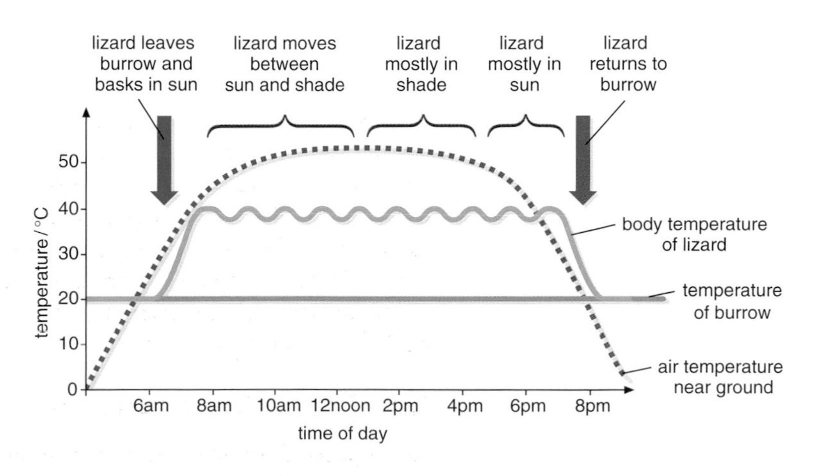


Figure 9.15
Comparison between the external temperature and the internal temperature of a lizard as it moves from sun to shade

## Temperature regulation in endotherms

Endothermic animals maintain a stable core body temperature using both physiological and behavioural means. Mammals and birds are endotherms.

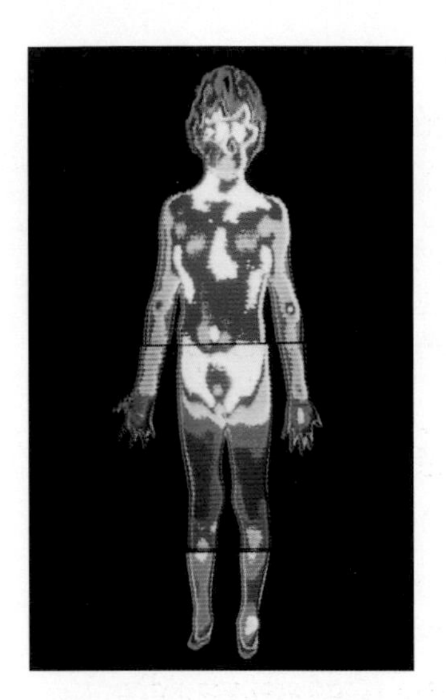

Figure 9.16
A thermal image of a person. Areas of high temperature are white, red or orange. Cooler areas are green, blue or purple

Mammals rely largely on physiological methods. Humans, for example, maintain a temperature within the range 36.1–37.8 °C. If the temperature rises too high, various internal mechanisms bring about loss of heat. If it falls too low, other mechanisms serve to increase heat production and reduce its loss from the body.

When we refer to body temperature, we actually mean **core body temperature.** In a human, the limbs and other extremities may be substantially cooler than 37 °C, but the core temperature usually remains constant (Figure 9.16).

### The role of behaviour

Just as ectotherms modify their behaviour to stay comfortable, so do endotherms. When we encounter a particularly warm or cold environmental temperature, temperature receptors in the skin send nerve impulses to the **voluntary** centre of the brain. So if we feel hot or cold we can decide to do something about it like changing position, changing clothing or turning the heat up or down.

### The role of physiology

If behaviour cannot deal with the problem, blood temperature starts to rise or fall. The **hypothalamus** of the brain detects the change in the temperature of the blood flowing through it and acts via the autonomic nervous system to initiate the appropriate response.

Temperature control in a mammal is achieved by negative feedback.

- When the core temperature rises too high, the resulting increase in blood temperature is detected by receptors in the hypothalamus. As a result the **heat loss centre**, which is also in the hypothalamus, sends nerve impulses to structures such as arterioles and sweat glands in the skin whose actions bring about the necessary fall in temperature.
- When the core temperature falls too low, this is also detected by receptors in the hypothalamus. This time, the **heat conservation centre** triggers mechanisms which conserve the body's heat or even generates more heat by actions such as shivering. The outcome, once again, is a return to the normal level (Figure 9.17).

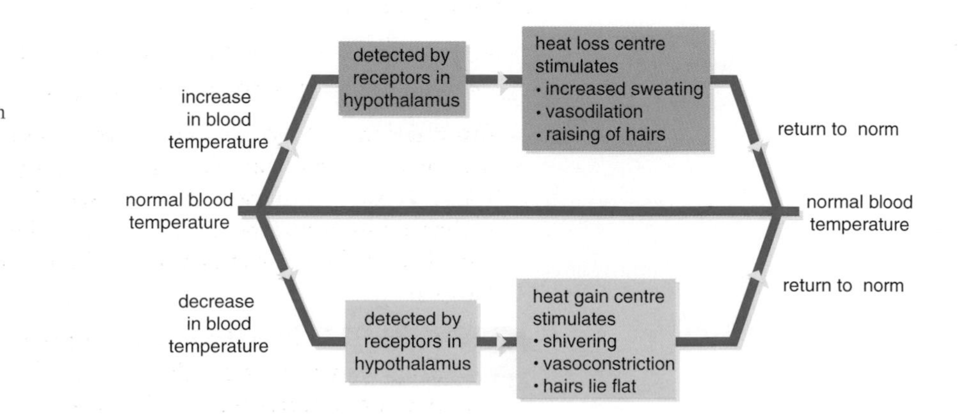


Figure 9.17
Summary of negative feedback in temperature control

### The role of temperature receptors in the skin

If the hypothalamus detects changes in temperature of the blood and initiates appropriate responses, what do the temperature receptors in the skin do?

While the hypothalamus detects temperature fluctuations inside the body, the skin receptors detect temperature changes at the surface. They enable you to feel whether the environmental temperature immediately outside your body is hot or cold. This information is sent by nerves to the voluntary centres of the brain and initiates voluntary activities such as jogging in severe cold or moving into the shade if conditions are hot.

### Physiological response to cold

There are four main physiological responses that occur when the hypothalamus detects a drop in blood temperature.

- **Shivering** – This occurs when muscles contract and relax rapidly. Shivering muscles give out four or five times as much heat as resting muscles.
- **Vasoconstriction** – This occurs when the arterioles leading to the capillaries in the surface layers of the skin constrict (narrow) so reducing the blood flow through the capillaries (Figure 9.19(a)). This cuts down the amount of heat lost through the skin by radiation and conduction. Remember, the very thin walls of the capillaries do not contain muscle, so they cannot constrict.
- **Hair raising (piloerection)** – The erection of hairs is a reflex more commonly used by the hairier mammals. In most mammals, raising the hair makes the fur 'thicker' so that it traps more air. This provides extra insulation. In humans, the erector pili muscles in the skin pull our tiny hairs upright, but only succeed in creating 'goose pimples'.
- **Increased metabolic rate** – The body secretes the hormone adrenaline in response to cold. This raises metabolic rate and therefore increases heat production. Mammals which live in cold conditions for a period of several weeks or months, show a more permanent increase in metabolic rate due to the secretion of the hormone thyroxine.

### Physiological response to heat

There are two main physiological responses that occur when the hypothalamus detects a rise in body temperature.

- **Vasodilation** – This occurs when arterioles leading to capillaries in the skin, **dilate**. At the same time **shunt vessels**, which carry blood deep below the surface of the skin are closed off. This results in a greatly increased blood flow near the surface of the skin resulting in more heat being lost to the environment by radiation and conduction (Figure 9.18(b)).
- **Sweating** – Sweat is a salty solution made by sweat glands. Evaporation of sweat from the skin's surface cools it. The evaporation of each gram of water requires 2.5 kJ of energy. Being furless, humans

have sweat glands over the whole of the body, making cooling by this mechanism very efficient. Animals with fur generally have sweat glands confined to areas of the skin where fur is absent, e.g. pads of the feet. In humans, sweat glands can lose up to 1000 $cm^3$ $hour^{-1}$ of sweat. The efficiency of sweating depends on the humidity in the environment.

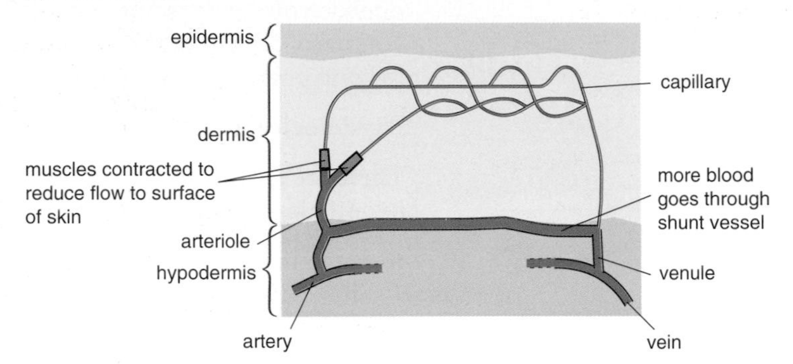


**Figure 9.18**
(a) In cold conditions blood flow to the surface of the skin when the arterioles dilate during vasodilation (b) In warm conditions blood flow is now mainly in the shunt vessel away from the surface of the skin

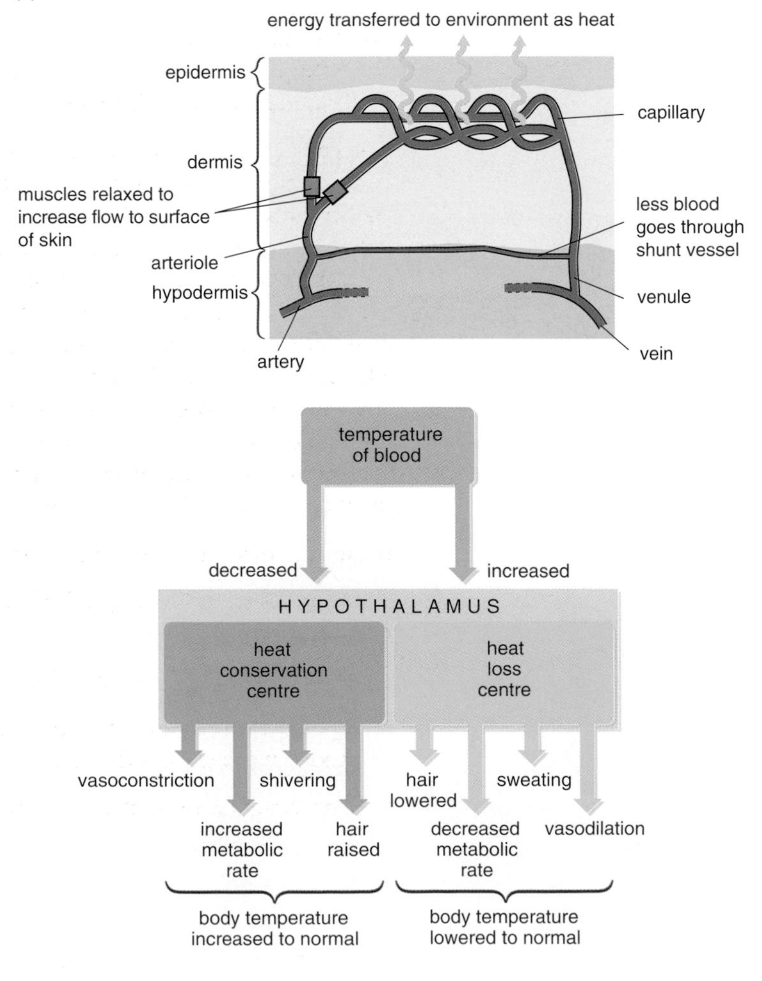


**Figure 9.19**
Summary of body temperature control by the hypothalamus

### The structure and role of the skin in temperature regulation

In terms of area, the skin is the largest organ in the body. It is in direct contact with the external environment and so plays a central role in maximising and minimising heat loss (Figure 9.20).

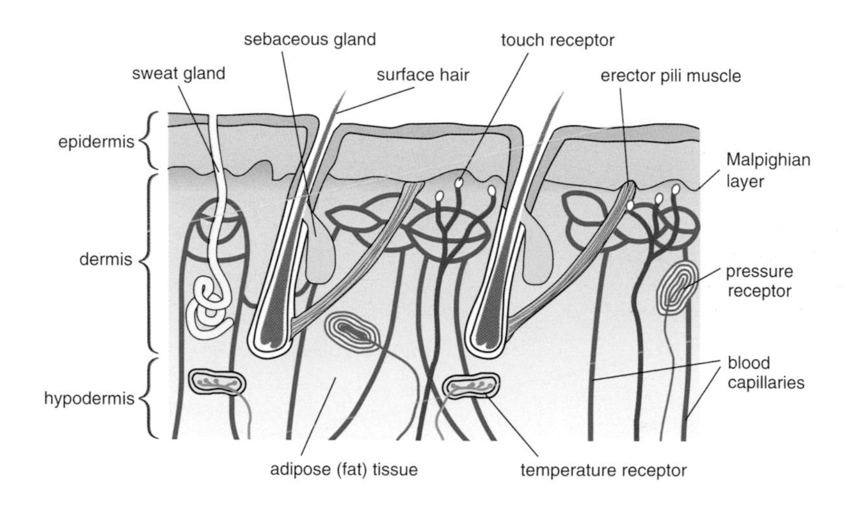


Figure 9.20
The structure of human skin

Structurally the skin is divided into two layers, the outer **epidermis** and the inner **dermis**. Forming the boundary between the two is the **Malpighian layer,** the bottom-most layer of epidermal cells.

The cells of the Malpighian layer divide repeatedly by mitosis. As new cells are formed, the older ones get pushed outwards towards the surface, flattening as they do. After a time, their cytoplasm becomes full of granules and the cells die. Finally they become converted into scales of **keratin**. Keratin is a tough fibrous protein and gives the skin its protective properties and makes it waterproof. We are constantly losing epidermal cells. The dust in our houses is mainly made up of lost epidermal cells. The dermis is much thicker than the epidermis and contains structures such as nerve endings and blood vessels held together with connective tissue. Beneath the dermis is the **hypodermis**, which usually contains at least some **subcutaneous** fat. This fat storage tissue, called **adipose** tissue, provides vital insulation in mammals.

Mammalian skin has several functions

- It detects stimuli using cells that are sensitive to heat, cold, pressure and pain.
- It prevents excessive water loss or gain.
- It plays a role in temperature regulation by adjusting heat loss according to the conditions.
- It prevents entry of microorganisms.

## Extension box 3

### Hypothermia

Hypothermia is the condition in which the body temperature falls dangerously below normal. It happens if heat energy is lost from the body more rapidly than it can be produced, for example when a person wearing inadequate clothing is subjected to prolonged cold.

As the body temperature falls, one of the first organs to be affected is the brain. This results in the person becoming clumsy and mentally sluggish. Since brain function is impaired, the victim may not realise that anything is wrong and so does nothing about it such as putting on more clothes. Indeed there have been examples when people even take off clothes.

As the body temperature falls, the metabolic rate falls too. That makes the body temperature fall even further, a case of positive feedback (Figure 9. 21).

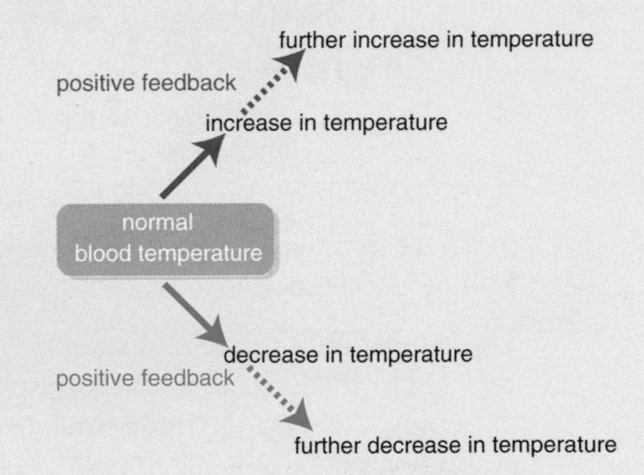


**Figure 9.21**
Positive feedback showing that after a deviation from the norm, the response is to take it further away from the norm

Death usually occurs when the body temperature drops to about 25 °C, though people have been known to survive lower body temperatures than this. The cause of death is usually ventricular fibrillation, a condition in which the normal beating of the heart is replaced by uncoordinated tremors.

Most at risk from hypothermia are babies and the elderly; babies because of their high surface area to volume ratio and undeveloped temperature regulation mechanisms, the elderly because their thermoregulatory mechanisms may have deteriorated through old age.

Deliberate hypothermia is sometimes used in surgical operations on the heart. The patient is cooled either by circulating the blood through a cooling machine or by placing ice packs in contact with the body.

By cooling the patient the metabolic rate is reduced and the demand for oxygen by the brain and other vital tissues is lowered. This allows the heart to be stopped without risk of the patient suffering brain damage through lack of oxygen. But the patient must not be cooled for too long or the tissues may be permanently damaged. Lowering the body temperature to 25 °C allows about ten minutes for the operation.

## Summary

- Homeostasis provides a constant internal environment and independence from fluctuating external conditions.
- 'Negative feedback' tends to restore systems to their original level.
- There are separate mechanisms that must be coordinated when there are departures from the original state in opposite directions.
- Temperature control in the body of an ectothermic reptile is achieved by its behaviour.
- In an endothermic mammal there are mechanisms that cause heat to be produced, conserved and lost.
- The hypothalamus monitors blood temperature and the autonomic nervous system coordinates temperature control in a mammal.
- Blood glucose concentration changes all the time depending on absorption of glucose from the digestive system and removal by active cells for respiration.
- Insulin and glucagon are hormones that control blood glucose concentration by activating enzymes involved in the interconversion of glucose and glycogen.
- Diabetes can be controlled using insulin or by manipulating carbohydrate intake.

## Assignment

### Frogs, toads and homeostasis

Figure 9.22
It is the ability to regulate their internal environments which allows organisms to survive in a wide range of environmental conditions. This lizard (a) lives in the deserts of Patagonia where temperatures range between −20 °C and +30 °C. Flowers like crocuses (b) can even flower when there is snow on the ground. Marine animals such as these barnacles (c) can survive in the changing conditions between the tide lines on a rocky shore

Most of the material in the chapter which you have just been studying was concerned with mammals. Although we usually associate the idea of homeostasis with mammals we must remember that there are many other organisms that live in conditions which, at first sight, appear unlikely to be able to support them (Figure 9.22). They often do this by keeping their internal environment remarkably constant.

In this assignment we shall look at a group of animals which can survive in a wide range of environmental conditions. This group is the amphibia – frogs and toads.

One of the skills you need to master as part of your A2 biology course is the ability to link together material from different parts of the specification. In AQA specification A, one way in which this ability is tested is with questions involving a comprehension passage, like the one in this assignment.

Read the following passage. Remember that a lot of the information here will be new to you. Read it carefully because the answers to many of the questions can be found in the passage.

With their moist skins, frogs and toads might seem best suited to living in or near ponds and streams. Surprisingly, they are also found in dry regions. They can survive in such places as they have evolved various behavioural and physiological mechanisms which allow
5 them to conserve water and keep cool.

A mammal has a skin which protects it against desiccation. The outer part, called the stratum corneum, is made up of many layers of flattened, dead epidermal cells. This part reduces water loss. Amphibia, however, generally have a stratum corneum consisting of only a single
10 layer of cells. This layer is very permeable and has some benefits. Frogs and toads do not drink but they can absorb moisture through the skin. Oxygen and carbon dioxide also pass readily through it.

Frogs and toads cannot make urine that is more concentrated than their blood. They therefore need relatively more water than
15 mammals to eliminate their waste. They have a number of characteristics which help to offset the handicaps of permeable skins and inefficient kidneys. When water is unavailable, these animals stop producing urine and allow wastes to accumulate in the body fluids. Dehydration also results in beneficial changes in the skin
20 and bladder. Increased permeability of the bladder permits the animals to make up for water lost by evaporation. In addition they absorb water through the skin much more readily than when they are hydrated. These responses are controlled by a hormone produced by the pituitary gland. Some desert species can store up to 25% of their body mass in
25 the form of dilute urine so, as water is lost through evaporation, water from urine is recycled back into the body.

Spadefoot toads show a further refinement of the excretory system. They normally survive a dry spell by burrowing deep into the soil. The drier the soil, the more urea they make. Accumulation of soluble urea in
30 the spadefoot's body fluids enables the animal to take up moisture in the dry conditions that exist just before the summer rains.

Frogs and toads achieve a degree of thermoregulation. Some species raise their internal temperature by basking in the sun. Juvenile green toads,

for example, basking in the sun have an internal temperature 10 to 15 °C
35 higher than animals in the shade. It has been suggested, however, that
this mechanism would be very inefficient because evaporative water
loss would dissipate essentially all of the solar heat gain. Some species
of frog, however, have a skin which is covered in a waterproof waxy
layer. The graph in Figure 9.23 shows how these species regulate their
40 temperatures.

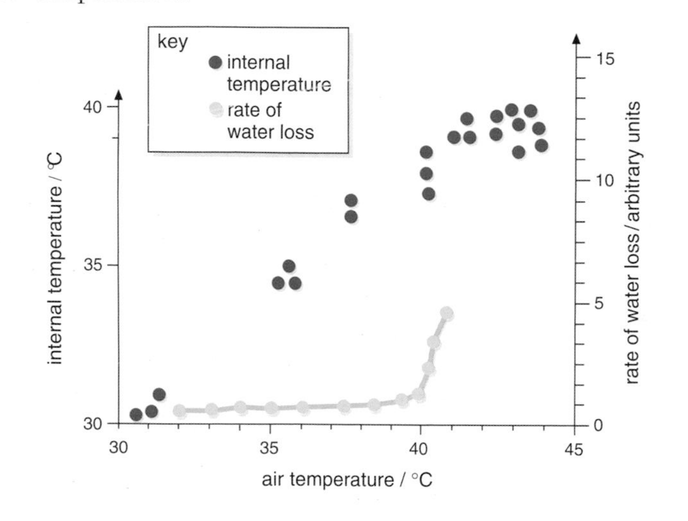


**Figure 9.23**
Graph showing how internal temperature and the rate of water loss vary with air temperature in the waterproof frog, *Phyllomedusa sauvagei*

Now use your knowledge and the information in this passage to answer the questions which follow.

1 Using an example from the passage to illustrate your answer in each case, explain what is meant by:

(a) a behavioural mechanism (line 4);
(b) a physiological mechanism (line 4).

*(2 marks)*

2 (a) Oxygen and carbon dioxide pass readily through the skin of a frog (lines 11–12). Explain how this is linked to the fact that the skin of a frog is moist.

*(3 marks)*

(b) The Titicaca toad lives in water at very high altitudes in the mountains of South America. Its skin hangs in loose folds. Explain how this might be an advantage to an amphibian living in this environment.

*(3 marks)*

3 Explain how increased permeability of the bladder 'permits the animals to make up for water lost by evaporation' (lines 20–21).

*(1 mark)*

4 (a) What happens to the water potential of the soil as the ground dries before the summer rains (line 31)?

*(1 mark)*

(b) Explain how accumulation of urea in the spadefoot's body fluids enables the animal to take up water in the dry conditions that exist just before the summer rains (lines 29–31).

*(3 marks)*

**5** Explain in your own words why basking in the sun would be a very inefficient method of raising the body temperature of a species such as the green toad (lines 35–37).

*(2 marks)*

**6** (a) Describe how the body temperature of the waterproof frog changes with an increase in air temperature. Explain what causes this change.

*(3 marks)*

(b) Suggest why there is a greater rate of water loss from the skin of this frog at an air temperature of about 40 °C.

*(2 marks)*

# Examination questions

**1** The graph shows the rate of glucose absorption in and excretion from a mammalian kidney in relation to the glucose concentration in the plasma.

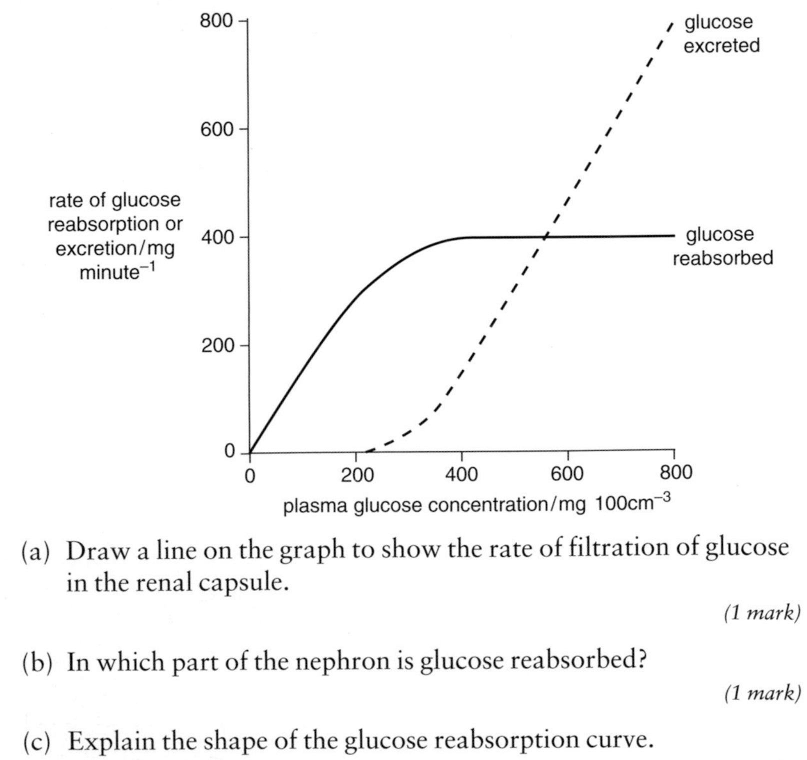


(a) Draw a line on the graph to show the rate of filtration of glucose in the renal capsule.

*(1 mark)*

(b) In which part of the nephron is glucose reabsorbed?

*(1 mark)*

(c) Explain the shape of the glucose reabsorption curve.

*(3 marks)*

2 (a) Small desert animals can survive without drinking. They meet their water requirements mainly from metabolic water.

(i) What is metabolic water?

*(1 mark)*

(ii) Describe one other way in which these animals are able to gain water.

*(1 mark)*

(b) The fat-tailed mouse has a characteristic store of fat within its tail. It has been suggested that stores of fat are ways of providing desert-living animals, like the fat-tailed mouse, with water. The table gives some figures relating to the metabolism of fat over a period of time.

| | |
|---|---|
| Mass of fat respired (g) | 1.06 |
| Mass of water formed on oxidation (g) | 1.13 |
| Volume of oxygen consumed ($cm^3$) | 2.13 |
| Mass of water evaporated from lungs (g) | 1.80 |

Use the figures in this table to explain why:

(i) one gram of fat can produce more than one gram of water on oxidation

*(1 mark)*

(ii) the suggestion that stores of fat are ways of providing desert-living animals with water cannot be correct.

*(2 marks)*

# Gas Exchange

Figure 10.1
(a) This whale may not breathe again for an hour (b) When a whale returns to the surface after a dive, it clears its 'blow-hole' before inhaling

**As a species we are not very good at swimming under water without special equipment. Most people cannot manage to hold their breath for more than a minute. However, diving mammals such as dolphins, whales and seals can stay under water for much longer periods. Sperm whales for example can dive for over an hour. During this time the animal is not breathing and therefore cannot take in any more oxygen from the air. This means that the oxygen breathed in before the whale dives has to last a long time. How do they do this? There are a number of ways which include:**

- **The muscles of a diving mammal contain a high concentration of myoglobin so can store more oxygen than those of non-diving mammals.**
- **These mammals can also move oxygenated blood away from organs that can function without oxygen for a while, such as the skin and direct it to those organs that need a constant supply, such as the brain and heart. The decreased flow to the skin is so effective that cuts sustained during a dive do not bleed until the animal surfaces.**
- **Sperm whales and other diving mammals breathe out just before a dive. This makes them less buoyant so they use less energy to swim down. It also means that there is less gas in the lungs to be forced into the blood at high pressure. Gas dissolved in the blood may form bubbles as an animal surfaces, causing a potentially life-threatening condition known as 'the bends' or decompression sickness.**

All living cells require a source of energy in order to survive. Energy is obtained by the process of respiration. Respiration occurs in every cell, generating ATP, using oxygen and producing carbon dioxide as a waste product. To keep the process going, a living organism needs to obtain oxygen and expel waste carbon dioxide continuously. The diffusion of oxygen and carbon dioxide in opposite directions across the surface of a cell is called **gas exchange**. In humans a ventilation mechanism, breathing, brings a continuous supply of air into the lungs. Special gas exchange surfaces, the alveoli, are adapted for efficient gas exchange.

Usually, small simple organisms, such as those made up of only one cell, have a relatively small oxygen demand. All the oxygen they need diffuses into their body through their surface. They remove carbon dioxide in the same way. But this strategy does not work in larger animals.

## Surface area to volume problems

Although the amount of respiratory gases an organism needs to exchange is proportional to its volume, the amount it can exchange is proportional to its surface area. When an organism gets larger, its volume increases at a

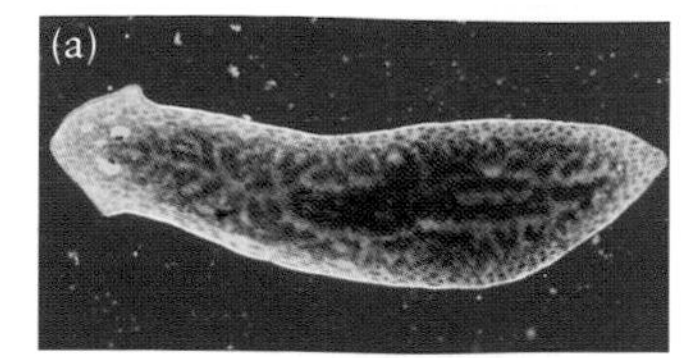

**Figure 10.2**
(a) This flatworm, a platyhelminth, is flat and ribbon-like, providing a large surface area to volume ratio so gas exchange can occur through its body surface (b) Larger and more complex organisms such as annelids like this ragworm are elongated, but this does not provide a large enough surface area to meet their gas exchange needs. Structures called gills at the sides of the body further increase their surface area (c) A sea slug is a mollusc. Its external gills increase its surface area (d) Tadpoles, the larvae of amphibians, also have gills. In young tadpoles they are outside the body. In older tadpoles, they are inside.

faster rate than its surface area. So the need to exchange materials could quickly outstrip an organism's ability to do so.

Organisms with a diameter larger than a few millimetres can survive only because they have evolved ways of increasing their surface area to volume ratio. Many species of worm, for example, have a flattened body shape. However, this sort of adaptation allows only a limited size increase. Much larger organisms have developed specialised organs which increase the surface area used for exchanging materials. Once absorbed, substances have to travel through the organism to reach every cell and thus the evolution of specialised exchange organs is linked to the development of transport systems.

## Gas exchange organs

Oxygen passes into the body of an organism by diffusion, and carbon dioxide leaves in the same way. Gas exchange organs increase the rate of gaseous exchange. They maximise the efficiency of diffusion.

## Fick's Law

Fick's Law states that the rate of diffusion is directly proportional to the area of the diffusion surface and to the difference in concentration of the substances concerned. It is inversely proportional to the thickness of the exchange surface. This is shown more simply in the formula:

$$\text{Rate of diffusion} = \frac{\text{surface area} \times \text{difference in concentration}}{\text{thickness of surface}}$$

This law provides an effective framework for considering how the maximum rate of diffusion of oxygen and carbon dioxide in an organism is achieved. To obtain the maximum rate of diffusion, a gas exchange system must have:

- the largest possible surface area
- a large difference in concentration
- the smallest possible diffusion pathway.

Later in this chapter, we will consider how some exchange surfaces are adapted to achieve the maximum rate of diffusion. The outer surface can be used as an exchange surface only if it is permeable. However, any surface that is permeable to oxygen and carbon dioxide will also be permeable to water, which is a smaller molecule. Therefore all gas exchange surfaces will be moist because water diffuses out of the organism.

## The problems of gas exchange in air and water

Animals obtain oxygen from the air or water that surrounds them. It is no coincidence that mammals and birds, the organisms with the highest metabolic rates, are air breathers. They could not obtain the amount of oxygen they need from water. However, a major drawback associated with breathing air is the loss of water. The combination of a large surface area and moist membranes mean that exhaled air is saturated with water vapour.

## Extension box 1

### Water or air?

The table shows a number of key properties of water and air.

| | Water | Air |
|---|---|---|
| Density | High | Low |
| Viscosity | High | Low |
| Oxygen concentration | Low | High |
| Diffusion rate | Low | High |

- Water is about 770 times denser than air and about 100 times as viscous. It therefore requires a lot more energy to move a current of fresh water over a gas exchange surface than to move the same volume of air.
- Oxygen is only sparingly soluble in water. At 20 °C air contains 27 times as much oxygen as an equal volume of water saturated with air.
- Diffusion of gases in solution is many thousands of times slower than diffusion in air. This is important because no matter how well ventilated, there is always a very thin layer of stationary water adjacent to the gill surface through which oxygen and carbon dioxide must move by diffusion. For this reason ventilation of a gill needs to be more or less continuous.

When we consider these facts together, the picture looks even worse.

- A kilogram of air contains 209 000 $cm^3$ of oxygen, over 30 000 times as much as the same mass of water. A fish thus needs to move a huge mass of water to obtain a small quantity of oxygen. Taking into account the greater density and viscosity of water, breathing is much more energy-expensive for fish than for a mammal. A trout uses about 20% of its energy in breathing, whereas a mammal uses less than 5%
- A further disadvantage of water-breathing is the fact that oxygen becomes less soluble with a rise in temperature, being only half as soluble at 35 °C as it is at 0 °C. The amount of oxygen in air varies little with temperature. Oxygen availability is thus more variable in water than it is on land.

However, although gas exchange in air seems to have all the advantages, this is not as one-sided as it appears.

Water has a lot of buoyancy and the gills can therefore protrude into it without sticking together as they would on land so a large surface area can be exposed to the water. To conserve water, lungs are ventilated through a single opening. On the other hand, gills can be ventilated by a one-way flow of water, in through the mouth and out through the operculum. As we have seen, this makes it possible for water to flow in the opposite direction to that of the blood so increasing the efficiency of gas exchange.

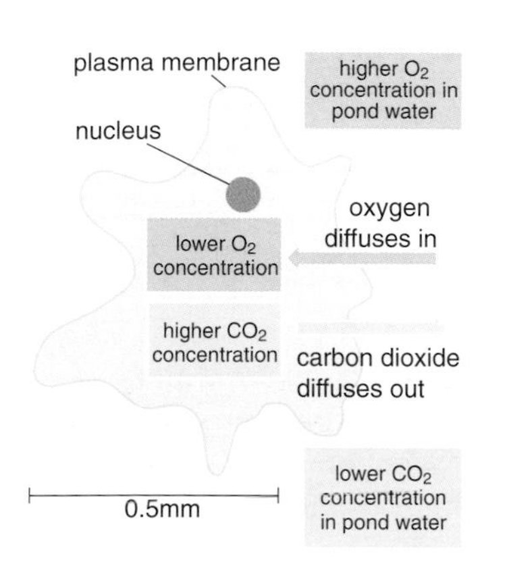


**Figure 10.3**
This amoeba is small enough for the plasma membrane to have a large enough surface area for sufficient gas exchange

## Gas exchange in a protoctist

The kingdom Protoctista (see Chapter 3) is a diverse collection of eukaryotic organisms including *Amoeba*, *Paramecium* and *Euglena* and even some multicellular forms like *Fucus* (brown seaweed, see Chapter 3). If you look at a drop of water containing amoebae you can just about see the organisms with the naked eye. They appear no bigger than dust particles. In single-celled organisms oxygen and carbon dioxide travel the entire distance between mitochondria and the external environment by diffusion alone. As the organisms are so small they have a very large surface area to volume ratio. Their external surface can thus be used to exchange gases, as the large surface area is able to supply sufficient oxygen to the small volume. No extra gas exchange surface is necessary.

*Amoeba* lives in water surrounded by the materials it needs. The water has a higher concentration of oxygen than is found in the organism. Oxygen diffuses through the plasma membrane into the cell where oxygen is at a lower concentration and can reach every part of its small volume. In the same way carbon dioxide produced by *Amoeba* diffuses out into the pond where its concentration is lower.

## How protoctists achieve efficient gas exchange – summary

Large surface area to volume ratio

- Though the actual surface area is very small its surface area to volume ratio is very large.

Small diffusion pathway

- This is because the oxygen and carbon dioxide have only to travel across the plasma membrane, which is 7 nm thick and the distance to the middle of the cell is small.

Large concentration difference

- Oxygen is being used in the cell by the process of respiration and carbon dioxide is being produced as a waste product. Thus there is always a diffusion gradient between the cell and its surroundings.

Larger animals, that have many cells, can also have a very large surface area to volume ratio. This is achieved by having a flattened body. A flattened body shape also means that no cell is far from the exchange surface.

Diffusion can provide every cell with oxygen and similarly carbon dioxide can be removed.

Oxygen diffuses through the body surface. It then diffuses through the body itself, towards the middle. Not only is diffusion a slow process but also some of the oxygen is used as it moves inwards. So there is a maximum thickness for any organism which relies on diffusion through the body surface. As a rule of thumb, the distance from the cell needing oxygen to the surface should be no more than about 1 mm.

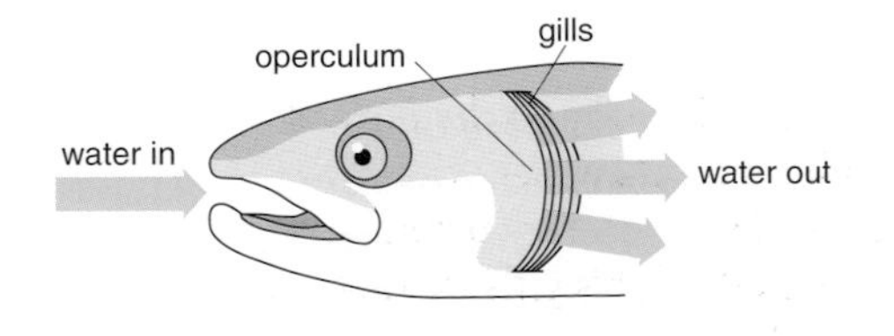


Figure 10.4
The movement of water through a fish and the position of its gills

Larger animals that are not flat have a small surface area to volume ratio. The large volume creates a demand for oxygen, which the relatively small surface area is unable to supply. Larger animals have solved this problem by developing systems to ensure that sufficient oxygen is supplied to every cell.

## Gas exchange in fish

Bony fish are relatively large animals so have a small surface area to volume ratio. Diffusion of gases through their outer surface would not provide sufficient oxygen for their large volume, so they need specialised gas exchange surfaces. These are the **gills**.

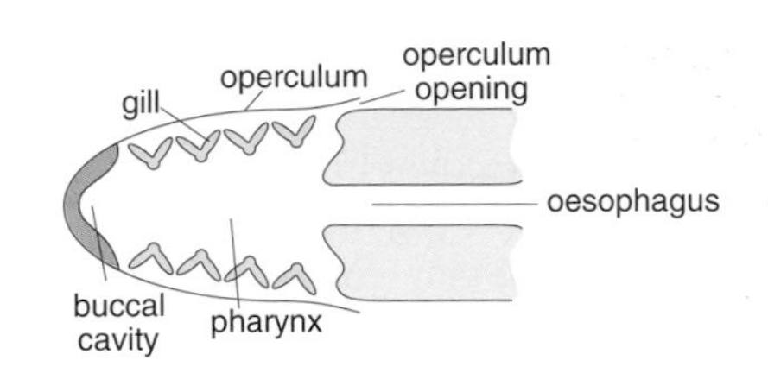


Figure 10.5
The position of the gills in the pharynx

Fish are water breathers, using oxygen in solution from the water surrounding them and passing carbon dioxide back into the water. Water contains a low concentration of dissolved oxygen, about 1%, when compared to air which contains 21%. To obtain enough oxygen, fish need to pass large quantities of water over their gills (Figure 10.4). This is achieved by a good ventilation system. Living in water means there is no risk of dehydration so the gills can be exposed more freely to the water, as long as they are adequately protected from damage.

### Structure of gills

The gills are fragile, complex structures. They lie in the pharynx, the area between the buccal cavity and the oesophagus (Figure 10.5).

Each gill is made up of two piles of soft thin plates called lamellae attached to a bony gill arch. These lamellae lie on top of each other (Figure 10.6). If not kept apart by water, they stick together, and their exposed surface area becomes too small for efficient gas exchange.

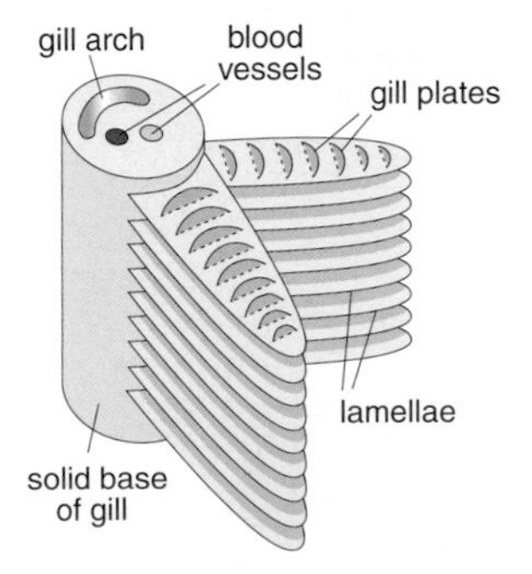


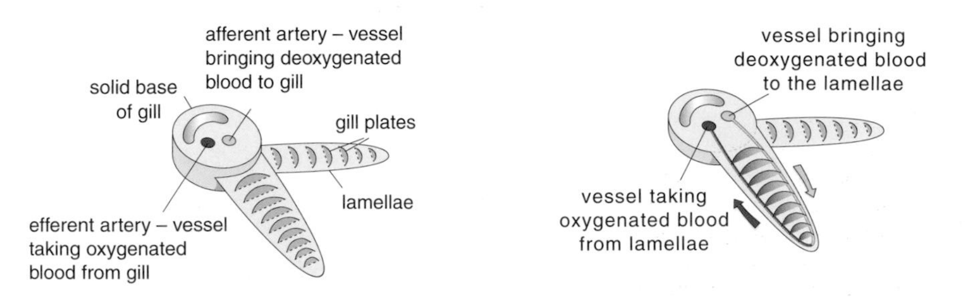


Figure 10.6 (left)
The arrangement of lamellae in a gill

Figure 10.7 (centre)
The position of the gill plates on the lamellae

Figure 10.8 (right)
The blood supply to and across the gill plate

As a way of increasing the surface area even further, each lamella is covered with flaps called gill plates (Figure 10.7).

### Blood supply

If you look at the gills of a fish you will see that they are pink. This is because they have a rich supply of blood. Blood flows from an afferent vessel in each gill arch down the inner side of each lamella, across each gill plate, and back to an efferent vessel (Figure 10.8). The great number of capillaries in the gills provide a large surface area for diffusion.

## Water flow

Water passes over the gill plates in the opposite direction to the flow of blood so that, across the whole gill plate, blood meets water with a higher concentration of oxygen than its own. This counter current mechanism increases the efficiency with which the fish's gills extract oxygen from water Figure 10.9.

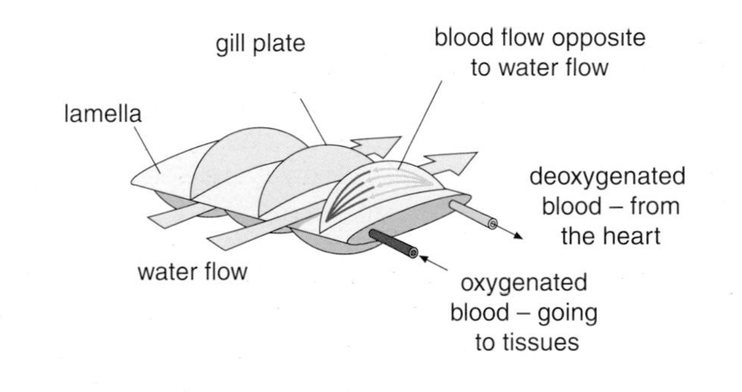


Figure 10.9
Blood flows through the gill plates in the opposite direction to the flow of water – a counter current mechanism

## The countercurrent system of the fish gill

A countercurrent system occurs when two substances flow through the same body part in opposite directions. Figure 10.10 shows the countercurrent system in the gill lamellae of a fish. To understand the advantage of blood and water flowing in opposite directions, consider what would happen if they both flowed in the same direction (Figure 10.10(a)). The two fluids would quickly reach equilibrium so the blood would extract much less of the oxygen available in the water. A countercurrent system ensures that there is a diffusion gradient along the whole of the gill plate (Figure 10.10(b)).

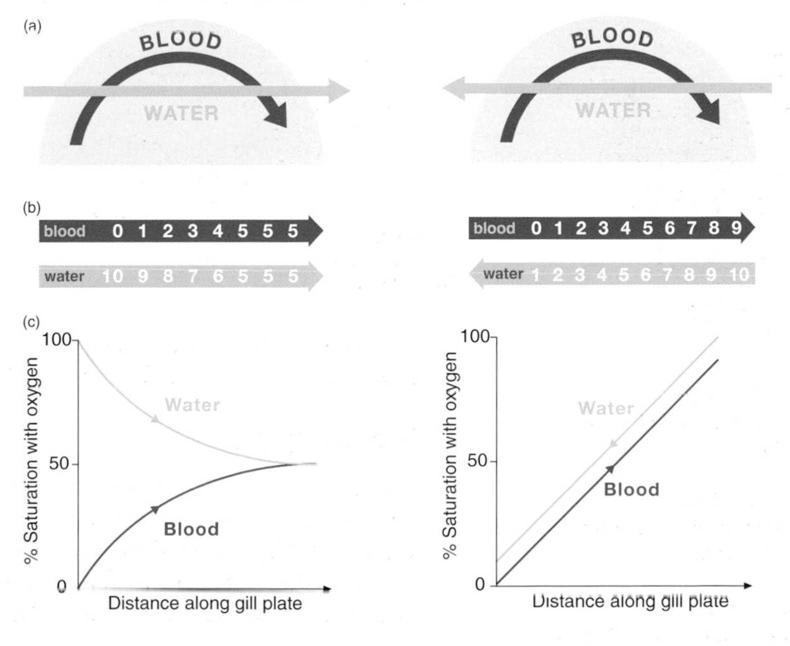


Figure 10.10
The advantage of blood and water flowing in opposite directions
(a) Direction of water and blood as they flow across a gill plate
(b) Oxygen concentration in water and blood (c) Percentage saturation of blood across a gill plate

**Q 1 Why would fish gills not be effective on land?**

## Ventilation

Bony fish achieve a continuous flow of water over their gills using a pumping system. A ventilation mechanism creates the pressure differences needed to provide a constant stream of water over the gills. Water is drawn through the mouth into the buccal cavity by muscular contractions, passes over the gills and leaves through the operculum. Thus a fresh supply of aerated water constantly flows over the gills.

The bulk movement of water depends on pressure differences. When the fish opens its mouth, the volume inside the buccal cavity increases. This decreases the pressure and water flows in. The fish then closes its mouth and squeezes the water out of the buccal cavity, over the gills and out of the operculum on each side.

### Extension box 2

## Ventilation in bony fish

Forcing water over the gills requires energy. There are two different ways of providing this. The simplest is for a fish to open its mouth and swim forwards (Figure 10.11). The faster the fish swims, and the wider open the mouth, the faster the water flows across the gills. Energy is provided indirectly, by the muscles used in swimming. This is called **ram ventilation** and is common in fast-swimming fish such as mackerel.

**Figure 10.11**
Mackerel swim with their mouths open – ram ventilation

## Ram ventilation

Most fish do not swim fast enough for ram ventilation to be effective. Some fish spend much of their time stationary, either hiding or lying in wait for their prey.

During inspiration in these fish the mouth is opened and the buccal cavity expands. This decreases the pressure within, causing water to enter through the mouth. Water does not come from the gill region because the outside water pressure closes the valve at the posterior end of the operculum. At the same time the opercular muscles contract causing the opercular cavity to be enlarged. The pressure in the opercular cavity is lower than that in the buccal cavity and hence water is drawn from the buccal cavity over the gills into the opercular cavity.

When expiration takes place, the mouth closes and the floor of the buccal cavity is raised. The entrance to the oesophagus also closes to stop water being swallowed. The raised pressure forces water in the buccal cavity to flow over the gills, to the outside via the now open operculum. The coordinated activity of the muscles of the buccal cavity and the opercular muscles ensures that an almost continual flow of water passes over the gills.

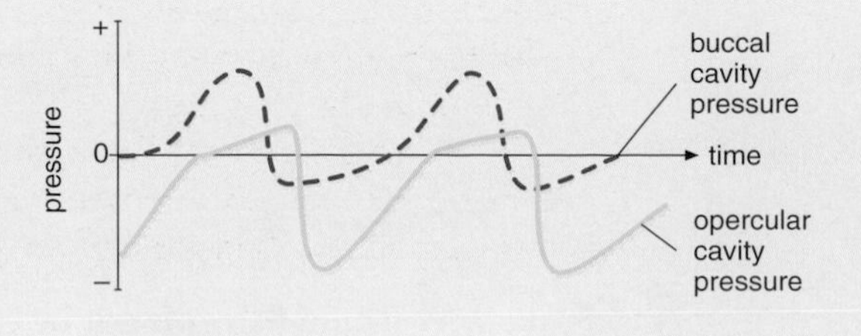


**Figure 10.12**
Recording of pressures in the buccal cavity and opercular cavity of a trout

**Q 2** Look at Figure 10.12: it shows the pressure changes measured inside the buccal and opercular cavities of a bony fish during normal breathing. Identify, with reasons, the part of the graph in which:

(a) water enters the mouth

(b) water is pushed from the mouth to the pharynx

(c) the mouth closes

(d) water is pushed over the gills from the pharynx

(e) the opercular valve is open.

**Q 3** The pressure differences are such that a continual flow of water from the buccal cavity to the opercular cavity is maintained throughout most of the ventilation cycle. At what stage of the cycle is this not so?

## How fish achieve efficient gas exchange – summary

Large surface area

- Each gill has two rows of gill lamellae.
- Each gill lamella has a further series of gill plates sticking out from it at right angles.
- Running through each gill plate is a mass of capillaries.

Short diffusion distance

- Lamellae have walls only one cell thick.
- The capillary wall is only one cell thick.
- Blood is close to the surface of the plate. There are therefore only two complete layers of cells between the water surrounding the gills and blood in the capillaries.

Maintaining a diffusion gradient

- A ventilation mechanism ensures that there is a continual flow of aerated water across the gills.
- The circulation of blood through the gills ensures that blood saturated with oxygen and low in carbon dioxide is replaced with blood low in oxygen and high in carbon dioxide.
- Water always passes over the gills in the opposite direction to blood flowing through the gills. This countercurrent mechanism ensures that the blood is always next to water containing a slightly higher oxygen concentration. Because of this, the oxygen concentration in the blood never reaches equilibrium with the oxygen in the gills, so absorption continues across the whole of the gill plate.

**Figure 10.13**
A locust's abdomen. One pair of spiracles is present on each of the segments

## Gas exchange in insects

Insects have a segmented body with a rigid exoskeleton made of a rigid substance called chitin. The outside of the exoskeleton is covered with a layer of wax, making it impermeable to water, and gases (Figure 10.13). Insects have solved the problems of gas exchange in a unique way. Instead of absorbing oxygen into a blood transport system as fish and mammals do, insects have tubes taking air close to all the body cells.

### Tracheal system

An insect has a **tracheal** system consisting of tubes leading from the outside to the inside of the body.

On the outside of the insect's body there are small holes on each side of the segments, through which gases can diffuse. The holes, called spiracles, can open and close to control the level of ventilation.

The spiracles lead into a system of large tubes, called the tracheae. Each trachea has rings of chitin in its wall and is impermeable to gases. The tracheae branch into smaller tubes called tracheoles. The tiniest tracheoles are less than 1μm in diameter and can pass between and even into the insect's cells. The tracheoles contain little or no chitin and are permeable to gases and therefore also to water. Carbon dioxide can diffuse from the cells into the tracheoles and oxygen can diffuse from the tracheoles directly into the cells. The finest tracheoles are filled with fluid because water diffuses into them from the surrounding cells.

**Q** 4 **Suggest a function for the chitin rings around the tracheae.**

### Gas exchange

The air-filled tracheal system penetrating into all the tissues of an insect allows gases to diffuse directly between air and tissues. Movement of air in the insect tracheal system depends mainly on the diffusion of oxygen from, and carbon dioxide to, the outside air. In the tissues, the partial pressure of oxygen is lower than that in the tracheoles, so oxygen diffuses from the tracheole into the tissue. This creates a diffusion gradient between tracheole and the air surrounding the insect. Oxygen diffuses through the spiracles and along the tracheae. Since tracheae are gas filled, diffusion is rapid. Diffusion from the tracheoles to mitochondria is in fluid, and therefore much slower. But the diffusion path from tracheoles to mitochondria is very short.

### Ventilation

In small insects, in which the total diffusion path is no more than a millimetre, gas transport is entirely by diffusion. In larger insects, such as locusts, and during periods of high activity, the tracheal system is actively ventilated by pumping movements of the abdomen. These insects compress their abdomen and squeeze air from their tracheal tubes. Fresh air moves into the tubes when the body returns to its normal size.

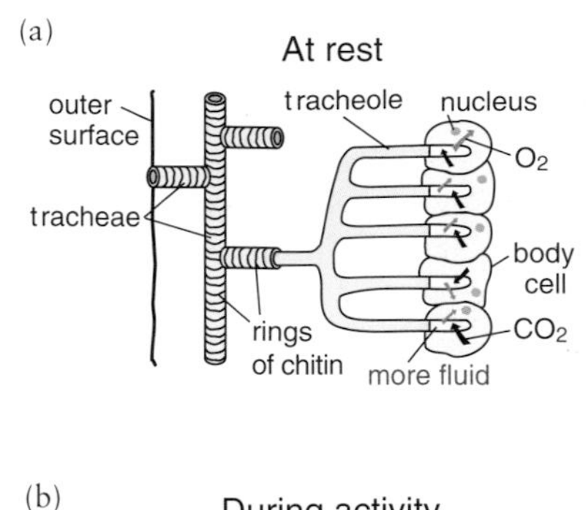


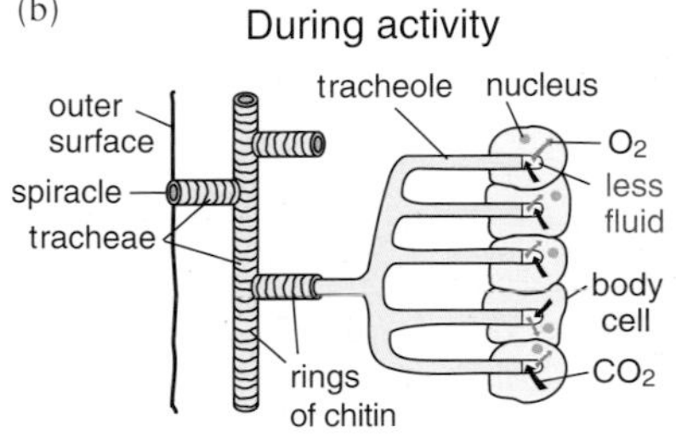


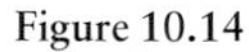

**Figure 10.14**
Gas exchange in insects when at rest (a) and when active (b)

When an insect is very active, lactic acid accumulates in the cells. This decreases their water potential. By osmosis, water in the tracheoles is drawn into the cells, causing more air to enter the tracheoles. This means that more oxygen comes into close contact with the tissues just at the time when it is required (Figures 10.14(a) and (b)).

## Coping with dry conditions

Some of the moisture in the tracheoles will inevitably evaporate and diffuse out through the spiracles. In hot, dry, windy conditions insects can reduce water loss through their spiracles by closing them by means of a muscular valve. Hairs around the spiracles can trap moist air, also reducing water loss.

## How insects achieve efficient gas exchange – summary

Large surface area

- extensive system of branching tracheoles
- withdrawal of fluid from tracheoles during vigorous activity, further increasing surface area for gas exchange

Maintaining a large diffusion gradient

- use of oxygen by cells in respiration keeps the concentration of oxygen low in cells
- production of carbon dioxide by cells increases the concentration aiding its diffusion from the cells and thus out of the insect
- some insects have a ventilation system which increases the concentration gradient
- continual flow of air through tracheae

Short diffusion distance

- thin tracheole walls
- withdrawal of fluid from the tracheoles enables air to get closer to the cells, so reducing the diffusion distance.

# Gas exchange in plants

## What gases do plants exchange?

Like animals, plants exchange gases with the atmosphere. It is easy to get the idea that gas exchange in plants is the opposite to that in animals, but that is not correct.

Like animals, plants respire aerobically which enables them to obtain energy from organic substances such as carbohydrates. This process uses oxygen and produces carbon dioxide as a waste product. Plants are less active than most animals and consequently their oxygen requirement is relatively low. During daylight, they also photosynthesise to make organic substances. This process uses carbon dioxide and releases oxygen as a waste product. In daylight, the rate of photosynthesis is usually greater than the rate of respiration and so the green tissues of a plant take in carbon dioxide from the atmosphere and allow excess oxygen to escape. Most photosynthesis happens in the leaves and so most gas exchange takes place here too (Figure 10.15).

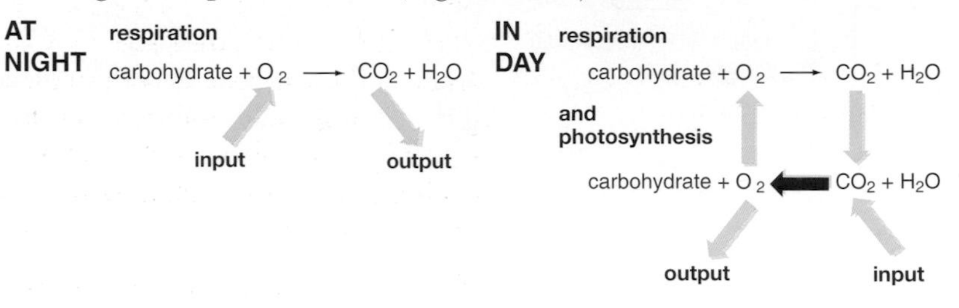


Figure 10.15
Gas exchanged by a plant in different quantities at night and during the day

Figure 10.16
Photograph of a leaf showing its flat shape

## Where are gases exchanged?

The overall shape of a leaf is one with which we are all familiar. They are for the most part flat. The flat shape means that no cell is far from the air. Gases enter the leaves through the stomata by diffusion and spread between the cells. There are air spaces between the spongy mesophyll cells, which are also in contact with the palisade mesophyll (Figure 10.16).

Both types of mesophyll cell contain chloroplasts, the organelles in which photosynthesis takes place and so they will take carbon dioxide, from the air inside the leaf, by diffusion. These cells are also active and thus must have oxygen for aerobic respiration, again obtained by diffusion from the air inside the leaf.

## Gas exchange and water loss

The main gas exchange surface within the leaf is the spongy mesophyll. (Figure 10.17) This area of cells is specially adapted for this function. The cells are loosely packed, creating many air spaces between them, thus increasing the surface area for gas exchange. As mentioned before, to be permeable to gas the cell is also permeable to water. As the cells absorb carbon dioxide for photosynthesis and oxygen for respiration they will lose water into the air spaces inside the leaf.

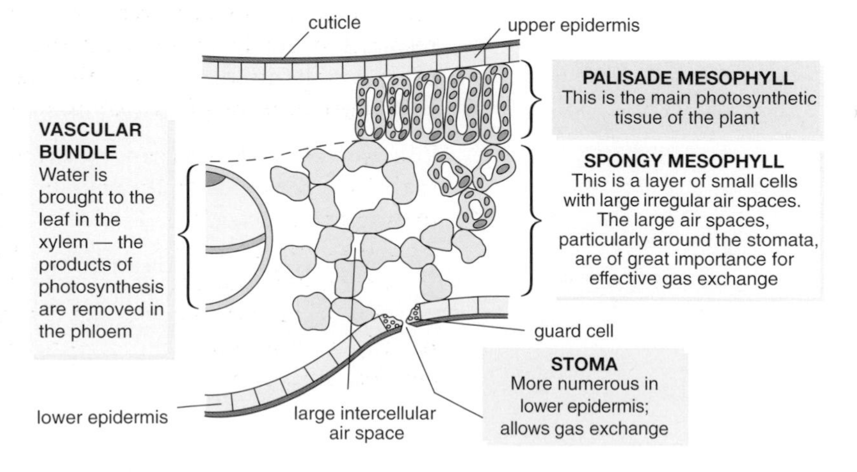


Figure 10.17
Cross section through a leaf showing the important features involved in gas exchange

Plants, just like animals, are exposed to the drying action of the air. To stop the loss of too much water a waterproof waxy cuticle covers them. Water can therefore only leave the leaf through the stomata down a water potential gradient, if the air outside the leaf contains less water than the air inside.

Some of the moisture on the gas exchange evaporates and diffuses out of the leaf through the stomata in the process of transpiration (see Chapter 7). The spacing and small size of the stomata allow sufficient gas exchange yet minimise water loss. Land plants have their gas exchange surface inside the leaf to reduce the water lost. If they were to exchange gases through the surface of the leaf, the water loss would be so great that the plant would dehydrate and die.

Plants are able to close their stomata if water loss becomes excessive, which also reduces the entry and exit of air for gas exchange. However, the large volume of air spaces in the leaf provides a sufficient reservoir, and gas exchange into the cells is rarely affected.

## When are gases exchanged?

The answer to this question is easy: gases are exchanged all the time.

Air enters through the stomata by diffusion; there is no ventilation process. As leaves are thin, and up to 40% of their volume is air, gases can diffuse quickly through to the mesophyll from the stomata. The walls of these cells are thin, so reducing the distance for diffusion of gases into the cell.

The direction of gas movement relies on differences in concentration of carbon dioxide and oxygen. That in turn relies on the processes which use and produce those gases.

### During the night

Plants, unlike animals, are able to photosynthesise but to do that they must absorb light. Therefore at night the only process that occurs which requires gas exchange is respiration. Oxygen readily passes through the walls of the mesophyll cells and is transported through the cytoplasm to the mitochondria where respiration occurs. Oxygen is used during respiration and this maintains a lower concentration of oxygen than is present in the air in the leaf. This difference in concentration ensures continued diffusion of oxygen into the mesophyll cells.

The numerous small holes of the stomata allow the entry and exit of a sufficient quantity of fresh air to keep the concentration of oxygen high in the air spaces. Carbon dioxide produced during respiration diffuses from the mesophyll cells, where it is present at a higher concentration, to the air spaces in the leaf where carbon dioxide is at a lower concentration. It then passes out through the stomata by diffusion, as there is an even lower concentration of carbon dioxide in the environment.

## During the day

Both respiration and photosynthesis take place during the day. The plant will therefore require oxygen for respiration and carbon dioxide for photosynthesis. The odd thing about this is that photosynthesis produces oxygen and respiration produces carbon dioxide as waste products. We have to consider the rates of these reactions and thus the amounts of the waste gas produced by one process, which could be used by the other process. The plant may not have to absorb either gas from the air. Plants are not very active and therefore the rate of respiration is always very low. The rate of photosynthesis is thus the major factor that determines which gas is absorbed and which is released.

In bright light, the rate of photosynthesis is greater than the rate of respiration. Therefore carbon dioxide for photosynthesis is required in larger quantities than it is produced by respiration. As carbon dioxide is being used for photosynthesis, a lower concentration exists in the mesophyll cells, compared with the leaf air spaces. Therefore carbon dioxide diffuses into the mesophyll cells and through the cytoplasm to the chloroplasts.

Constant use of the carbon dioxide by photosynthesis will maintain this difference in concentration and diffusion of carbon dioxide into the cells will continue. Oxygen will be produced by photosynthesis in larger quantities than is required for respiration. The excess oxygen will build up in the cell and diffuse into the leaf air space where it is in a lower concentration. It diffuses across the narrow leaf and out through the stomata, down a concentration gradient.

## How plants achieve efficient gas exchange and cope with water loss – summary

Large surface area

- the mesophyll cells are in contact with many air spaces between them
- the leaf has a large internal surface area in relation to its volume

Maintaining a large concentration gradient

- the diffusion of oxygen and carbon dioxide into and out of the cells
- air spaces inside leaf are continuous with external air through stomata
- there are huge numbers of stomata, several thousand per $cm^2$. Their size and density ensure an adequate diffusion of gas
- leaves held out into the air. Breeze renews air outside stomata

Short diffusion distance

- the cell walls are thin and permeable to gases
- all mesophyll cells are close to stomata.

### Reducing water loss

As in any gas exchange surface, water escapes from the cells, so the surface is moist. However, excessive loss of water into the atmosphere is avoided because

- the exchange surface is inside the leaf
- each stoma is surrounded by two guard cells that can vary the width of the stoma, allowing more or less water vapour to escape. The action of stomata is covered in more detail in Chapter 7.

**Q** 5 **What are the main gases that diffuse from a leaf:**

**(a) at noon on a bright summer's day**

**(b) at midnight in summer?**

## Summary

- Fick's law can be used as a framework to understand the essential features of a gas exchange structure to ensure the maximum rate of diffusion of respiratory gases.
- The body surface of a protoctistan, a fish gill, the tracheal system of insects and the spongy mesophyll of a leaf and a lung all have a large surface area; a small diffusion pathway and mechanisms maintain the maximum diffusion gradient.
- As terrestrial insects and mesophytic plants exchange gases they tend to lose water and so have developed structures to allow both processes to occur without causing harm.

## Assignment

### Water and gas exchange

You may have come across the saying 'There is no such thing as a free lunch'. What it means is that sooner or later you have to pay for everything. It is an idea which applies to the gas exchange surfaces of organisms which live on land. An efficient gas exchange surface is one over which oxygen will readily diffuse. If oxygen diffuses in, however, water will be able to diffuse out and water is in short supply in many land-living organisms. In this assignment, we will explore some of the links between water and gas exchange surfaces in a little more detail.

Table 10.2 compares the rates of diffusion of water, oxygen and carbon dioxide

| Substance | Molecular mass | Rate of diffusion/arbitrary units | |
|---|---|---|---|
| | | in air | in water |
| Water | 18 | $2.42 \times 10^{-5}$ | |
| Oxygen | 32 | $1.95 \times 10^{-5}$ | $2.5 \times 10^{-9}$ |
| Carbon dioxide | 44 | $1.51 \times 10^{-5}$ | $1.7 \times 10^{-9}$ |

1 (a) The figures in this table all refer to a temperature of 20 °C. How would you expect these figures differ if the temperature were increased to 30 °C? Give a reason for your answer.

*(2 marks)*

(b) The mantle cavity is the gas exchange organ of a snail. Figure 10.18 shows a section through the body of a snail. Use information from the table and your knowledge of diffusion to help explain what causes the floor of the mantle cavity of the snail to be moist.

*(3 marks)*

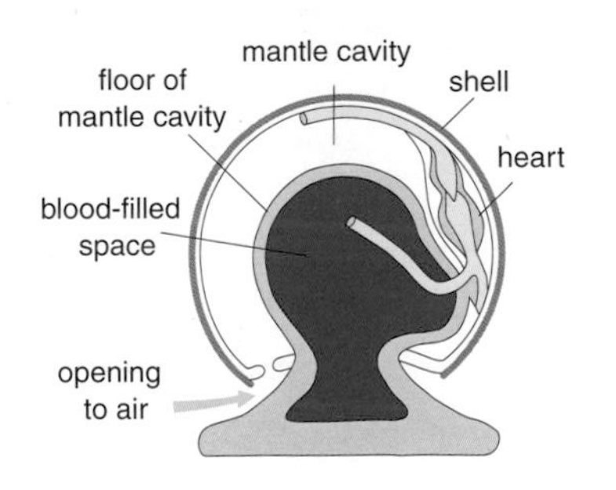


**Figure 10.18**
This diagram shows a section cut through a snail. The gas exchange organ of a snail is the mantle cavity

(c) It is sometimes suggested that a moist surface makes gas exchange more efficient. Use information from the table to explain why this suggestion is incorrect.

*(2 marks)*

One property of water molecules is that they stick to each other as a result of hydrogen bonds formed between molecules. This results in surface tension. Surface tension is the reason that wet surfaces stick to each other. Table 10.3 shows the surface tension of a number of liquids.

2 Cells in the alveoli of mammalian lungs secrete surfactants. These are phospholipids combined with proteins, and form a layer two to three molecules thick lining the alveoli. Explain how a surfactant increases the efficiency of breathing.

*(3 marks)*

| Liquid | Surface tension/arbitrary units |
| --- | --- |
| Water at 0 °C | 77 |
| Water at 20 °C | 73 |
| Water at 40 °C | 70 |
| Detergent solution | 25 |
| Surface liquid in mammalian lung | 1 |

In the last part of this assignment we will look at the relationship between the loss of water and the uptake of oxygen in a little more detail. Figure 10.19 shows sections through two different animals, an earthworm and an insect.

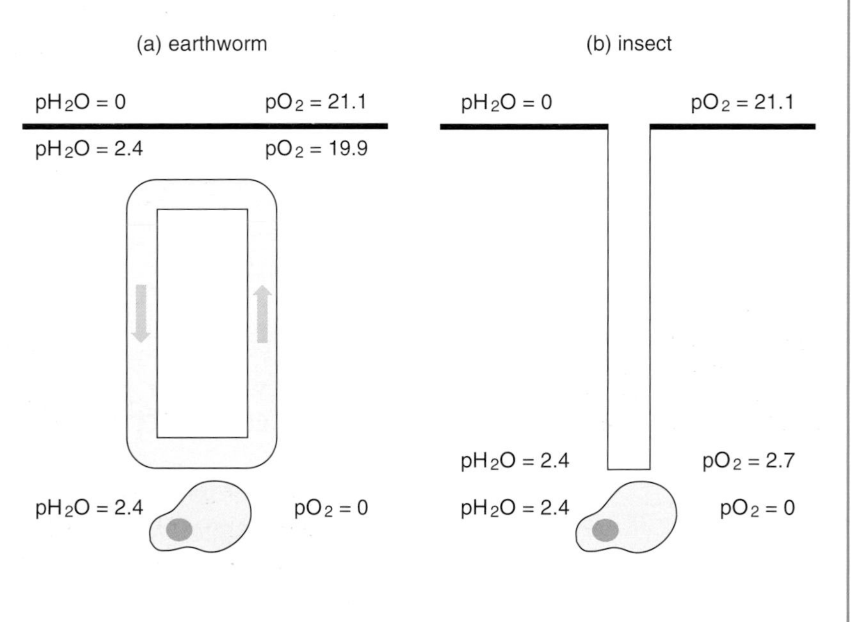


**Figure 10.19**
Gas exchange in (a) an earthworm and (b) an insect

3 Use the diagram showing the earthworm to describe how a muscle cell inside the animal gets oxygen.

*(2 marks)*

The figures on this diagram show the partial pressure of water molecules ($pH_2O$) and the partial pressure of oxygen molecules ($pO_2$) at a number of different points. You can assume that the partial pressure of a particular substance is a measure of its concentration.

4 The partial pressure of water has the same value at the surface of an earthworm as in one of its cells but the partial pressure of oxygen is much lower in the cell than it is at the earthworm's surface. Explain why.

*(2 marks)*

The difference in partial pressure across the exchange surface of the earthworm gives us a measure of the rate of loss or gain by diffusion. We will use the symbol $\Delta H_2O$ to represent the change in partial pressure of water and to represent the change in partial pressure of oxygen.

5 If we calculate the ratio $\frac{\Delta H_2O}{\Delta O_2}$ it has a value of 2. Show how this value has been calculated.

*(1 mark)*

6 Now look at the diagram representing the insect. Use this diagram to explain why the partial pressure of oxygen at the gas exchange surface of an insect is lower than it is at the gas exchange surface of an earthworm.

*(2 marks)*

7 (a) Calculate the ratio $\frac{\Delta H_2O}{\Delta O_2}$ for an insect.

*(1 mark)*

(b) Use the $\frac{\Delta H_2O}{\Delta O_2}$ ratios for an earthworm and an insect to suggest why insects are able to survive in drier conditions than earthworms.

*(2 marks)*

## Examination questions

1 **Diagram A** shows part of a gill filament from a fish. **Diagram B** shows a transverse section through one of the gill plates.

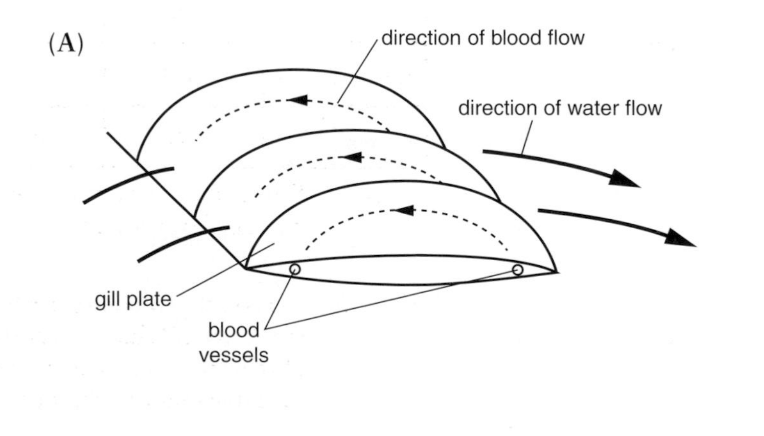

(B)

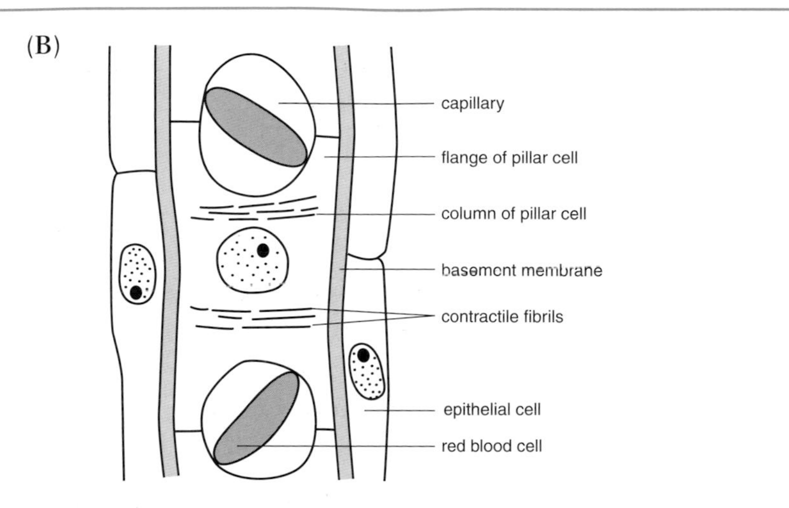


(a) Explain why the direction of blood flow in the gill filament results in efficient gas exchange.

*(2 marks)*

(b) Although fish do not maintain a constant body temperature, they have mechanisms which conserve some of the heat generated by metabolism.

(i) Describe how contraction of the fibrils in the pillar cells will affect blood flow through the gills.

*(1 mark)*

(ii) Suggest how this mechanism might enable an inactive fish to conserve heat.

*(1 mark)*

| Species | Gill area ($mm^2\ g^{-1}$) | Thickness (μm) | | |
|---|---|---|---|---|
| | | epithelium | basement membrane | pillar cell flange |
| Mackerel | 1158 | 0.17 | 0.07 | 0.03 |
| Bonito | 595 | 0.01 | 0.08 | 0.02 |
| Dab 188 | 188 | n/a | n/a | n/a |
| Skaten/a | n/a | 0.51 | 0.13 | 0.03 |

(c) (i) Why is it necessary to give the surface area of the gills *per gram of body mass?*

*(1 mark)*

(ii) Mackerel and bonito are both active species living in the open sea. Dab and skate are bottom-dwelling fish that spend much of their time inactive on the sea bed. Explain how the data in the table relate to the way of life of mackerel and bonito.

*(3 marks)*

# Transporting Respiratory Gases

Figure 11.1
The environmental conditions encountered by a climber on Everest are very different to those found at sea level. Given time, the human body is able to adapt to these conditions. Many of these adaptations involve the blood system and the transport of oxygen to the tissues

**Whichever way you look at it, 9 000 metres or 29 000 feet, Everest is a very impressive mountain. Although it has been climbed many times, getting to the top is still a challenge. Not only is it bitterly cold, but the atmospheric pressure is far lower than at sea level. This brings real problems for the climber. The air at these high altitudes contains the same proportion of oxygen – around 21% – as it does at sea level but, because of the low pressure, there are far fewer molecules present. This means that there will be less oxygen going into the lungs, less oxygen being transported by the red cells in the blood and less oxygen delivered to the muscles and other tissues in the body.**

**One of the most remarkable features of the human body is that it can adapt to different environmental conditions, even when they are as demanding as those faced by mountaineers. These changes, however, do not occur immediately (Table 11.1) and mountaineers spend time at high altitude to acclimatise. This helps the body to adjust to the demanding environmental conditions.**

| Change | Time scale |
|---|---|
| Increase in resting heart rate<br>Increase in resting breathing rate | These are the first responses to living at high altitude. They start to occur after a few minutes. |
| Increase in concentration of blood plasma | A medium term change which takes a week or so. |
| Increase in red blood cell production<br>Increase in number of blood capillaries | Long term changes beginning to become noticeable after 1–2 weeks at high altitude. |

Table 11.1
The changes which take place as a mountaineer acclimatises

A red blood cell differs from other cells in the human body. It does not have a nucleus and there are no mitochondria in its cytoplasm. Instead, it is packed with around 280 million molecules of **haemoglobin**. In this chapter we shall look at the remarkable properties of haemoglobin which allow a mere five litres of blood to transport all the oxygen the human body needs. The haemoglobin found in different organisms differs very slightly in its chemical structure. We shall consider how these chemical adaptations enable different organisms to live in different environments. Finally, we shall look at the other side of the respiratory equation. How is the carbon dioxide produced during respiration removed from the body? You may find it useful to read Chapter 7 in the AS book again before you start work on this chapter.

## Haemoglobin and the transport of oxygen

A molecule of haemoglobin from the blood of an adult human is made up of four subunits (Figure 11.2). Each of these subunits consists of two parts – **haem** and **globin**. The haem part consists of a ring of atoms linked to iron ($Fe^{2+}$). The globin part is a polypeptide chain.

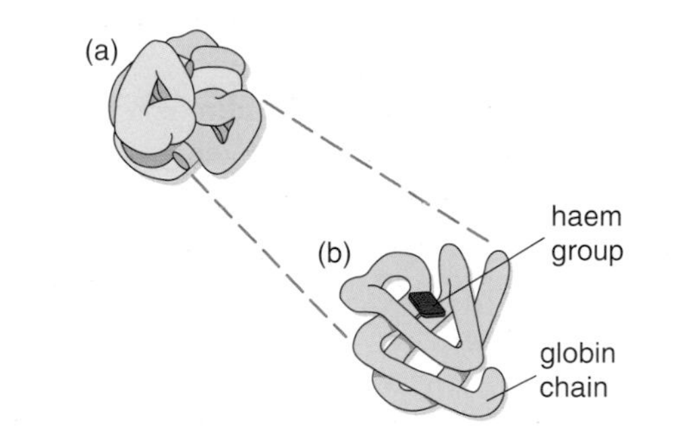


**Figure 11.2**
(a) There are four subunits in a molecule of haemoglobin from an adult human. The two blue ones are the α subunits. The two yellow ones are the β subunits. The subunits differ from each other in having slight differences in the sequences of amino acids which form the globin chains. (b) shows one of these subunits enlarged. The haem and globin parts of the subunit have been labelled

**Q 1 What is the evidence from Figure 11.2 that haemoglobin has a quaternary structure?**

As you will find out later in this chapter, there are many different sorts of haemoglobin. One way in which they differ from each other is in the sequences of amino acids which form the globin chains. These differences affect the oxygen-carrying properties of the molecule (Figure 11.3).

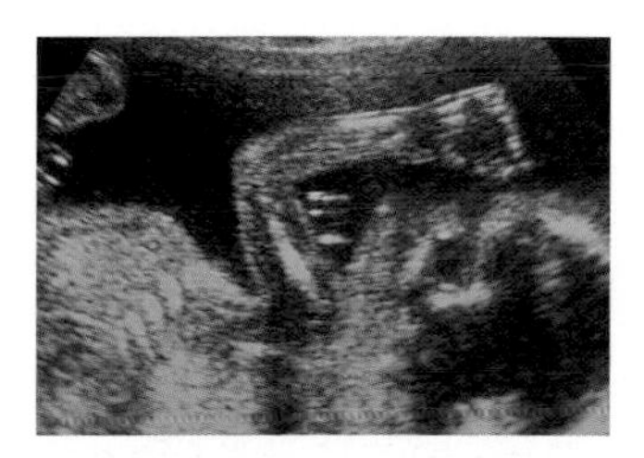

**Figure 11.3**
An ultrasound scan showing a human fetus inside the uterus of its mother. It has a slightly different type of haemoglobin from that found in an adult. This means that its haemoglobin is able to pick up oxygen from the mother's blood in the placenta

It is haemoglobin that enables blood to be so efficient at transporting oxygen. In the lungs, haemoglobin combines with oxygen to form **oxyhaemoglobin**. Oxyhaemoglobin is transported in the blood to the respiring tissues where it dissociates. It gives up oxygen which diffuses into the cells of the body while the haemoglobin is transported back to the lungs. To understand this process in more detail, we need to look at the **dissociation curve** shown in Figure 11.4. This is a graph showing the percentage saturation of a sample of haemoglobin plotted against the partial pressure of oxygen. The **partial pressure** of oxygen, usually abbreviated to $pO_2$, is simply a measure of the amount of oxygen present.

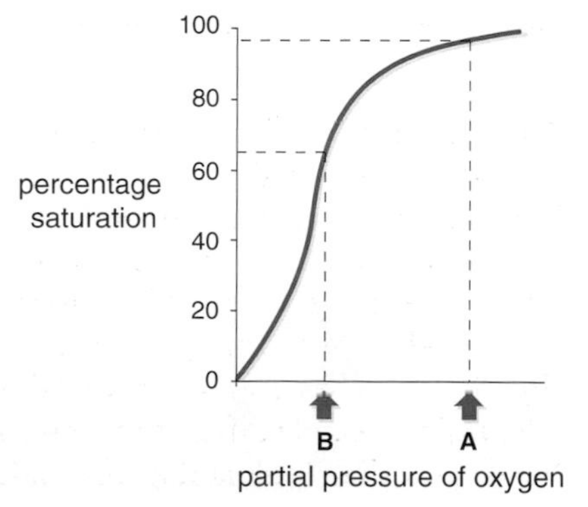


**Figure 11.4**
A dissociation curve for human haemoglobin. This graph shows how the percentage saturation of haemoglobin with oxygen varies with the amount of oxygen present

We will look at this graph in a little more detail. Find the point marked **A** on the horizontal axis. This point represents the partial pressure of oxygen in the lungs. You can see from the curve that haemoglobin is almost totally saturated at this high partial pressure. Now find the point marked **B**. This corresponds to a muscle. The muscle is respiring so it is using up oxygen. There will obviously be a lower partial pressure of oxygen in an actively respiring muscle than there is in the lungs. The curve shows that the percentage saturation of haemoglobin is lower at this lower partial pressure of oxygen, so a lot of the oxygen that was being carried by the haemoglobin has been given up. So, here is the basic idea. The haemoglobin in the blood is almost saturated with oxygen in the lungs. This oxygen is transported by the blood to the tissues of the body where it is unloaded.

**Q** 2 **(a) What will happen to the partial pressure of oxygen in a muscle if its rate of respiration increases?**

**(b) What effect will this change in partial pressure have on the amount of oxygen supplied to the muscle?**

Another feature that you should note from Figure 11.4 is that the curve is S-shaped. The explanation for this lies in the fact that each haemoglobin molecule can carry up to four molecules of oxygen. The first oxygen molecule only binds with some difficulty but, as it does, it brings about a change in the molecular shape of the haemoglobin allowing it to bind more easily with the remaining oxygen molecules. This means that the last oxygen molecule binds to the haemoglobin several hundred times faster than the first one.

The graph you have just been looking at shows what happens when there is very little carbon dioxide present, in other words, when there is a low partial pressure of carbon dioxide. Look at Figure 11.5. This shows you how a change in the partial pressure of carbon dioxide will affect the dissociation curve for haemoglobin. Increasing the partial pressure of carbon dioxide causes the curve to move over to the right. We refer to this effect as the **Bohr effect**.

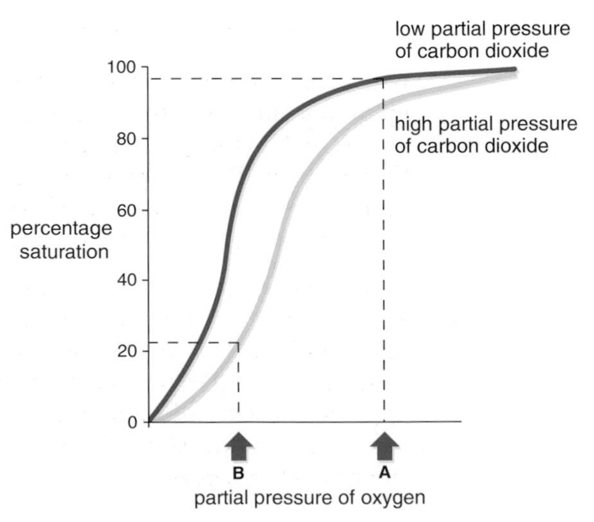


**Figure 11.5**
The effect of changing the partial pressure of carbon dioxide on the dissociation curve for human haemoglobin

It is important to appreciate that Figure 11.4 represents a simplification. In the body, not only does the partial pressure of oxygen vary but so does the partial pressure of carbon dioxide. Bearing this in mind, we will now look at the graph in Figure 11.5 in more detail.

**Q** 3 **Would you expect the highest partial pressure of carbon dioxide and the lowest partial pressure of oxygen to be in the lungs or in a respiring muscle? Explain your answer.**

Find the point marked **A** on the horizontal axis of Figure 11.5. As we saw in Figure 11.4, this represents the partial pressure of oxygen in the lungs. But now we have the extra complication of carbon dioxide to think about. In the lungs we will have a low partial pressure of carbon dioxide because it is being removed from the body. We are dealing with the upper curve and you can see from this that haemoglobin is almost totally saturated.

Now we will look at what is happening in the respiring muscle at the point marked **B**. The muscle is using up oxygen so there will obviously be a low partial pressure of oxygen. It is also producing carbon dioxide as a result of respiration so there will be a high partial pressure of carbon dioxide. We need to look at the lower curve, which represents the situation with a high partial pressure of carbon dioxide. The effect of increasing the partial pressure of carbon dioxide therefore results in the haemoglobin giving up even more of the oxygen it is carrying to the tissues.

If you think about your own activity pattern, you will realise how important it is to match the amount of oxygen the blood supplies to the needs of the body. Going from standing still to walking quickly will result in a faster rate of respiration and the need for more oxygen to be supplied to the tissues. Two very simple mechanisms help haemoglobin to meet this need by releasing more of the oxygen it is carrying. As the rate of respiration increases, more oxygen is used up and more carbon dioxide is produced. The resulting fall in the partial pressure of oxygen and rise in the partial pressure of carbon dioxide both lead to the release of more oxygen.

## Extension box 1 Mutations and malaria

As you have just seen, a molecule of adult human haemoglobin is made up of four subunits – two α subunits and 2 β subunits. The sequence of amino acids in the globin chain of the α subunit is different from that in the β subunit. The two sequences are coded for by different genes. The gene coding for the α subunit globin is on chromosome 16 while that coding for the β subunit globin is on chromosome 11.

Mutations sometimes occur in these genes and we now know of over 400 different ones which affect the globin chains of human haemoglobin. Most of these involve substitutions of single DNA bases which affect single amino acids. Although this might not seem important, some of these mutations have very serious consequences. We will look at one of them – the one that gives rise to a condition known as **sickle cell anaemia**.

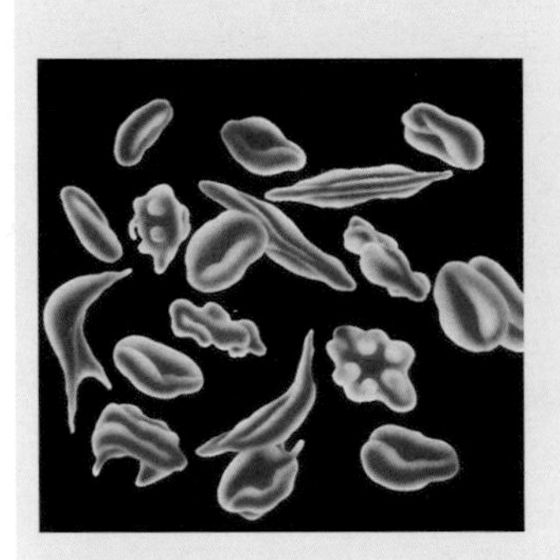

Figure 11.6
The blood cells in this photograph were taken from a patient with sickle-cell anaemia. Note their characteristic sickle shape

Sickle-cell haemoglobin differs from normal adult haemoglobin by just one amino acid. In sickle-cell haemoglobin the sixth amino acid in the β subunit globin is valine, instead of glutamic acid. This results in sickle-cell haemoglobin being less soluble in conditions where there is a low concentration of oxygen, such as in the blood which is being carried back to the heart in the veins. The abnormal haemoglobin molecules stick together to form long fibres which alter the shape of the red blood cells, making them sickle shaped (Figure 11.6)

The body rapidly destroys these abnormal cells so an affected person develops severe anaemia. In addition, sickle cells clump together and may block smaller blood vessels. This reduces the supply of oxygen to the organs of the body. Not surprisingly, a person who is homozygous and has two alleles for the condition, is likely to die in early childhood unless he or she receives the necessary medical treatment. You might expect, as a result of natural selection, that the allele which causes sickle-cell anaemia would be very rare. This is not the case, however. In Africa, the condition is much more common than might be expected. This is particularly true of areas in which malaria is found (Figure 11.7). How can we explain this?

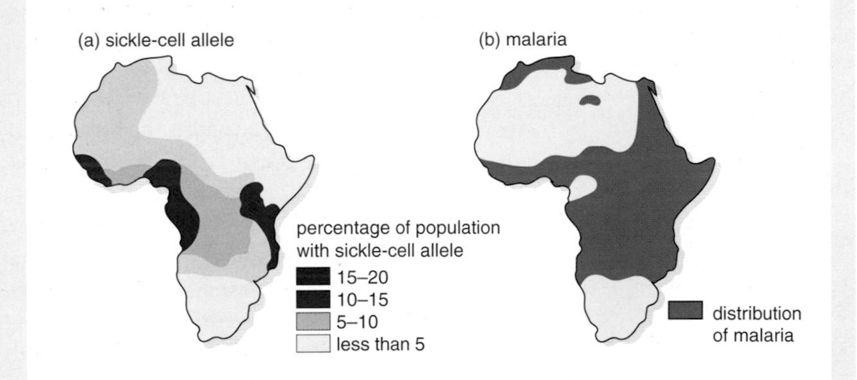


Figure 11.7
Maps of Africa showing the distribution of (a) the sickle-cell allele and (b) malaria. Note how high frequency of the allele that produces sickle-cell haemoglobin occurs where malaria is common

In order to understand the link between the sickle-cell allele and malaria we need to look at what happens in people who are heterozygous and only have a single copy of the defective allele. They make some sickle-cell haemoglobin and some normal haemoglobin. As a result, they rarely experience illness due to the sickle-cell condition, although in some conditions, such as at high altitude and while carrying out vigorous exercise, some of the red blood cells become sickle shaped.

**Q 4 Explain why vigorous exercise causes some of the red blood cells in a heterozygote to become sickle shaped.**

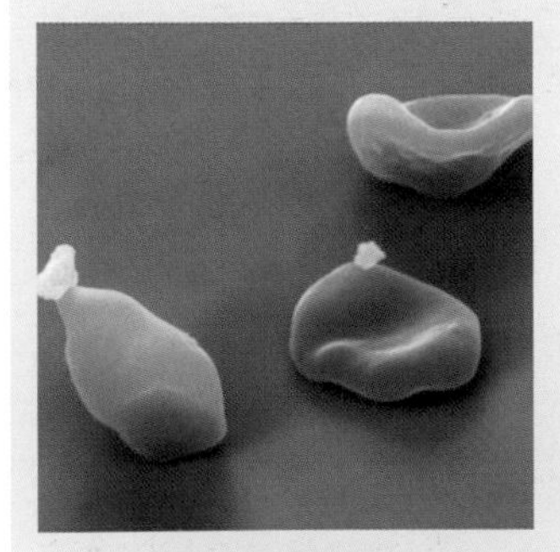

Figure 11.8
Malaria is a disease which affects the blood. This photograph shows red blood cells infected with malarial parasites. Eventually the cell will burst. A new generation of parasites will be released and infect more red blood cells

People who are heterozygous have an advantage if they get malaria. The malarial parasite (Figure 11.8) causes the red blood cell it has infected to become sickle shaped. As a result, this blood cell is destroyed by the body and the parasites it contains are unable to complete their life cycles.

Figure 11.9
This is the end of the Earth! It is a photograph of the dramatic scenery at the southern tip of South America. In an area such as this, the amount of oxygen available to different animals is very variable. The partial pressure of oxygen is particularly low on the top of the mountains in the background and in the burrows of the many small mammals that escape the extreme conditions by living underground. In addition, the whales and seals that live in the sea also have adaptations that enable them to remain underwater for up to an hour

Figure 11.10
A bull elephant seal. Tracking devices attached to the skin of elephant seals show that they frequently dive to depths of 1000 metres and remain under water for up to an hour at a time

## Different sorts of haemoglobin

Different animals live in different places with different environments. These environments are characterised by features such as temperature and humidity. A desert, for example, usually has a much greater temperature range and a lower relative humidity than a tropical forest. Environments also vary in the amount of oxygen that is available to the organisms living in them. The partial pressure of oxygen will be low, for example in deep underground burrows, on high mountains, and in lakes full of decomposing vegetation. Animals that live in these places have a number of adaptations which allow them to survive in conditions where the partial pressure of oxygen is low. To do this they often have haemoglobin which is adapted to these conditions.

There are many thousands of different species which have haemoglobin in their blood – mammals, birds and fish, and invertebrates such as worms. In this next section we will look at some of these types of haemoglobin and consider how they are linked to the different environments in which organisms live.

### Providing an oxygen store

Seals, such as the one shown in Figure 11.10 are aquatic mammals. Like all mammals, they have lungs. This means that they need to come to the surface of the water to breathe. Despite this, they are able to remain under water for long periods of time. They have many adaptations which enable them to remain under water. One of them concerns a pigment found in their muscles, a pigment called **myoglobin.**

Myoglobin is very similar to haemoglobin although there is one big difference in its chemical structure. Myoglobin molecules are made up of a single subunit, not the four we have in a molecule of haemoglobin. However, it still functions as a respiratory pigment and is able to combine with and release oxygen. Look at Figure 11.11. It shows dissociation curves for myoglobin and human haemoglobin.

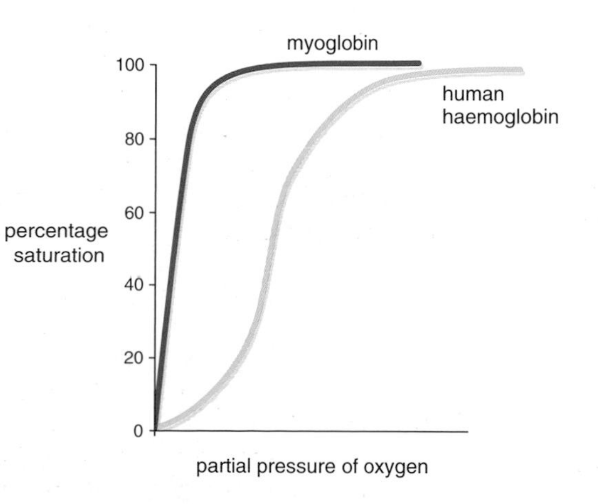


Figure 11.11
This graph shows a dissociation curve for myoglobin. Note how it differs in shape to the curve for human haemoglobin. Myoglotrin is found in the muscles of many mammals. It is particularly abundant in the muscles of diving mammals

Myoglobin has two properties which make it very useful for its function as an oxygen store. It picks up oxygen very readily and, as the graph shows, it is saturated at relatively low partial pressures of oxygen. The second feature is that as the partial pressure of oxygen falls to very low levels, myoglobin readily releases all the oxygen it is carrying.

**Q 5 Explain how the properties of myoglobin enable it to keep the muscles of a diving seal supplied with oxygen.**

In some animals, haemoglobin also acts as an oxygen store. Extension box 2 gives some information about lugworms, which use their haemoglobin in this way.

## Extension box 2

## How lugworms use their haemoglobin

A lugworm (Figure 11.12) is a large worm which is found on sandy sea shores. It lives in a U-shaped burrow. When the tide is in, the area in which the lugworm lives is covered with sea water. The worm is able to suck in a mixture of sand particles and tiny bits of organic matter. This organic matter is digested by enzymes in the animal's gut. The undigested sand passes out through the lugworm's anus forming a worm cast. While submerged, there is a constant flow of water through the burrow. The oxygen it contains diffuses into the worm, mainly through the animal's gills but also through the rest of its body surface. It dissolves in the blood and is then distributed to the cells of the body.

What happens when the tide is out? The sand remains moist but there is no current of water through the burrow to bring a supply of oxygen. This is where the lugworm's haemoglobin is important. You can see from the graph in Figure 11.13 that the shape of the dissociation curve for lugworm haemoglobin is very different from that for human haemoglobin. It is more like myoglobin. Lugworm haemoglobin acts as an oxygen store.

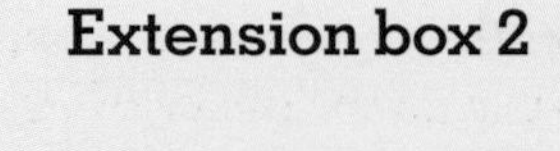


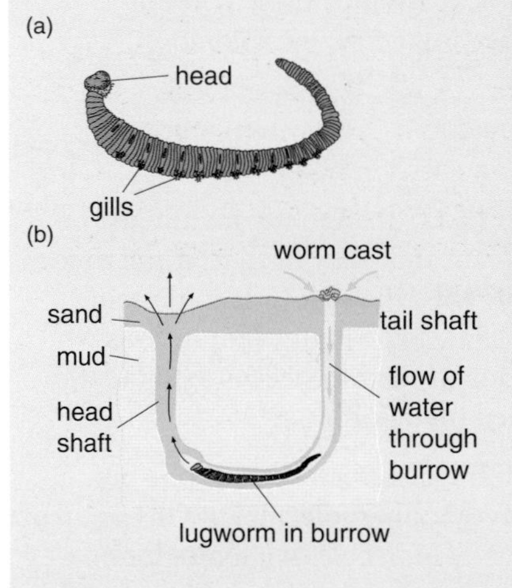


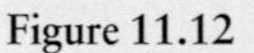

**Figure 11.12**
(a) A lugworm is a large worm approximately 15 cm in length and as thick as a person's little finger. (b) A lugworm in its U-shaped burrow

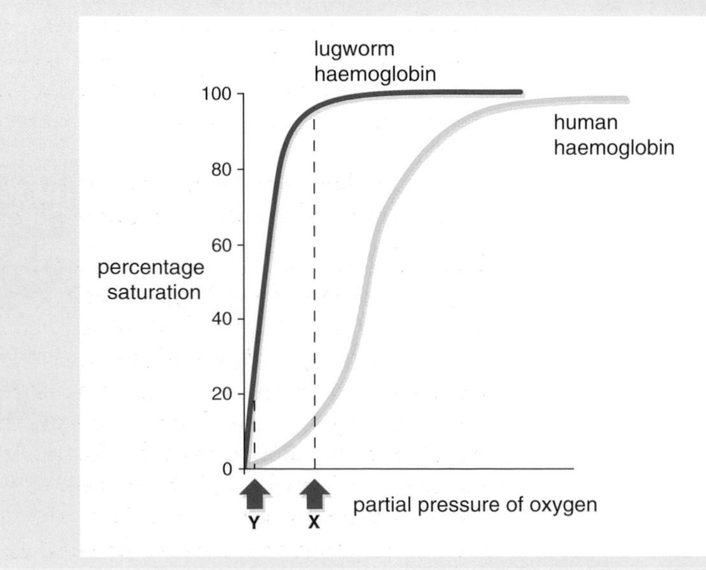


**Figure 11.13**
Dissociation curves for human haemoglobin and lugworm haemoglobin

## Getting rid of carbon dioxide

As a result of respiration, cells produce carbon dioxide and this has to be taken from the tissues to the gas exchange surface where it is removed from the body. Carbon dioxide dissolves in water. It also reacts chemically with water to form carbonic acid. Carbonic acid is a weak acid but, under the conditions found in the body, it splits into hydrogen ($H^+$) ions and hydrogencarbonate ($HCO_3^-$) ions. This process can be summarised with an equation.

$$CO_2 + H_2O \rightleftharpoons H_2CO_3 \rightleftharpoons H^+ + HCO_3^-$$

carbon dioxide, water, carbonic acid, hydrogen ion, hydrogencarbonate ion

We will look at what happens in respiring tissues. Carbon dioxide is being produced. This diffuses from the cells into the blood plasma. A small amount, about 10%, reacts with the water in the plasma and produces hydrogen ions and hydrogencarbonate ions. Most of the rest diffuses straight into the red blood cells where it also reacts with water. Red blood cells, however, contain an enzyme called **carbonic anhydrase**. This enzyme catalyses the reaction so it goes much faster in red blood cells than in the plasma. (Figure 11.18).

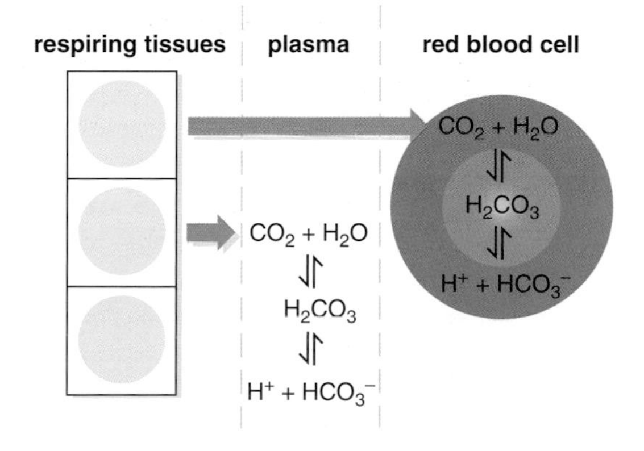


Figure 11.18
Carbon dioxide produced as a result of respiration reacts with water in the plasma and in the cytoplasm of red blood cells. As a result, hydrogen ions and hydrogencarbonate ions are formed

### Transporting hydrogencarbonate ions

Because the reaction between carbon dioxide and water is much faster in the red blood cells than in the plasma, there is a much higher concentration of hydrogencarbonate ions in a red blood cell. As a result, they diffuse out through the plasma membrane into the blood plasma. Ions, however, are charged particles and movement of all these negatively charged hydrogencarbonate ions into the blood plasma will upset the equilibrium. There is a mechanism which corrects this. The plasma contains a higher concentration of chloride ($Cl^-$) ions than the red blood cells. The hydrogencarbonate ions diffusing out are matched by chloride ions diffusing in. This mechanism is sometimes called the **chloride shift**.

Although most of the carbon dioxide is transported in the blood as hydrogencarbonate ions, a small amount – again, about 10% – reacts directly with haemoglobin to form **carbamino-haemoglobin**. The amount of carbon dioxide that haemoglobin can carry in this way depends on the amount of oxygen that it is carrying. The smaller the amount of oxygen, the more carbon dioxide it can carry.

**Q 7 Would you expect more carbon dioxide to be carried by a carbamino-haemoglobin in an artery or in a vein? Explain your answer.**

## Taking up the hydrogen ions

The more carbon dioxide that diffuses into the blood, the more hydrogen ions are formed. If the concentration of hydrogen ions were allowed to go up, the pH of the blood would fall. This would have severe effects on the biochemical reactions which take place in the body by affecting proteins such as enzymes, and carrier molecules found in plasma membranes.

You may remember that when you investigated the various factors which influenced the rate of enzyme-controlled reactions, you made use of **buffer solutions**. A buffer solution is one that keeps the pH of a solution constant by binding to hydrogen ions when their concentration is high and releasing them when it is low. The blood contains a number of substances which act as buffers and take up the hydrogen ions. In the plasma this function is carried out by phosphates and by plasma proteins. In the red blood cell the most important buffer is haemoglobin.

The faster a tissue respires, the more carbon dioxide it produces. As we have seen, this carbon dioxide combines with water to produce hydrogencarbonate ions and hydrogen ions. So, clearly, there are more hydrogen ions produced at greater rates of respiration. In red blood cells, most of these hydrogen ions are taken up by the haemoglobin. This, in turn, causes the haemoglobin to release the oxygen that it is carrying. So, the faster the rate of respiration, the more carbon dioxide produced and the greater the amount of oxygen supplied to the tissues. The way in which haemoglobin transports oxygen and carbon dioxide is summarised in Figure 11.19.

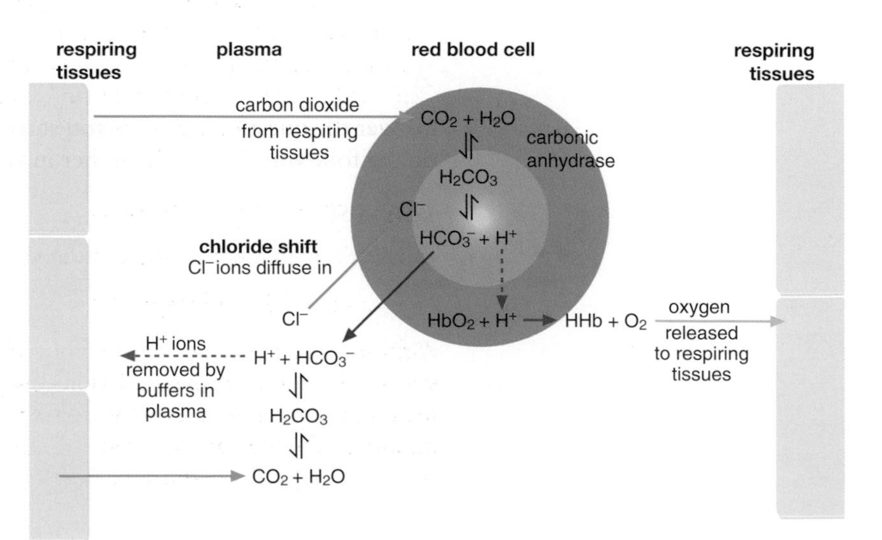


**Figure 11.19**
A summary of the way in which oxygen and carbon dioxide are transported in the blood

## Summary

- In mammals and many other animals, oxygen is transported from the gas exchange surface to the respiring tissues mainly by haemoglobin. When the concentration of oxygen is high, such as in the lungs, haemoglobin combines chemically with oxygen to form oxyhaemoglobin. When the oxygen concentration is low, such as in actively respiring tissues, the oxyhaemoglobin dissociates and oxygen is released.
- An oxygen dissociation curve is a graph showing the relationship between the percentage saturation of haemoglobin and the partial pressure of oxygen. The presence of carbon dioxide alters the position of the curve on the graph. This is known as the Bohr effect.
- Different organisms have different sorts of haemoglobin which have different oxygen-carrying properties. There is a link between the type of haemoglobin an organism possesses and the environment in which it lives.
- Most of the carbon dioxide produced in respiration is transported to the gas exchange surface as hydrogencarbonate ions.
- Haemoglobin is an important buffer. It absorbs the hydrogen ions produced by the reaction of carbon dioxide and water. As a result the pH of the blood remains more or less constant.

## Assignment

### Size and oxygen supply

You have seen in this chapter that different animals have different sorts of haemoglobin. These different types of haemoglobin differ in their oxygen-carrying properties. In this assignment we will look at how the size of a mammal is related to the properties of its haemoglobin. You will need to use material from various parts of the specification to answer the questions. Before you start, it would be a good idea to look up the following key topics either in your notes or in your textbook:

- surface area to volume ratio
- metabolism and metabolic rate
- mitochondria and respiration

Although the mammals are a group of animals that are similar in many ways – their skin is covered in hair, they possess sweat glands and they feed their young on milk – they differ considerably in size. The largest mammal, the blue whale, weighs about a hundred tonnes. This is about two hundred million times heavier than the smallest shrew! Look at the graph in Figure 11.20. It shows the oxygen dissociation curves for some different mammals.

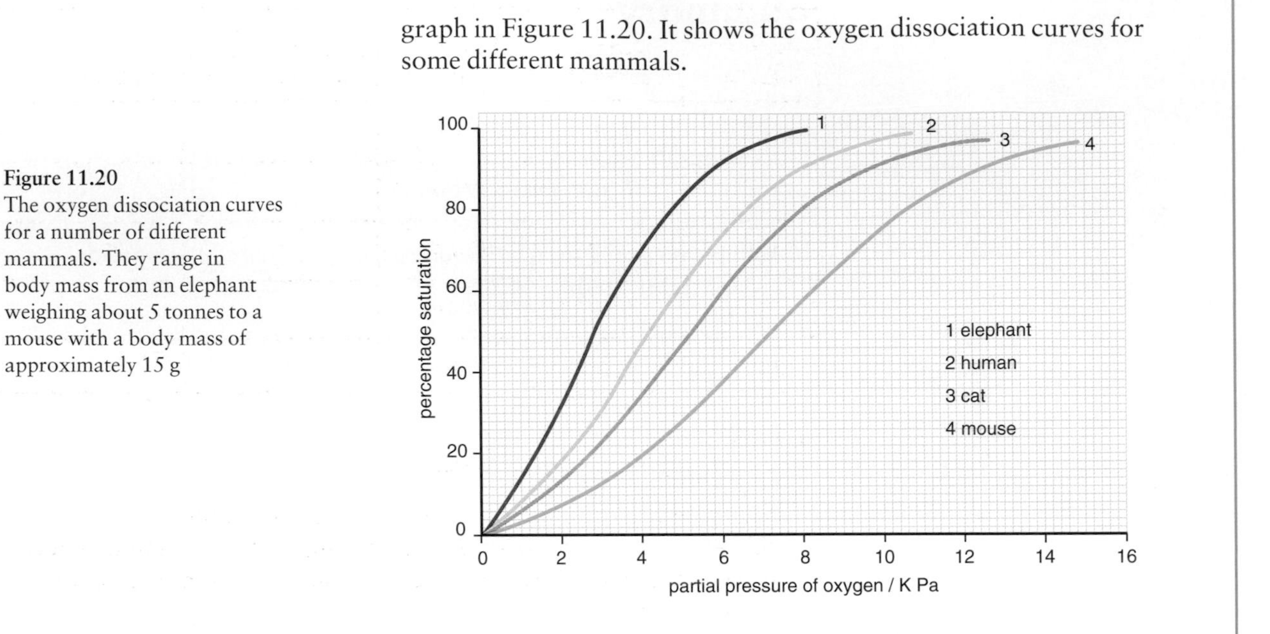


**Figure 11.20**
The oxygen dissociation curves for a number of different mammals. They range in body mass from an elephant weighing about 5 tonnes to a mouse with a body mass of approximately 15 g

When we compare oxygen dissociation curves, it is often useful to look at what is called their **unloading pressure**. This is the partial pressure of oxygen at which the haemoglobin is 50% saturated with oxygen.

1 Use Figure 11.20 to find the unloading pressure of the haemoglobin from each of the four mammals shown in the graph. Write your answers in a suitable table.

*(2 marks)*

The unloading pressure gives us a measure of how readily a sample of haemoglobin gives up the oxygen that it is carrying. The higher the unloading pressure, the more readily the haemoglobin gives up the oxygen it is carrying to the tissues.

2 Describe the relationship between the size of a mammal and the ease with which oxygen is unloaded from its haemoglobin to its tissues.

*(1 mark)*

Now, how can we explain this? How do small mammals benefit from haemoglobin that gives up its oxygen very readily? In order to answer this question, we need to look at another aspect of mammalian physiology. Look at the data in Table 11.2. This shows some of the results of an investigation carried out over a hundred years ago into heat lost by dogs with different body sizes.

| Body mass/kg | Surface area/$m^2$ | Heat loss/ kJ $hour^{-1}$ | Relative heat loss/ | |
|---|---|---|---|---|
| | | | kJ $hour^{-1}$ $kg^{-1}$ | kJ $hour^{-1}$ $m^{-2}$ |
| 31.2 | 1.08 | 191.1 | | |
| 24.0 | 0.88 | 171.0 | | |
| 18.2 | 0.77 | 146.4 | | |
| 9.6 | 0.53 | 108.9 | | |
| 3.2 | 0.24 | 49.1 | | |

Table 11.12
Heat lost by dogs with different body sizes

3 Copy out this table and complete the two blank columns. The first of these columns shows the heat loss compared to the animal's body mass. The second shows the heat loss compared to the surface area of the animal.

*(2 marks)*

4 Which do you think is more important in determining heat loss, the animal's mass or its surface area? Explain the evidence from the table that supports your answer.

*(? marks)*

5 Mammals produce the heat that results in a high body temperature from the metabolic reactions which take place in the body. Use the information in this assignment to explain why:

(a) when compared to their size, small mammals lose more body heat than large mammals
(b) small mammals have a higher oxygen consumption relative to their body size than large mammals
(c) small mammals benefit from haemoglobin that gives up its oxygen very readily.

*(5 marks)*

6 Since the early work on heat loss in dogs, studies have been carried out linking the body size of mammals with other factors. Explain the link between each of the following and your findings about the metabolic rates of different sized mammals:

(a) the number of mitochondria per gram of liver is higher in small mammals than in large mammals
(b) the Etruscan shrew is the smallest of all mammals. Its mitochondria have many more cristae than the mitochondria of larger mammals
(c) the tissues of small mammals have a higher density of capillaries than the tissues of large mammals.

*(5 marks)*

## Examination questions

1 The graph shows an oxygen dissociation curve for human haemoglobin.

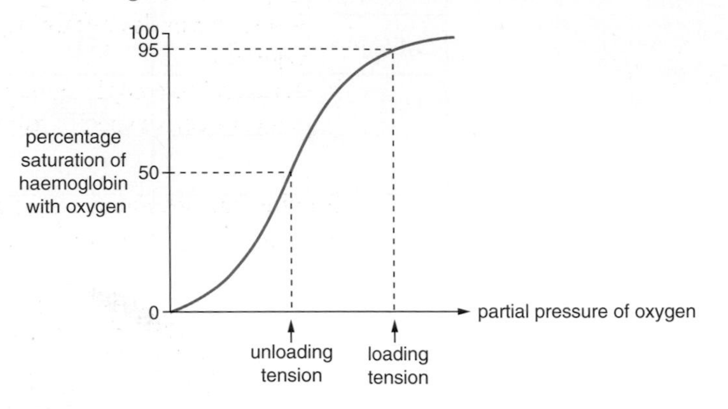


The loading tension is the partial pressure at which the haemoglobin is 95% saturated with oxygen. The unloading tension is the partial pressure at which the haemoglobin is 50% saturated.

(a) (i) What would be the effect on the unloading tension of an increase in the partial pressure of carbon dioxide?

*(1 mark)*

(ii) Explain how this may be of value in supplying tissues with oxygen.

*(2 marks)*

(b) The prairie dog is a small mammal that spends much of its life in an extensive system of burrows where the air may have a low partial pressure of oxygen.

(i) Sketch a curve on the graph which would represent an oxygen dissociation curve for prairie dog haemoglobin.

*(1 mark)*

(ii) Explain why you have drawn the curve in this position.

*(2 marks)*

2 The graph shows the oxygen dissociation curves for two species of fish, the carp and the mackerel.

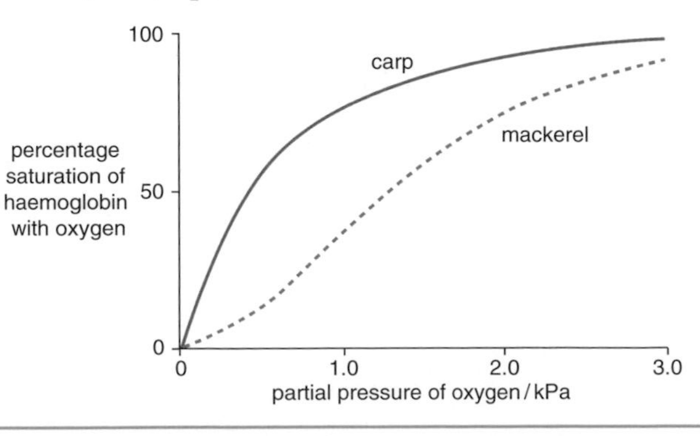

(a) Explain what is meant by the partial pressure of oxygen.

*(1 mark)*

(b) Explain how the oxygen dissociation curve of:

(i) carp haemoglobin is related to the fact that carp may be found in ponds containing large amounts of decomposing vegetation.

(ii) mackerel haemoglobin is related to the fact that the mackerel is a very active species found in the surface waters of the sea.

*(2 marks)*

3 The drawing has been made from an electronmicrograph. It shows three red blood cells in a capillary.

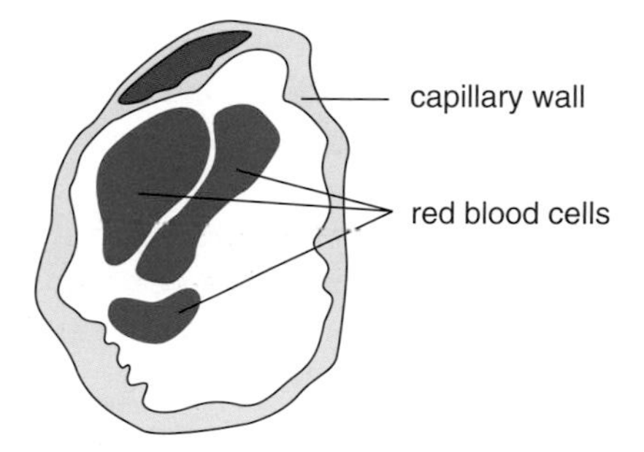


(a) Explain why the red blood cells in the drawing differ in shape.

*(1 mark)*

(b) Carbon dioxide and water react with each other.

(i) What is the product of this reaction?

*(1 mark)*

(ii) Explain why the reaction takes place much faster in the red blood cells than it does in the blood plasma.

*(1 mark)*

(c) Blackwater fever is a rare disease in which large numbers of red blood cells burst. Suggest why the urine of patients with blackwater fever contains haemoglobin but the urine of healthy individuals does not.

*(2 marks)*

# 12 Digestion and Diet

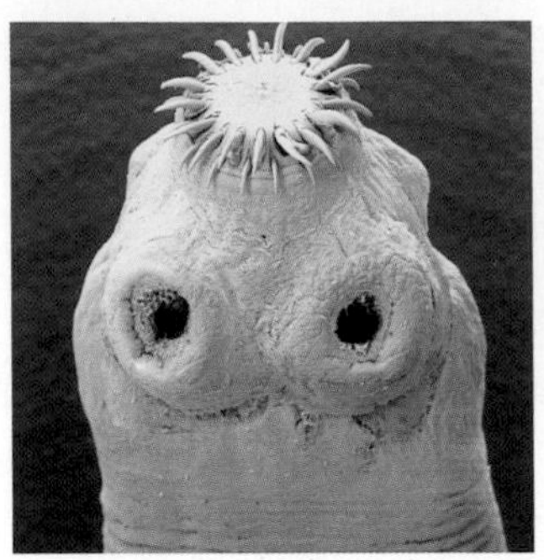

Figure 12.1
A tapeworm is a parasite which lives in the intestine of its host. It has a long, flattened body through which it absorbs digested food. This picture shows the hooks which the tapeworm uses to secure itself to the intestinal lining

Life must be very easy for a tapeworm (Figure 12.1). It is a parasite and spends its time lying in the intestine of its host, surrounded by digested food. All it has to do is to absorb the products of digestion. It does not even need a gut of its own. Tapeworms, however, do have a number of adaptations which make them very efficient at absorbing the products of their host's digestion.

We will look at the body surface of a tapeworm. It is covered in cells like those shown in Figure 12.2. The surface of each of these cells has a large number of tiny, hair-like processes called microtriches. Each of these processes is about 2 μm in length and ends in a spine. Not only does the long, flattened shape of the tapeworm provide the animal with a large surface area but these processes increase it even more. The spines may help to hold the parasite in place in the intestine.

Although the tapeworm only secretes few digestive enzymes, it is now known that it is able to capture enzymes such as pancreatic amylase produced by its host. It absorbs these enzymes onto its body surface. It gains an advantage from being able to do this. Food molecules are broken down on the tapeworm's surface. This will help to make absorption more efficient as the distance they have to travel to get into the body of the parasite is very small.

A tapeworm has another adaptation which allows it to compete very successfully with its host for the products of digestion. It secretes hydrogen ($H^+$) ions. This results in the pH of the intestine of an infected animal being lower than that of an uninfected animal. A lower pH makes the host intestine inefficient at absorbing molecules such as those of glucose, while glucose absorption by the parasite appears to be faster. The tapeworm's carrier molecules work better than those of the host at low pH values.

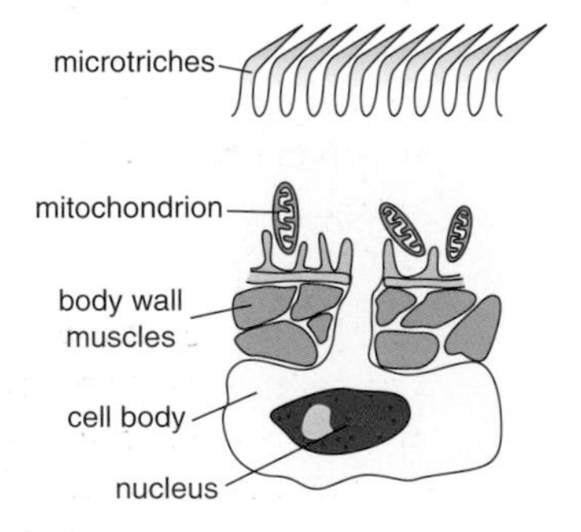


Figure 12.2
A cell from the body surface of a tapeworm. The microtriches provide a large surface area for absorption. Compare them with the microvilli shown in Figure 12.16. They have a similar function

The tapeworm and its human host are **heterotrophs**. They get their nutrients by breaking down large molecules such as carbohydrates and proteins into smaller ones. They do this, however, in different ways. In this chapter we shall concentrate mainly on mammalian digestion but even here we shall see how this system is closely related to the animal's way of life.

There are four basic stages in the way in which a mammal processes its food.

1. **Ingestion** – taking food in through the mouth.
2. **Digestion** – breaking down the large molecules which make up the food into smaller, soluble ones. This is the function of enzymes secreted in digestive juices produced by various glands.
3. **Absorption** – taking up the products of digestion. A variety of processes are involved here. They include diffusion and active transport.
4. **Egestion** – the removal of undigested food, bacteria and dead cells from the lining of the gut as faeces.

**Q** 1 **Excretion is the removal of waste products formed by biochemical reactions in the body. Why do we not refer to the removal of faeces as excretion?**

We shall look at the processes of digestion and absorption and consider how the flow of digestive juices is controlled so that they are only secreted when there is undigested food present in the gut. We shall complete the chapter by looking at nutrition and considering how the nutritional requirements of animals differ at different stages in their lives.

## Digestion

In order to function properly, the human body needs to be supplied with carbohydrates, triglycerides and proteins. These substances are all composed of large molecules, and large molecules, such as those of proteins and polysaccharides, are unable to cross plasma membranes. They cannot be taken up by cells lining the gut, so they cannot be taken directly into the body. Digestion involves breaking these large molecules down into their smaller components: proteins into amino acids; triglycerides into monoglycerides, glycerol and fatty acids; and starch into glucose. The chemical reaction involved is the same each time and involves breaking links between larger molecules by adding molecules of water. This reaction is hydrolysis and digestive enzymes are therefore all **hydrolases.**

The gut or **alimentary canal** of a mammal consists of a tube running through the body from the mouth to the anus. Food is swallowed and travels down the **oesophagus** to the **stomach**. From the stomach it passes first through the **small intestine** then through the **large intestine**. Various glands add their secretions to the contents of the gut. These include the **salivary glands**, the **pancreas** and the **liver**. The arrangement of the organs which make up the digestive system is shown in Figure 12.3.

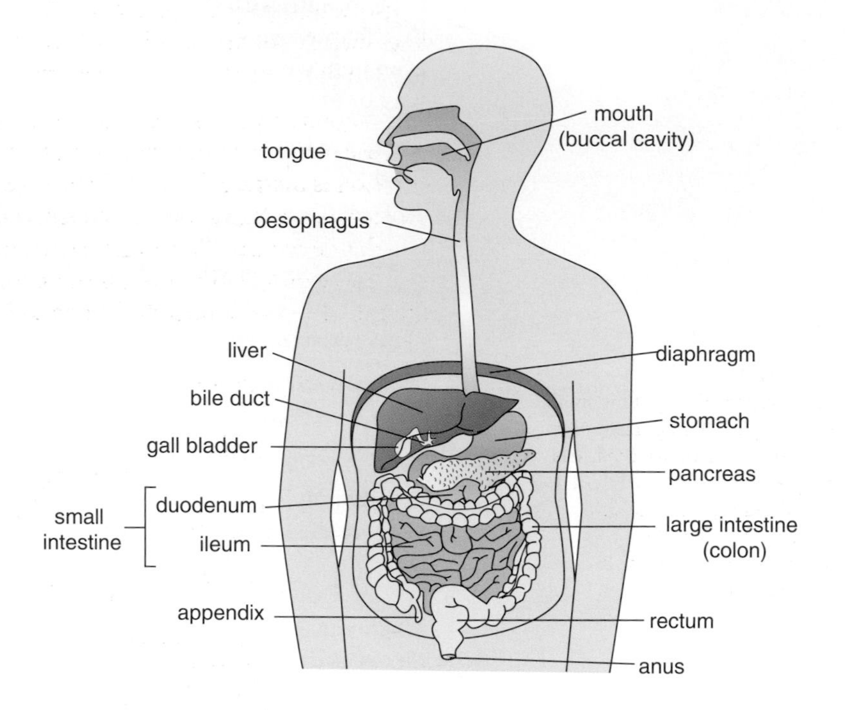


**Figure 12.3**
The digestive system of a human. The digestive system of most other mammals is very similar to this although the relative sizes of different organs may vary

We will now look in more detail at the way in which different substances are digested.

## Starch

Starch is a polysaccharide built up from a large number of glucose monomers. Digestion of starch involves two main steps:

1 breaking down starch to maltose

2 breaking down maltose to glucose.

The process starts in the mouth. Saliva contains **amylase.** Amylase is an enzyme which hydrolyses starch producing maltose as an end product. To be more accurate, salivary amylase is really a group of amylase enzymes, each of which hydrolyses starch by breaking the bonds at different places in the molecule.

**Q 2 Using your knowledge of enzymes, explain why a particular amylase hydrolyses starch by breaking bonds at a particular place in the molecule.**

Salivary amylase plays a relatively minor part in starch digestion. It works in slightly alkaline conditions. Food is rarely in the mouth for long and once it has been swallowed the pH changes rapidly as it encounters acid conditions in the stomach. In addition, much of the food we eat is hot. The temperature of a potato chip when it is eaten, for example, is likely to denature any amylase with which it comes into contact. Amylase is also secreted by the pancreas. This is a gland which opens through a duct into the small intestine. Most of the starch in the food we eat is broken down to maltose by pancreatic amylase.

The second stage in starch digestion – the hydrolysis of maltose to glucose – occurs in the small intestine. The enzyme that catalyses this reaction is **maltase.** Look at Figure 12.4 and you will see that the enzyme molecules are located in the plasma membrane of the epithelial cells which line the small intestine. Maltase breaks a molecule of maltose into two molecules of glucose. The glucose molecules are either absorbed into the cell or released into the lumen to be absorbed further along the small intestine.

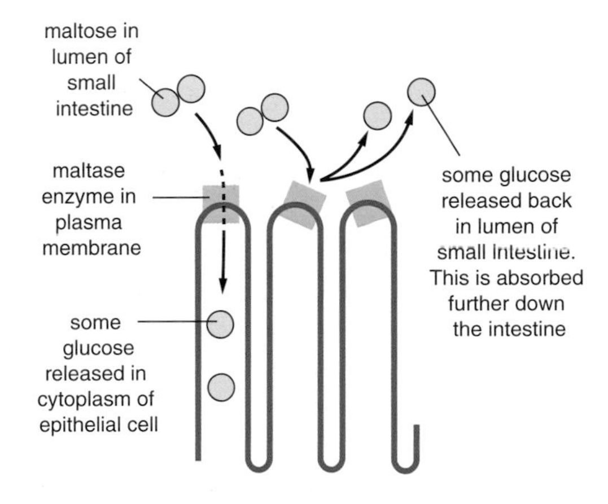


**Figure 12.4**
Cells from the lining of the small intestine play an important part in the digestion of disaccharides such as maltose. The enzymes which catalyse this process are found in the cell surface or plasma membrane of these epithelial cells

## Proteins

Protein digestion is similar to the digestion of polysaccharides because it also involves hydrolysis. A protein molecule consists of one or more polypeptide chains. Protein-digesting enzymes or **proteases** break the peptide bonds joining amino acids together. The human intestine secretes a number of different proteases but they can be divided into two main sorts (Figure 12.5).

- **Endopeptidases** break peptide bonds between specific amino acids in the middle of polypeptide chains. They break large polypeptides into smaller ones.
- **Exopeptidases** break peptide bonds between amino acids at the ends of polypeptide chains. They produce a mixture of amino acids and dipeptides.

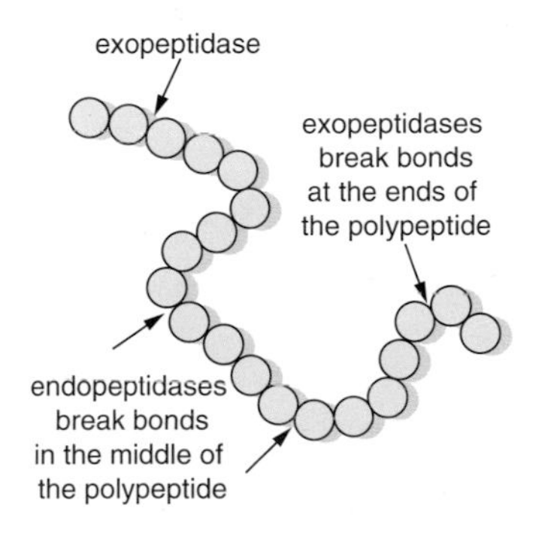


**Figure 12.5**
Polypeptides are long chains of amino acids linked by peptide bonds. Endopeptidases break bonds in the middle of the polypeptide while exopeptidases break bonds at the ends

**Q** 3 **As food passes along the gut, it is mixed first with endopeptidases and later with exopeptidases. Suggest why this arrangement is efficient at digesting proteins.**

Endopeptidases are secreted by glands, and glandular tissue contains a lot of protein ... so, why do you not digest yourself? Why do these enzymes not break down the proteins of which your cells are made? Part of the answer to this question is that endopeptidases are secreted in an inactive form. Only when they get into the alimentary canal are they activated.

**Pepsin** is an endopeptidase secreted by the gastric glands in the stomach. It is secreted in an inactive form called **pepsinogen**. The stomach also secretes hydrochloric acid and this acid, together with pepsin already formed, converts pepsinogen to pepsin. This can be summarised in a simple equation:

$$\text{pepsinogen} \xrightarrow[\text{pepsin}]{\text{hydrochloric acid}} \text{pepsin}$$

Another endopeptidase is **trypsin**. Trypsin is secreted by the pancreas, again in an inactive form. This inactive form, **trypsinogen**, is converted to trypsin by another enzyme, **enterokinase**, which is secreted by the wall of the small intestine. Again we can summarise this in a simple equation.

$$\text{trypsinogen} \xrightarrow{\text{enterokinase}} \text{trypsin}$$

As well as producing trypsin, the pancreas produces a number of exopeptidases. They split groups of one or two amino acids off the ends of the smaller polypeptides which result from the action of pepsin and trypsin. The final stage in protein digestion occurs in the epithelial cells lining the small intestine. These cells have enzymes in their plasma membranes and in their cytoplasm which are able to convert dipeptides to amino acids.

## Lipids

Triglycerides are an important part of the diet. They are digested by **lipase**, an enzyme secreted by the pancreas, which breaks them down to produce glycerol, monoglycerides and fatty acids.

**Q** 4 **How many molecules are produced when lipase digests a molecule of triglyceride and produces:**
**(a) glycerol and fatty acids**
**(b) monoglyceride and fatty acids?**

Vegetable oil is a triglyceride. If you shake a small quantity of vegetable oil vigorously with water, the oil breaks up into tiny droplets. These droplets are suspended in the water and form an **emulsion**. After a short time, however, they run back together to form a layer of oil. Lipase breaks down triglycerides much more effectively if they are dispersed to form an emulsion. With lots of very small droplets, there is a much larger surface area on which the enzyme can act. In the small intestine an emulsion of triglycerides is produced by the action of muscles and the presence of bile. Muscles in the wall of the stomach and the small intestine are continually mixing and squeezing the food. This produces an emulsion. The droplets do not run together again because of the presence of bile. Bile is made in the liver and stored in the gall bladder before being secreted into the small intestine. It contains a number of different substances including **bile salts**. It is these bile salts that make sure that the triglycerides stay as an emulsion.

| Part of gut | Secretion | Enzyme | Substrate | Products |
|---|---|---|---|---|
| Mouth cavity | Saliva | Amylase | Starch | Maltose |
| Stomach | Gastric juice | Pepsin (endopeptidase) | Polypeptides | Small polypeptides |
| Pancreas | Pancreatic juice | Trypsin (endopeptidase)<br>Exopeptidases<br>Amylase<br>Lipase | Polypeptides<br>Polypeptides<br>Starch<br>Triglycerides | Small polypeptides<br>Amino acids and dipeptides<br>Maltose<br>Glycerol, monoglycerides and fatty acids |
| Wall of small intestine | | Maltase<br>Dipeptidase | Maltose<br>Dipeptides | Glucose<br>Amino acids |

**Table 12.1**
Summary of the main enzymes involved in digestion

## Extension box 1

## With a little bit of help

The problem with plant tissues as a source of food is that their cells are surrounded by cell walls. As you learned during your AS-course, plant cell walls consist mainly of cellulose. The cell walls of woody tissues such as xylem have another polymer present. This is lignin. Both cellulose and lignin are difficult to break down and very few animals produce the necessary cellulase and lignase enzymes. Not only does this mean that they miss out on two very common substances as sources of nutrients but, if they cannot break cell walls open, they cannot make use of the molecules of starch, protein and lipids that plant cells contain.

**Figure 12.6**
Termites are colonial insects and many species live in huge nests. A nest like this may contain over a million termites

Termites are small insects found throughout the tropics (Figure 12.6). They feed on various sorts of plant material. Some species feed almost entirely on wood while others feed mainly on grass and dead plants. Although they can cause a lot of damage, they play a very important part in the cycling of the carbon locked up in cellulose and lignin.

If they do not produce the necessary enzymes to digest either cellulose or lignin, how do termites survive on their plant diet? The answer is with a little bit of help from various microorganisms. The hind gut of many termites contains large numbers of protozoa. These microorganisms engulf small particles of food material from the termite gut by phagocytosis. Enzymes secreted by the protozoa hydrolyse the cellulose and produce glucose. The protozoa then use this glucose as a respiratory substrate. They respire anaerobically and release fatty acids and carbon dioxide as waste products. The fatty acids are absorbed and used by the termite.

**Q 5 The relationship between termites and the protozoa which live in their guts may be described as mutualistic. Explain why.**

One group of termites has a relationship with a species of fungus which is able to produce the ligninases which break down lignin. These fungi grow inside the termite nest on 'gardens' (Figure 12.7) made from termite faeces. The fungi break down the undigested lignin in the termite faeces into simpler substances. The termites eat the fungus and digest these simpler substances, absorbing the products.

**Figure 12.7**
The corky-looking material that makes up this fungus 'garden' is made from termite faeces. The immature termites are feeding on the small white bodies produced by the fungus. This is a mutualistic relationship. The fungus is provided with a lignin-rich medium, while the termite gains substances which it can digest and absorb

## Getting food from plants – ruminant digestion

The commonest of all organic substances is cellulose. Many mammals feed entirely on plants but they do not produce cellulase enzymes so how do they manage to gain their nutrients? The answer is that they, like the termites described in Extension box 1, rely on a mutualistic relationship with other organisms, in this case the bacteria and protozoa living in their guts.

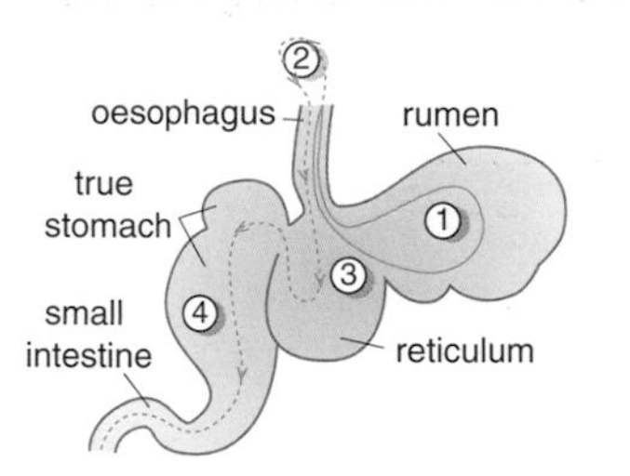


**Figure 12.8**
The path of food through the stomach of a cow. Food is swallowed and enters the rumen (1). After being thoroughly mixed with saliva and microorganisms, small amounts are brought back up into the mouth and chewed once more (2). These are swallowed again (3). This may be repeated many times as the animal ruminates or 'chews the cud'. Eventually the contents of the rumen pass into the true stomach (4) and on down the small intestine

One group of these mammals includes cattle, sheep, antelopes and deer. These are the ruminants and they are characterised by having large, four-chambered stomachs (Figure 12.8). The first two of these chambers, the **rumen** and the **reticulum**, form a huge fermenter packed with microorganisms. The second two chambers are the **omasum** and the **abomasum**. These make up the true stomach. When a cow eats, the food it takes in is swallowed and enters the rumen and reticulum. These organs have thick, muscular walls which churn the food with large amounts of saliva and microorganisms. Conditions are kept remarkably constant. The temperature stays between about 38 °C and 42 °C, the cow's internal temperature. The pH is prevented from changing significantly by the phosphates and hydrogencarbonate in the saliva. These act as buffers.

**Q 6 Describe three ways in which the rumen of a cow is like a fermenter used to produce enzymes from microorganisms.**

When the cow **ruminates** or 'chews the cud', small amounts of the rumen contents are brought back up the oesophagus into the mouth. Here they are chewed thoroughly and then swallowed again and return to the rumen. A considerable amount of time is spent ruminating – cattle can ruminate for 8 hours a day, about as long as they spend feeding. Eventually, the partially digested rumen contents, along with enormous numbers of microorganisms, pass into the true stomach and on along the small intestine. Here they are exposed to the cow's enzymes and the process of digestion continues.

## Ruminants and cellulose

Cellulose is a polysaccharide. Figure 12.9 summarises the way in which microorganisms break down the cellulose in the plants which a cow has eaten. The first step involves the breakdown of cellulose into the glucose monomers from which it is formed. Very little of this glucose is available to the animal, however, as it is used by the microorganisms for their own metabolism. It is converted to pyruvate in a process similar to glycolysis (Chapter 5) and the pyruvate is further metabolised. The waste products of this metabolism are carbon dioxide, methane and various fatty acids. Methane and carbon dioxide escape from the cow's body when the animal belches. In a cow, belching is not simply bad manners; it is the only way these waste gases can be removed. The fatty acids produced by the microorganisms have very short carbon chains. They diffuse out of the microorganism and can be absorbed directly through the wall of the rumen into the blood of the cow. They form an important energy source in the animal's respiration.

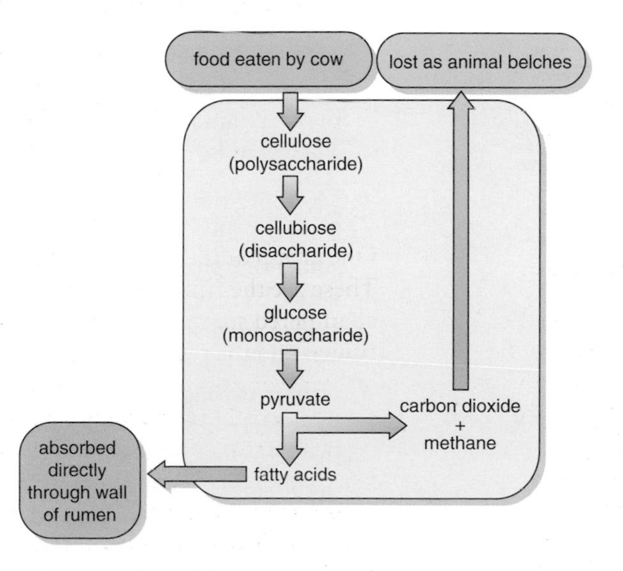


**Figure 12.9**
The microorganisms living in the rumen of a cow carry out a series of biochemical reactions in which they break down cellulose and make fatty acids available to the animal

## Ruminants and protein

The microorganisms living in the rumen also play an important role in making protein available. They do this in several ways (Figure 12.10).

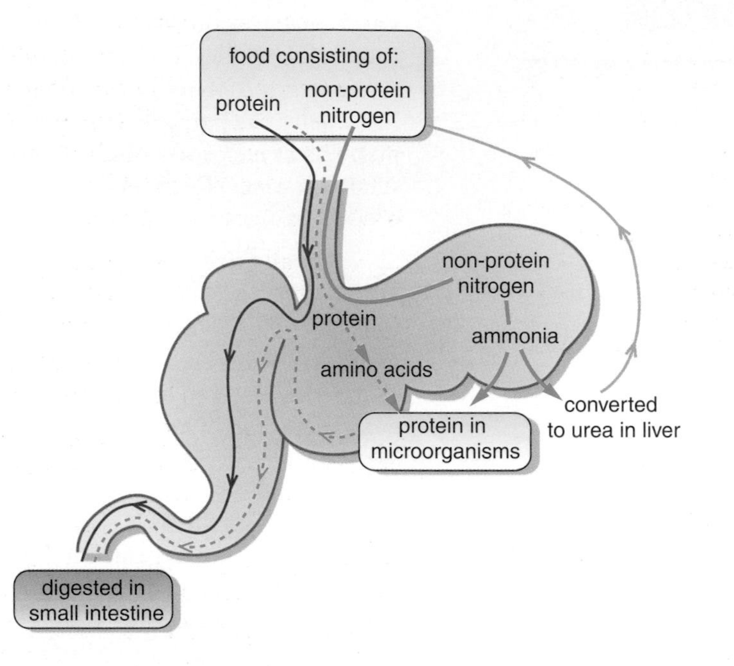


**Figure 12.10**
Some of the ways in which microorganisms help ruminants to obtain protein

- Plant cells are surrounded by cellulose walls. By breaking down these walls, the proteins and other nutrients contained in the cells are made available.
- The microorganisms themselves are an important source of nutrients. Their cells are made of cytoplasm and a large part of that cytoplasm is protein. As they pass through the animal's true stomach and small intestine, these bacteria and protozoa are digested. The proteins which they contain are broken down by the cow's enzymes to amino acids. These can be absorbed.
- Microorganisms have a number of unusual biochemical pathways. They are able to convert **non-protein nitrogen** into proteins, something that a mammal such as a cow is unable to do. The proteins which they produce in this way are available to the cow as the microorganisms are digested. A waste product produced by the microorganisms in the rumen is ammonia. Ammonia is absorbed rapidly through the wall of the rumen and transported to the cow's liver. Here it is converted to urea. Some of this urea leaves the body through the kidneys but a lot goes to the salivary glands and is secreted in the saliva. Urea is a non-protein nitrogen source, so the rumen microorganisms are able to convert this into protein.

**Q** 7 **Describe how urea gets from the liver to the salivary glands.**

### Extension box 2

## Problems at the back

Figure 12.11
A zebra relies on bacteria in its colon to hydrolyse cellulose. It cannot, however, gain proteins from digesting these bacteria. Consequently when food is scarce, members of the horse family do not do as well as ruminants

Cattle, sheep, deer and antelope are all ruminants. Horses and zebras (Figure 12.11) are not but they also feed on plant material. How do they manage without a rumen? The answer is that they rely instead on a greatly enlarged large intestine. Just like the rumen of a cow, this organ contains huge numbers of bacteria which hydrolyse cellulose and produce a variety of substances including carbon dioxide, methane and fatty acids. These fatty acids are absorbed directly through the wall of the colon and are used by the animal as an important respiratory substrate.

However, all this takes place at the rear of the gut and this means that these bacteria are not passed through the stomach and intestines where they would be digested. Consequently, members of the horse family get the products of cellulose digestion but they do not gain the extra proteins from digesting the bacteria.

Figure 12.12
A rabbit produces soft mucus-covered faecal pellets which it immediately eats. It digests the bacteria contained in these pellets and gains nutrients from them

Rabbits (Figure 12.12) also depend on bacteria in the large intestine to break down cellulose but they have a different strategy. They produce two sorts of faecal pellet. One is the familiar hard, dark-coloured rabbit 'currant'; the other is soft, pale and mucus covered. Over 50% of the dry mass of these soft pellets is made up of bacteria from the large intestine. The soft pellets are eaten by the rabbit and the bacteria they contain are digested as they pass through the stomach and intestines. Unlike a zebra, a rabbit is able to gain both the products of cellulose digestion and the amino acids which result from breaking down the bacteria which live in the large intestine.

## Absorbing digested food

Different parts of the alimentary canal have different functions. The role of the oesophagus is to squeeze food that has been chewed and mixed with saliva into the stomach. Once there, more mixing and churning take place and gastric juice starts the digestion of proteins. The small intestine is concerned with further digestion. It is also the region where most of the digested food is absorbed. Undigested food is stored in the large intestine before being egested from the body as faeces.

If you examine a section through any of these parts of the gut with a microscope, you can see that they all have a similar basic structure. They are hollow organs with a layer of epithelial cells surrounding the **lumen**, and their walls contain muscles and blood vessels. There are, however, slight differences in the structure of each region and these differences can be linked to function. Figure 12.13 shows the appearance of a cross-section through the small intestine. Study this figure and you will see how the organ is adapted for its function of absorbing digested food.

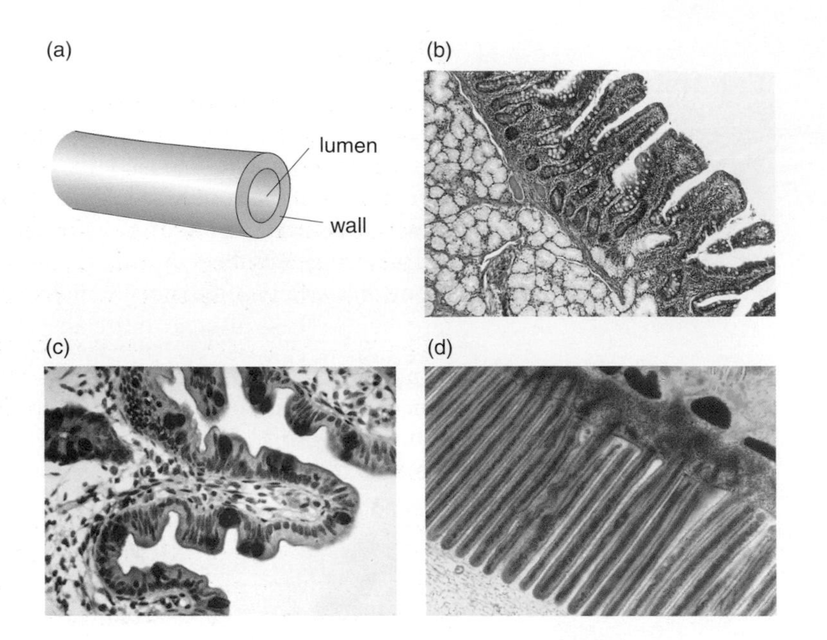


**Figure 12.13**
The structure of the small intestine. The cross section (a) shows that the intestine is a tube with a thick wall surrounding a hollow lumen. When part of the wall is seen through a light microscope (b), the tissues that make up the organ are clearly visible. Higher magnification shows the epithelium in more detail (c). Photograph (d) is an electron micrograph of part of these epithelial cells

**Q 8 List the features which result in the small intestine having a large surface area.**

## Salts and water

By the time that the partly digested food is half way along the small intestine, it is rather like soup in consistency. It has been mixed with digestive juices from various glands and with sodium and chloride ions from the intestinal epithelium. These ions result in a water potential gradient. The water potential in the lumen of the first part of the small intestine is lower than that in the epithelial cells. Water therefore moves from these cells into the lumen by osmosis.

**Q 9 Which glands will have secreted digestive juices onto food by the time it is half way along the small intestine?**

Further along the small intestine, these ions are pumped back out again by the epithelial cells. This involves active transport. The result is that the water potential in the cells is now less than that in the lumen of the gut, so water moves back out by osmosis (Figure 12.14).

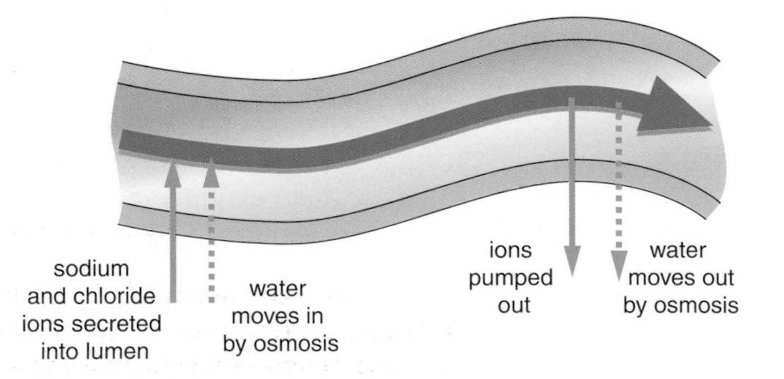


**Figure 12.14**
Sodium and chloride ions are secreted into the first part of the small intestine. This results in osmosis and water is added to the gut contents. The processes of digestion and absorption are much more efficient in the resulting soupy mixture. Further along the gut, the ions and the water are reabsorbed

## Glucose

The digestion of carbohydrates results in the formation of monosaccharides such as glucose. Glucose is absorbed from the gut by active transport. There are specific carrier molecules in the plasma membranes of the epithelial cells of the small intestine and these take the molecules into the cytoplasm of the cell. Uptake requires ATP and is linked to the transport of sodium ions. Glucose passes from the inside of the cell into the capillaries by facilitated diffusion (Figure 12.15).

Other monosaccharides, such as fructose, are not absorbed by active transport but by facilitated diffusion and move from a higher concentration in the gut to a lower one in the cells without using ATP.

## Amino acids

The uptake of amino acids works in a slightly different way from that in which glucose is absorbed. This time, amino acids pass into the epithelial cells by facilitated diffusion. Different sorts of carrier molecule in the plasma membrane transport different types of amino acid. The carrier molecules only work, however, in the presence of sodium ions and each time an amino acid is transported into the cell, so is a sodium ion. A diffusion gradient is maintained across the plasma membrane by pumping the sodium ions actively out into the blood. Carriers also transport some dipeptides across the membrane. Enzymes break these dipeptides down to amino acids in the cytoplasm (Figure 12.15).

Some proteins can be absorbed directly by the epithelial cells of the small intestine by pinocytosis. One example in which this occurs is in new-born mammals. They can absorb antibodies from their mother's milk in this way.

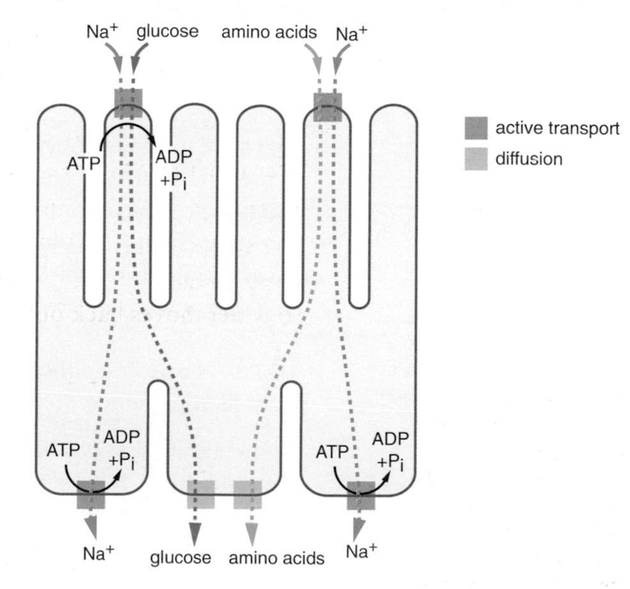


**Figure 12.15**
The ways in which monosaccharides, such as glucose and fructose, and amino acids are absorbed by the epithelial cells lining the small intestine. Note the importance of sodium ions in these processes

**Q 10 New-born mammals do not produce trypsin and their stomachs do not secrete hydrochloric acid. Explain how these features enable the intestines of new-born mammals to absorb proteins.**

## Lipids

We saw that triglycerides were digested by lipase enzymes to produce a mixture of glycerol, monoglycerides and fatty acids. The monoglycerides combine in the small intestine with bile salts to form tiny droplets called **micelles**. Each of these micelles is about 5 μm in diameter and also contains other molecules such as fatty acids and glycerol. The micelles transport their contents to the plasma membranes of the epithelial cells (Figure 12.16). They do not go into the cell, but their contents do. The glycerol, monoglycerides and fatty acids they contain dissolve readily in the phospholipid bilayer and enter the cytoplasm. Triglycerides are resynthesised and pass out through the sides and base of the cells into lymph capillaries called **lacteals**.

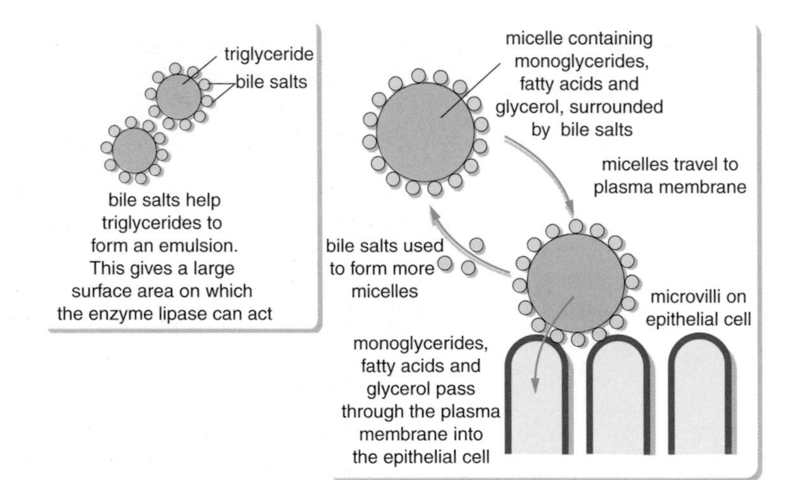


**Figure 12.16**
Bile salts have two important functions in digestion. They help triglycerides to form an emulsion and they also help in transporting the products of triglyceride digestion to the epithelial cells lining the small intestine

## Controlling digestion

Digestive juices are only secreted in large amounts when food is present in the gut. There are good reasons for this. Think, for example, about gastric juice. It contains hydrochloric acid and pepsin which is a protein-digesting enzyme. Prolonged contact with gastric juice would damage the tissues which form the stomach wall. In addition, secreting digestive enzymes when they are not required wastes the body's resources.

**Q** 11 **Name two substances which would be wasted if digestive juices were secreted when no food was present in the gut.**

**Figure 12.17**
A feeding baboon. Food spends longer in the small intestine than it does in the mouth. It also takes longer to get to the intestine. These differences are reflected in the way that digestive juices are controlled

Figure 12.17 shows a baboon searching for tasty morsels in a pile of elephant dung. It does not know precisely when it is going to find an undigested seed or a maggot. Once it finds an item of food, the baboon puts it in its mouth, chews it and swallows it. It is necessary to have a control system which ensures that the animal will have saliva in its mouth as soon as the food is put in. Food does not remain in the mouth for all that long, however, so the flow of saliva need only be short term.

The situation is rather different in the small intestine. This part of the gut is rarely empty and there is usually plenty of time before food put in the mouth arrives in the small intestine. A different sort of control system, one which takes these factors into account, is therefore needed for pancreatic juice.

A mammal has two main communication systems – the nervous system and the endocrine system. When you read Chapter 13 you will see that nervous communication relies on electrical impulses passing along nerve cells. Because of this, it is an ideal system for controlling processes which are short-lived and need to be switched on or off rapidly. The endocrine system is based on hormones travelling in the blood, from the glands or cells which secrete them, to the target organ. Hormones generally have long-lasting effects and take more time to bring about changes.

**Q** 12 **Which of the nervous system or the endocrine system would you expect to control:**
**(a) a change in the diameter of the pupil when a light is shone at the eye**
**(b) the calcium concentration in the blood?**

There is a change of emphasis from nervous control to endocrine control as you move down the gut. Saliva is needed as soon as, or even just before, food enters the mouth. This is an ideal situation for nervous control. On the other hand, food takes a considerable time to reach the small intestine and is present there for some hours. Not surprisingly, secretion of bile and of pancreatic juice are mainly controlled by hormones.

There are three basic ways in which the nervous and endocrine systems control the secretion of digestive juices.

- **Nervous reflexes** – A reflex is a nerve pathway involving a small number of nerve cells (often only two or three). The response is automatic as it does not involve the front part of the brain. This means that a particular stimulus will always have the same effect. The other important feature of a reflex is that, as it only involves a small number of neurones, it is very rapid. You can find out more about reflexes on page 265.
- **Conditioned reflexes** – Has your mouth ever watered at the smell of your favourite meal cooking? This is an example of a conditioned reflex. The stimulus responsible for increasing the secretion of saliva is contact between food in your mouth and tastebuds on your tongue. But there are many sights, sounds and smells associated with meals. In time, one of these will make you salivate, even if you have no food in your mouth. We describe a conditioned reflex as when an unrelated stimulus, such as the sight or smell of food, produces the response which was originally associated with an appropriate stimulus.
- **Hormones** – Hormones are secreted in response to the presence of food in a particular region of the gut. These hormones are carried in the blood to glands where they stimulate the secretion of digestive juices.

| Part of gut | Stimulus | Effect |
|---|---|---|
| Buccal cavity | Various stimuli associated with food such as the smell or sight of food, or sound of food being prepared | Trigger a conditioned reflex leading to the secretion of saliva before food arrives in the mouth. |
| | Contact of substances in food with taste buds on tongue | Reflex leading to secretion of saliva when food is in the mouth |
| Stomach | Various stimuli associated with food | Trigger a conditioned reflex leading to the secretion of gastric juice before food arrives in the stomach. |
| | Contact of substances in food with taste buds on tongue | Reflex leading to secretion of gastric juice as food arrives in the stomach |
| | Food in stomach | Secretion of the hormone gastrin. Gastrin travels in the blood to the glands in the stomach wall. These glands secrete gastric juice. |
| Pancreas and gall bladder | Food in the small intestine | Secretion of the hormone cholecystokinin-pancreozymin (CCKPZ). This hormone stimulates the release of digestive enzymes by the pancreas and bile by the gall bladder. |
| | | Secretion of the hormone secretin. Secretin stimulates the release of alkaline fluid by the pancreas. |

**Table 12.2**
Summary of the ways in which some digestive secretions are controlled in humans

## Diet and development

The food a human needs changes with age. It is not just the total amount that changes but the relative amounts of different nutrients.

**Table 12.3**
How age affects the mean daily requirements for some nutrients in females

| Age | Mean daily requirement for | | |
|---|---|---|---|
| | protein/g | iron/mg | calcium/mg |
| 0–3 months | 12.5 | 1.7 | 525 |
| 1–3 years | 14.5 | 6.9 | 350 |
| 7–10 years | 28.3 | 8.7 | 550 |
| 15–18 years | 55.5 | 14.8 | 800 |
| 19–50 years | 55.5 | 14.8 | 700 |
| over 50 years | 53.5 | 8.7 | 700 |

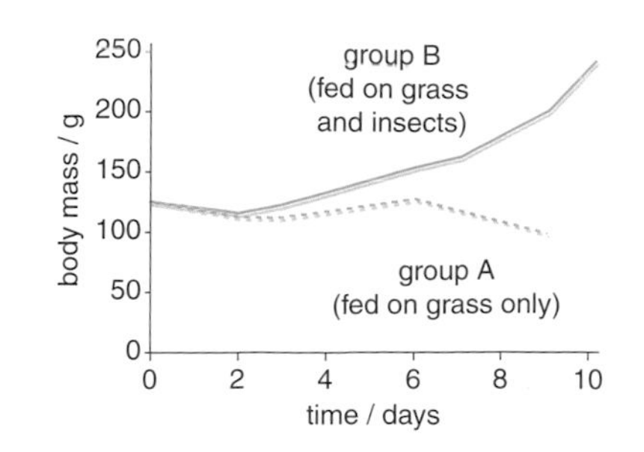


Figure 12.18
The result of an investigation into the effect of diet on growth in young geese. Group **B**, fed on a protein-rich diet containing insects, grew much faster than the control group, **A**, fed on grass

Look at this table carefully. The amount of protein required can be linked with growth. Variation in the dietary requirement for calcium can be explained in terms of the growth of bones. A lot of calcium is required as the skeleton develops immediately after birth and there is another surge around puberty.

**Q 13 Iron is required for the formation of red blood cells. Suggest an explanation for the figures in the Table 12.3 relating to the large amount of iron required by women in the 15–18 and 19–50 age groups.**

Diet also varies with development in other animals. Adult geese feed mainly on grass but young geese require insects in their diet. The graph in Figure 12.18 shows the result of feeding two groups of newly hatched geese on different foods. Group **A** were given the adult diet and grew very little as a result. Group **B** ate a mixture of grass and insects. They grew rapidly on their protein-rich diet.

The assignment at the end of this chapter contains some more information about diet and growth.

## Diet and development in insects

Figure 12.19
The larva, or caterpillar, of a puss moth is little more than an eating machine. The eggs of this species are laid on willow leaves. They hatch and larvae emerge and feed on the leaves

Many insects have life-cycles in which the young stages are very different in appearance from the adults. One such group includes butterflies and moths. The adults lay eggs on a suitable food plant. The eggs hatch to produce larvae (Figure 12.19).

An insect has an **exoskeleton**. In other words, its skeleton is on the outside of its body. This exoskeleton takes the form of a tough, hardened cuticle. Larvae are therefore only able to grow by shedding their cuticle. They can only increase in size after they have moulted this old cuticle and before the new one underneath has hardened.

**Q 14 The larva of a particular species of insect moults five times as it grows. Sketch a graph to show the pattern of growth of this insect.**

Figure 12.20
A newly-emerged swallowtail butterfly. Adult butterflies and moths differ from their larval stages in being sexually mature and in possessing wings. There are also differences in their mouthparts and digestive systems

Eventually, the larva reaches its maximum size and moults to become a pupa. Inside the cuticle of the pupa, the larval tissues are broken down and built up again to form a sexually mature, adult moth or butterfly (Figure 12.20). The term **metamorphosis** is used to describe the complete change of form which occurs as the larva develops into an adult.

Once it is an adult, an insect does not moult any more so it cannot grow. Growth is therefore confined to the larval stages. In order to grow and produce new tissue, a diet is needed that contains protein, lipid and carbohydrate. The larvae of most butterflies and moths feed mainly on leaves and these provide an adequate supply of all these nutrients.

All the body systems in an adult insect are fully formed. The adult does not grow any more and it is sexually mature. Adult moths and butterflies also spend a lot of time in flight. Much of this flight is associated with courtship and, if it is a female, laying eggs. Adult butterflies and moths therefore need very little protein or lipid but they do require large amounts of carbohydrate, which acts as a respiratory substrate to

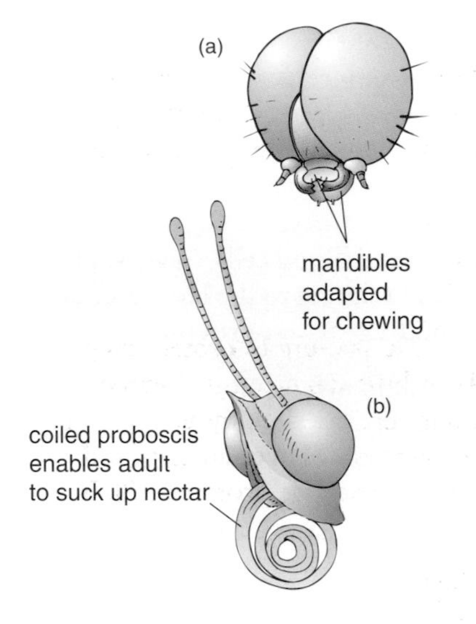


**Figure 12.21**
In butterflies and moths the mouthparts of the larvae (a) are adapted for chewing. This enables them to break down plant tissue. Adults, however feed on a liquid diet and have sucking mouthparts (b)

provide the energy they need. They feed on the nectar produced by flowers. Nectar is a dilute solution of sugars. Its constituents are mainly sucrose and the two monosaccharides, glucose and fructose, from which sucrose is formed.

Both larval and adult butterflies and moths have adaptations which allow them to exploit these different sources of food. Their mouthparts are very different (Figure 12.21) and they also produce different digestive enzymes at different stages in their life-cycles (Table 12.4).

| Stage in life-cycle | Food | Digestive enzyme | | | | | |
|---|---|---|---|---|---|---|---|
| | | Protease | Lipase | Amylase | Maltase | Sucrase | Cellulase |
| Larva | Leaves | ✓ | ✓ | ✓ | ✓ | ✓ | ✗ |
| Adult | Nectar | ✗ | ✗ | ✗ | ✗ | ✓ | ✗ |

✓ present ✗ absent

**Table 12.4**
Comparison of the foods eaten and digestive enzymes produced by larval and adult butterflies

## Summary

- Digestion involves the hydrolysis of large insoluble molecules by the enzymes present in digestive juices and in the epithelial cells which line the small intestine. Cellulose is digested by microorganisms living in the rumen of animals such as cattle and sheep.
- The small molecules resulting from digestion are absorbed through the wall of the small intestine. Absorption involves both diffusion and active transport.
- The nervous and endocrine systems are involved in controlling the secretion of gastric juices.
- The dietary requirements of organisms vary at different stages in their lives. In insects, changes in protein and energy requirements are associated with growth in the larva and with reproduction and dispersal in the adult.

## Assignment

### Protein makes your wool grow

You have seen in this chapter how the nutritional requirements of insects differ at different stages in their lives. In this assignment we will look at the protein requirements of sheep and how they differ. You will need to use material from various parts of the specification as well as from this chapter to answer the questions. Before you start, it would be a good idea to look up the following key topics either in your notes or in your textbook:

- proteins and protein structure
- genetic engineering.

It is important that farm animals are given the amount of protein they need. Too much would be wasteful and too little would slow down their growth. In order to be able to calculate the exact quantity required, we need a way of measuring the amount of protein in a food sample. One way of doing this is to calculate the crude protein (CP). It can be calculated from the equation:

$$CP = N \times \frac{1000}{160}$$

In this equation:

- N = the mass of nitrogen present in a 1 kg sample
- the ratio $\frac{1000}{160}$ is based on the fact that 1000 g of protein contains 160 g of nitrogen.

1 The calculation of CP makes two assumptions. Explain why each of these assumptions is not true and could lead to an inaccurate result.

(a) All the nitrogen present in the food sample is present as protein.

*(1 mark)*

(b) All proteins contain 160 g of nitrogen per kilogram.

*(2 marks)*

| Number of lambs in litter | Crude protein/g day$^{-1}$ | | | | | | |
|---|---|---|---|---|---|---|---|
| | Pregnancy/weeks before birth | | | | Lactation/weeks after birth | | |
| | 8 | 6 | 4 | 2 | 2 | 6 | 10 |
| 1 | 96 | 105 | 120 | 143 | 196 | 184 | 151 |
| 2 | 104 | 119 | 140 | 170 | 261 | 235 | 186 |

**Table 12.5**
The different amounts of crude protein required in the diet of a sheep during pregnancy and lactation

2 (a) Plot these data as a suitable graph.

*(4 marks)*

(b) Describe and explain the trend in the protein required in the diet during lactation.

*(2 marks)*

(c) The amount of protein required by the sheep changes during pregnancy. Suggest two functions for this extra protein other than providing protein for the growth of the lamb.

*(2 marks)*

3 Some of the protein in the diet of the sheep may be described as rumen degradable protein (RDP). Use your knowledge of digestion in ruminants to suggest the meaning of this term.

*(2 marks)*

Sheep's wool consists mainly of the protein keratin. The total mass of wool produced by a sheep in a year contains 3 kg of keratin.

4 (a) Assuming that wool grows evenly, a sheep will need to synthesise approximately 8 g of wool keratin per day. Explain how this figure is calculated.

*(1 mark)*

(b) For each gram of protein in its food, a sheep can digest and absorb 0.54 g. How much protein would it need to eat in order to synthesise 8 g of wool keratin? Show your working.

*(1 mark)*

(c) Keratin contains 120 g $kg^{-1}$ of sulphur-containing amino acids. Plant protein contains 30 g $kg^{-1}$ of these amino acids. Use these figures to make a better estimate of the amount of plant protein a sheep would need to eat in a day order to synthesise its wool keratin.

*(1 mark)*

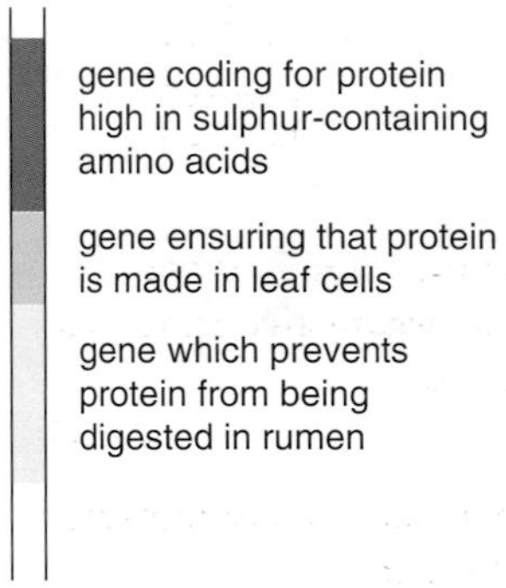

gene coding for protein high in sulphur-containing amino acids

gene ensuring that protein is made in leaf cells

gene which prevents protein from being digested in rumen

**Figure 12.22**
Piece of engineered DNA for insertion into clover

Australian scientists have recently produced genetically-engineered clover containing a protein high in sulphur-containing amino acids. It is thought that feeding sheep on this new strain of clover will improve the yield of wool. A piece of DNA was prepared with the sequences shown in Figure 12.22. This piece of DNA was inserted into a clover plant

5 Describe how each of the following enzymes may be used in preparing this piece of DNA:

(a) ligase

*(1 mark)*

(b) restriction endonuclease.

*(1 mark)*

6 Briefly explain why it is necessary to:

(a) ensure that the protein is made in leaf cells

*(1 mark)*

(b) prevent the protein from being digested in the rumen of the sheep.

*(2 marks)*

## Examination questions

1 A cow obtains most of its nutritional requirements from fermentaion by mutualistic (symbiotic) microorganisms in its rumen. The diagram summarises the biochemical processes involved.

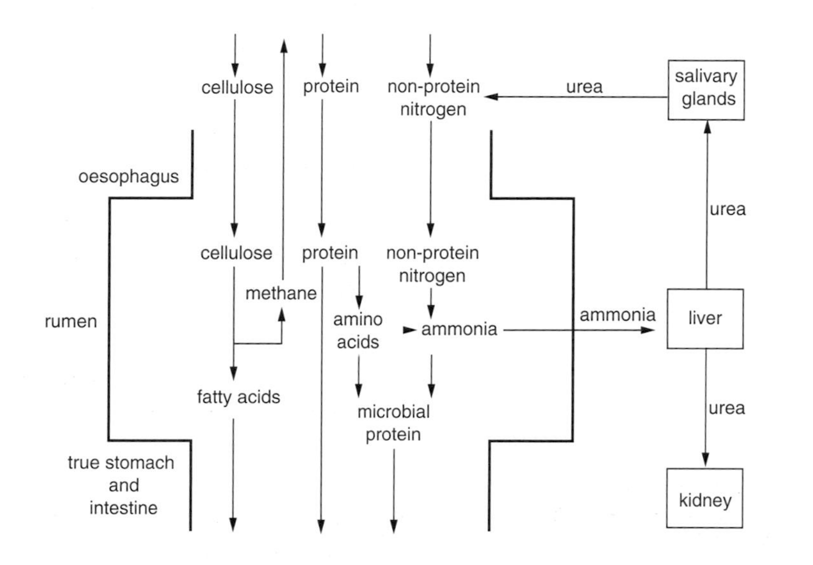


(a) With the aid of information in the diagram, explain why:

(i) the relation between the cow and the organisms living in its rumen may be described as mutualistic;

*(2 marks)*

(ii) it is possible for cattle to survive on a diet that contains no protein for a considerable period of time;

*(2 marks)*

(iii) ruminants such as the cow are less efficient than non-ruminent animals at converting energy in their food into energy in their tissues.

*(2 marks)*

(b) (i) What is likely to be the main respiratory substrate of the cow?

*(1 mark)*

(ii) How does the cow obtain ATP from this respiratory substrate? Details of biochemical pathways are **not** required.

*(2 marks)*

Fermentation in the rumen is sometimes likened to the process in an industrial fermenter. An industrial fermenter may be used for the continuous production of substances such as penicillin by the fungus *Penicillium*. Sterile medium is continuously added and the penicillin havested. Temperature and pH are carefully controlled.

(c) (i) Suggest **two** important differences between fermentation in an industrial fermenter and fermentation in the rumen.

*(2 marks)*

(ii) How are constant conditions of temperature achieved in the rumen?

*(1 mark)*

2 (a) Give **one** way in which the nutrition of:

(i) a heterotroph differs from that of an autotroph;

*(1 mark)*

(ii) a chemoautotroph differs from that of a photoautotroph.

*(1 mark)*

(b) (i) Different organisms absorb glucose at different rates. Suggest **two** features of the epithelial cells in the small intestine which help to determine the rate at which glucose is absorbed.

*(2 marks)*

(ii) The contents of the small intestine are constantly being mixed as a result of muscle action. What effect would you expect mixing to have on the rate of uptake of glucose? Give an explanation for your answer.

*(2 marks)*

(c) Cholera bacteria produce a toxin which causes the epithelial cells in the intestine to secrete chloride ions into the lumen of the intestine. Explain why diarrhoea is a symptom of cholera.

*(3 marks)*

The digestive enzymes produced by an insect at different stages in its life cycle are shown in the table.

| Stage in life cycle | Food | Enzyme | | | | |
|---|---|---|---|---|---|---|
| | | Exopeptidase | Lipase | Amylase | Sucrase | Maltase |
| Caterpillar (larva) | Leaves | ✓ | ✓ | ✓ | ✓ | ✓ |
| Butterfly (adult) | Nectar | | | | ✓ | |

(d) Describe the role of exopeptidases in the digestion of proteins.

*(2 marks)*

(e) Explain how the differences in digestive enzymes shown in the table can be related to the fact that:

(i) growth takes place only in the larval stage of the insect's life;

*(2 marks)*

(ii) the adult stage of the insect's life is associated with reproduction and dispersal.

*(2 marks)*

# 13 Nervous Systems and Receptors

Parkinson's disease results when a group of nerve cells in the brain release insufficient amounts of a neurotransmitter, called dopamine. Normally, dopamine regulates the flow of nerve impulses that stimulate the contraction of specific skeletal muscles. When insufficient dopamine is released, these skeletal muscles contract irregularly, causing muscle tremors and involuntary movements. This is accompanied by stiffness in the joints.

Sufferers of Parkinson's disease can be treated with drugs that are converted to dopamine in the brain. In the 1980s, a more controversial treatment was discovered. This involved transplanting fetal brain cells into the brains of Parkinson's disease patients. The hope was that the transplanted fetal cells would mature in the recipient's brain and secrete dopamine. Most of the fetal cells that were used came from aborted fetuses. In Britain, fetal transplant treatment was stopped following a public outcry over this use of aborted fetuses. In the USA, the treatment continued.

In March 2001, the results of the first full clinical trial of the fetal transplant technique were published by a team based in Denver and New York. They discovered that the treatment did not help patients over the age of sixty. Whilst some younger patients did appear to benefit, the treatment produced effects that were worse than the disease itself in about 15% of recipients of the transplants. In these recipients, it appears that the fetal cells had grown too well and had begun to produce too much dopamine. What scientists now needed was a way to switch off the transplanted cells once they had done their work.

This story illustrates two aspects of medical research. Firstly, in countries such as Britain, democratic processes determine the use of advances in scientific techniques. Secondly, as science students, we should be cautious when reading about new 'wonder cures' until such time as full scientific trials have been completed.

## Transmission of impulses through the nervous system

In this part of the chapter, we will consider nervous transmission. In doing so, we will learn about the structure of nerve cells (**neurones**), the way that they transmit impulses along their length and the way they pass impulses from one to another.

### Neurones

Neurones are cells that are adapted to transmitting information through the nervous system. All neurones do this by:

- transmitting nerve impulses along their length
- stimulating other cells.

We need to be familiar with three types of neurone, each of which has a different function. Table 13.1 shows the names of these neurones and summarises their functions.

| Name of neurone | Stimulated by: | Transmits impulse to: |
|---|---|---|
| Motor neurone (also called an effector neurone) | another neurone – either a relay neurone or a sensory neurone | an effector organ (gland or muscle) |
| Relay neurone | another relay neurone or a sensory neurone | another relay neurone or a motor neurone |
| Sensory neurone | a receptor, such as a Pacinian corpuscle, rod cell or cone cell, that is described later in this chapter | a relay neurone or a motor neurone |

**Table 13.1**
The three types of neurone that are found in the nervous system. Later in this chapter, you will see how sensory neurones are stimulated by receptors, such as Pacinian corpuscles, rod cells and cones cells, and how the three types of neurone are linked together in a reflex arc

**Q** 1 **Suggest why motor neurones are also called effector neurones.**

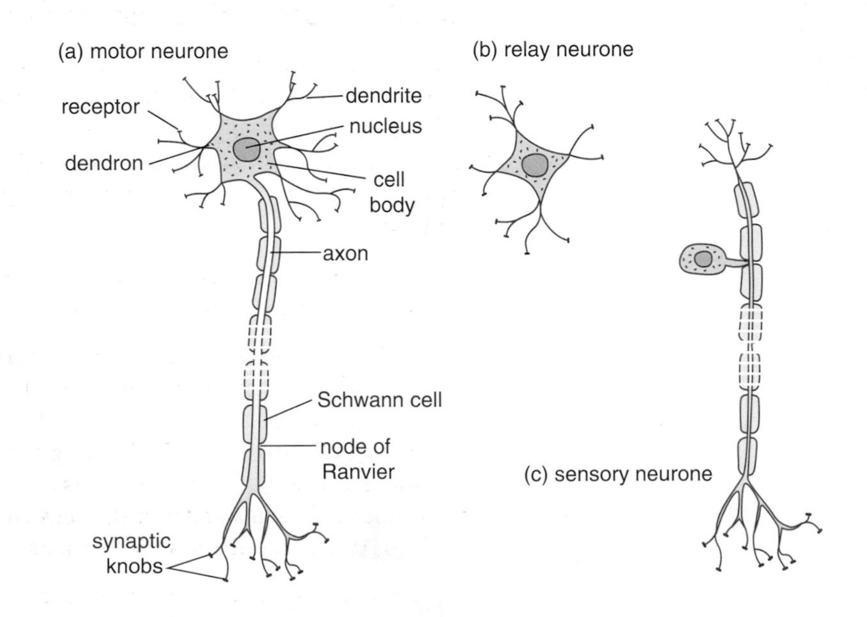


**Figure 13.1**
The structure of (a) a motor neurone, (b) a relay neurone and (c) a sensory neurone.

Figure 13.1 represents the structures of a motor neurone, a relay neurone and a sensory neurone. The structure of a motor neurone is shown in greater detail than the other two types of neurone, reflecting the level of recall that is expected of you in the AQA Specification A. The motor neurone has a large **cell body**, which contains the nucleus and large numbers of cell organelles. The cell body has two types of cytoplasmic extensions that are adaptations to transmitting impulses.

- **Dendrons** are short extensions of the cell body. Each dendron has large numbers of smaller extensions, called **dendrites**. These dendrites are stimulated by other neurones and are the points at which impulse transmission always starts in a motor neurone. They transmit nerve impulses towards the cell body
- The **axon** is a single extension, which can be up to one metre long in humans. The axon always transmits nerve impulses away from the cell body. The axon ends in a series of **synaptic knobs**, which are structures that stimulate the neurone's target organ (effector).

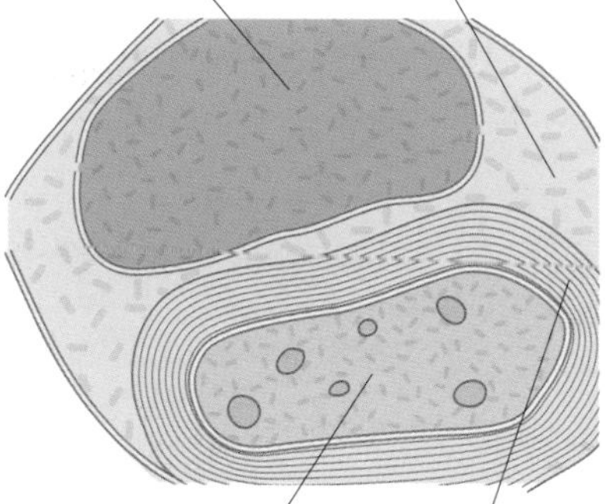

nucleus of Schwann cell
cytoplasm of Schwann cell
axon of neurone
layers of cell surface membrane

Figure 13.2
This cross-section through a myelinated axon shows how the Schwann cell has wrapped itself around the axon

**Q 2 Which of the structures shown in the motor neurone is also present in relay neurones and sensory neurones?**

The motor neurone shown in Fig 13.1 has one further feature of interest to us: its axon is surrounded by **Schwann cells** with small gaps, called **nodes of Ranvier**, between them. As Figure 13.2 shows, each Schwann cell wraps itself around an axon. As a result, many layers of the surface membranes of Schwann cells surround each axon. This is important because the membrane of the Schwann cell is rich in a lipid, called **myelin**. Neurones that have Schwann cells wrapped around their dendrons or axons are called **myelinated neurones**. As we shall see later in this chapter, myelinated neurones transmit impulses much faster than unmyelinated neurones.

## Resting potentials in neurones

Charged ions are present in the cytoplasm of a cell and in the fluid surrounding a cell. As you learned during your AS course, the movement of ions across any membrane is restricted. The way in which membranes restrict the movement of potassium ($K^+$) ions and sodium ($Na^+$) ions – two ions that are important in nerve impulses – is summarised below and in Figure 13.3.

- The phospholipid bilayer prevents the diffusion of ions, such as potassium ($K^+$) ions and sodium ($Na^+$) ions.
- Intrinsic proteins, which span membranes, form **ion channels**. Some types of ion channel are permanently open, enabling constant diffusion of ions, such as $K^+$ and $Na^+$. Other types of channel have **voltage-sensitive 'gates'**, allowing diffusion only when they are opened. The surface membranes of neurones contain separate voltage-sensitive $K^+$ gates and voltage-sensitive $Na^+$ gates, which are important in controlling the transmission of impulses.
- Active transport of ions also occurs across membranes. One type of intrinsic protein, called a **sodium-potassium pump**, uses energy to remove $Na^+$ ions from the cytoplasm and take up $K^+$ ions into the cytoplasm.

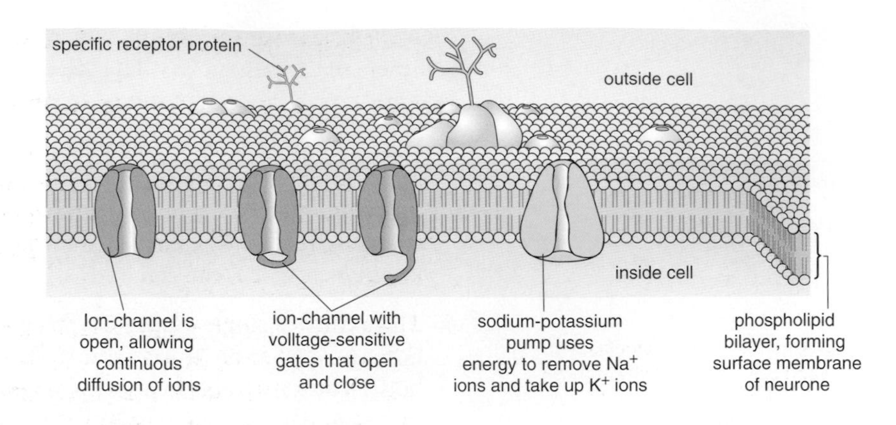


**Figure 13.3**
The methods by which the movement of potassium and sodium ions across membranes is restricted

Table 13.2 shows the concentration of three types of ion inside and outside a mammalian motor neurone when it is not transmitting an impulse ('resting' neurone). As we have seen above, $Na^+$ ions and $K^+$ ions can move across the surface membrane of the neurone through open ion channels. However, the movement of each type of ion is determined by its electrochemical gradient, which is made up of both the electrical gradient and concentration gradient of that ion across the membrane. Take the $K^+$ ions in Table 13.2, for example. You might expect $K^+$ ions to move out of the neurone down their concentration gradient. However, the build up of positive charges ($Na^+$ ions and $K^+$ ions) outside the membrane eventually repels the movement of any more $K^+$ ions, preventing their outward movement. An equilibrium is reached when these effects balance each other out and no net movement of $K^+$ ions occurs.

Table 13.2 also shows a concentration gradient of $Cl^-$ ions, from which you might expect $Cl^-$ ions to move into the neurone. However, the cytoplasm contains large, negatively charged protein molecules that cannot cross the surface membrane and which repel incoming $Cl^-$ ions, preventing their movement into the cell. The imbalance of ions shown in Table 13.2 causes a potential difference (or voltage) between the inside of the neurone and its surroundings. Since a neurone in this condition is not transmitting an impulse, this potential difference is called a **resting potential**. The value of this resting potential is different in different types of neurone, but in the mammalian motor neurone in Table 13.2 it is –70 mV. A membrane that has different charges on its inside and outside is **polarised**.

| Ion | Concentration inside cell/mmol $dm^{-3}$ | Concentration outside cell/mmol $dm^{-3}$ |
|---|---|---|
| $K^+$ | 150.0 | 2.5 |
| $N^+$ | 15.0 | 145.0 |
| $Cl^-$ | 9.0 | 101.0 |

**Table 13.2**
The concentration of three types of ion inside and outside a 'resting' mammalian motor neurone. The concentrations shown in the table result in a resting membrane potential of –70 mV

**Q** 3 Why is the membrane of a neurone said to be polarised?

## Action potentials in neurones

An **action potential** is an abrupt but short-lived reversal of the resting potential of a neurone. It occurs at a specific point of the neurone – we will see how a wave of action potentials passes along a neurone later in this chapter. Figure 13.4 shows the electrical changes that occur during a single action potential

1 The action potential begins when a particular stimulus causes the membrane at one part of the neurone suddenly to increase its permeability to $Na^+$ ions. This is labelled **A** in Figure 13.4 and happens because the stimulus has caused voltage-sensitive $Na^+$ gates in this part of the membrane to open. As a result, $Na^+$ ions diffuse rapidly into the neurone along their electrochemical gradient, reducing the negativity inside that part of the neurone. In a positive feedback loop, this causes more and more voltage-sensitive $Na^+$ gates to open until the voltage difference across the membrane reverses. The highest positive membrane potential is the action potential and this part of the membrane is now said to be **depolarised.**

2 At a certain point, labelled **B** in Figure 13.4, the depolarisation of the membrane causes the voltage-sensitive $Na^+$ gates to close.

3 Shortly after the voltage-sensitive $Na^+$ gates close, the voltage-sensitive $K^+$ gates open. This is labelled **C** in Figure 13.4. The voltage-sensitive $K^+$ gates open more slowly than the voltage-sensitive $Na^+$ gates and allow $K^+$ ions to flow out of the neurone, along their electrochemical gradient. This outflow of $K^+$ ions restores the resting potential in the neurone. In a negative feedback loop, the outward movement of $K^+$ ions results in the closing of the voltage-sensitive $K^+$ gates (labelled **D** in Figure 13.4). There is a slight 'overshoot' in the movement of $K^+$ ions, which causes the membrane potential to become slightly lower than its normal resting potential. This is called **hyperpolarisation** and is labelled region **E** in Figure 13.4.

4 After the resting potential has been restored, sodium-potassium pumps move $Na^+$ ions out of the neurone and move $K^+$ ions back into the neurone by active transport.

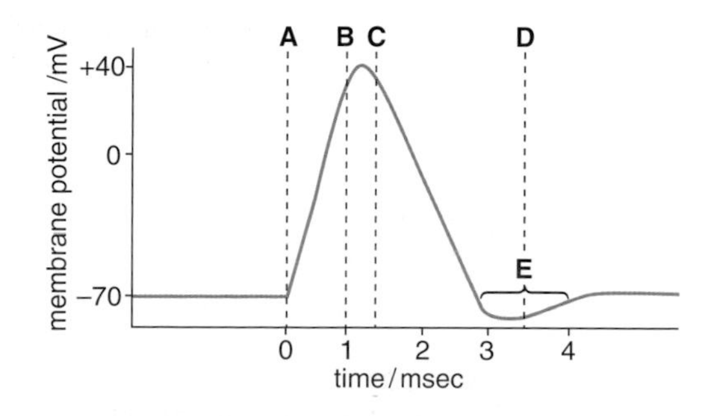


Figure 13.4
The changes in membrane potential associated with an action potential at one tiny part of the surface membrane of a neurone. Voltage-sensitive $Na^+$ gates open at point **A** and close at point **B**. Voltage-sensitive $K^+$ gates open at point **C** and close at point **D**

**Q** 4 Use Figure 13.4 to give the value of:
(a) the resting potential
(b) the action potential
(c) the length of time for which the action potential lasts.

**Extension box 1**

## The all or nothing principle

Any stimulus strong enough to initiate an impulse is called a **threshold stimulus**. If a stimulus is strong enough to cause an action potential, a nerve impulse is transmitted along the entire neurone at a constant and maximum strength. The transmission of the impulse is independent of any further intensity of the stimulus. This is the **all or nothing principle**: a stimulated neurone either fails to transmit an impulse at all or it transmits an impulse at a constant strength along its whole length.

A stimulus that is weaker than a threshold stimulus is called a **subthreshold stimulus** It is incapable of causing an action potential. However, if a second stimulus, or a series of subthreshold stimuli, is quickly applied to the neurone, the cumulative effect might be enough to cause an action potential. This is called **summation**. You will learn later in this chapter how rod cells work together in groups to produce a threshold stimulus on a single relay cell, thereby helping us to see in dim light conditions.

## Conduction of a nerve impulse along a single neurone

The above account explains what happens at a tiny part of the surface membrane of a neurone as it becomes depolarised. We now need to understand how this depolarisation enables a neurone to transmit an impulse all the way along its length towards its target cell.

1. First look at Figure 13.5(a). This represents a membrane in which all the voltage-sensitive $Na^+$ gates are closed. The entire membrane shown is polarised and the length of neurone shown has a resting potential.
2. In Figure 13.5(b), voltage-sensitive $Na^+$ gates have opened and the resulting influx of sodium ions has caused an action potential.
3. Figure 13.5(c) shows how the action potential has caused voltage-sensitive $Na^+$ gates to close and voltage-sensitive $K^+$ gates to open. Potassium ions flow out of the neurone, restoring the resting potential. However, the action potential has disturbed the adjacent membrane, causing its voltage-sensitive $Na^+$ gates to open. A new action potential is occurring in this part of the neurone membrane.
4. In Figure 13.5(d), the action of Figure 13.5(c) has been repeated and a third action potential has been produced. Although the resting potential has been restored, the balance of $Na^+$ ions and $K^+$ ions has not. This is done by sodium-potassium pumps, using the energy released by the breakdown of ATP, and is the only time that active transport is involved in this entire process.

In this way, a single action potential results in a step-by-step wave of action potentials along the entire length of the membrane. This is how a nerve impulse travels along a neurone.

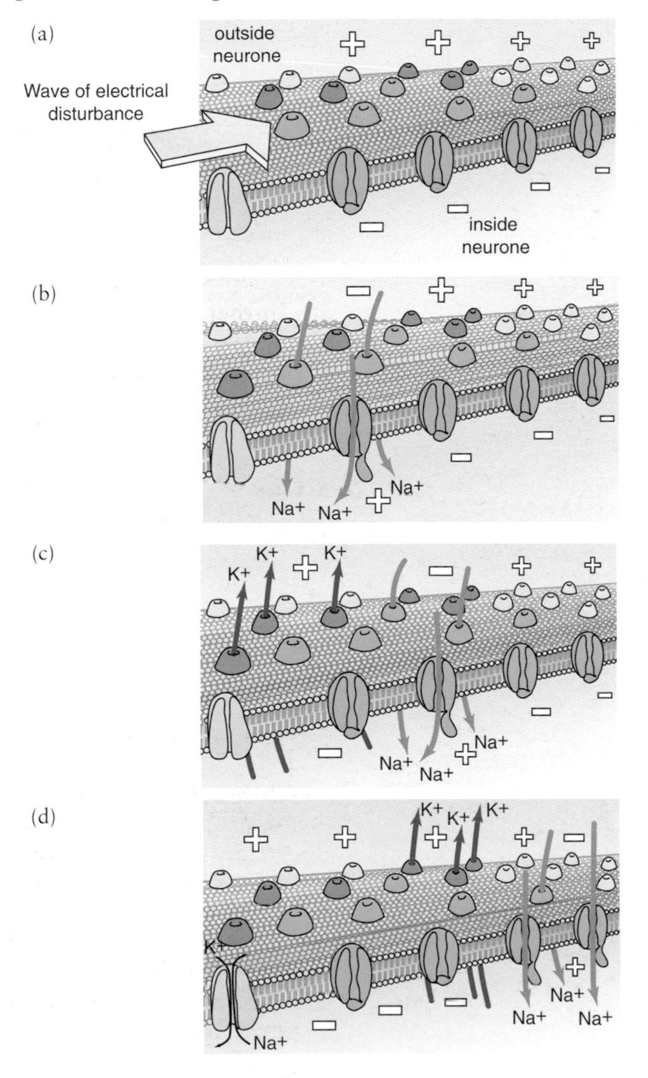


**Figure 13.5**
Conduction of a nerve impulse along a neurone. (a) The neurone is at rest. (b) A tiny part of the membrane on the left of the diagram has become depolarised and an action potential has occurred. (c) This action potential has depolarised another tiny part of the membrane immediately adjacent to it so that a second action potential occurs. (d) The second action has depolarised the next tiny section of membrane and a third action potential has occurred. In this way, the action potential moves steadily from left to right of the diagram

Whilst following Figure 13.5(c), you might have wondered why the second action potential depolarised only the membrane to its right and did not depolarise the membrane to its left as well. This is because the membrane to its left was recovering from the previous action potential in a short time, called the **refractory period**. During the refractory period of a section of membrane, its voltage-sensitive $Na^+$ gates cannot be opened. As a result, the membrane immediately 'behind' an action potential cannot be depolarised and so cannot set up a further action potential of its own. This has two important consequences. Firstly, it explains why nerve impulses can only travel in one direction along a neurone. An action potential can only depolarise the membrane 'in front' as the membrane 'behind' is in its refractory period and cannot be depolarised again. Secondly, because the refractory period lasts for up to 10 milliseconds, it limits the frequency with which neurones can transmit impulses.

## Speed of conduction of impulses

The above account of conduction of a nerve impulse is true for unmyelinated neurones. Each action potential triggers the next one, in a self-propagating wave. However, conduction of nerve impulses is affected by the myelin in the membranes of Schwann cells that surround myelinated neurones. Myelin almost stops the diffusion of $Na^+$ ions and $K^+$ ions. In other words, it insulates those parts of the neurone that it covers. As a result, the diffusion of $Na^+$ ions and $K^+$ ions can only occur at the nodes of Ranvier, shown in Figure 13.1. In a myelinated neurone, action potentials occur only at these nodes, so that the impulse 'jumps' from one node of Ranvier to the next. This is known as **saltatory conduction** (from the Latin *saltare*, meaning to jump) and results in impulses travelling much faster in myelinated neurones than in unmyelinated neurones.

Irrespective of whether they are myelinated or unmyelinated, temperature affects the speed at which neurones conduct impulses. As you learned in your AS Biology course, temperature affects the rate of diffusion. Provided they do not begin to denature cell proteins, high temperatures cause an increase in the rate of diffusion of ions. As a result, they will increase the rate of conduction of nerve impulses. Impulses are also faster in an axon with a large diameter. This is because there is less 'leakage' of ions in neurones with a large diameter than in neurones of small diameter. Since ion leakage weakens membrane potentials, the conduction of action potentials is often lost along unmyelinated neurones with a small diameter. As we have already seen, the presence of myelin stops ion leakage, so diameter is only an important consideration for unmyelinated neurones.

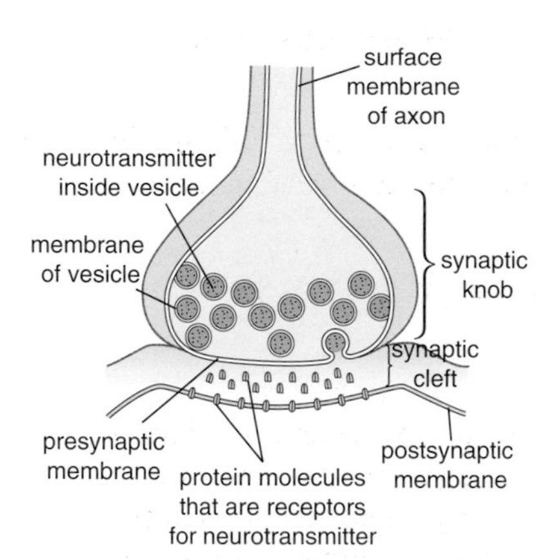


**Figure 13.6**
The structure of an excitatory synapse. When an impulse arrives at the end of a presynaptic neurone, it causes vesicles of neurotransmitter to fuse with the presynaptic membrane and diffuse through the synaptic cleft to the postsynaptic membrane. Here they bind with specific receptors and cause depolarisation of the post-synaptic membrane

## Transmission at synapses

Nervous pathways involve chains of at least two neurones that pass impulses from one to another. The junction between two neurones is called a **synapse**. The two neurones at a synapse do not touch each other. Instead, there is a gap of about 20 μm between one neurone and the next in the chain. This gap is called a **synaptic cleft**. On one side of the synaptic cleft is the **presynaptic membrane**, which is the surface membrane at the end of one neurone (the presynaptic neurone). On the other side of the synaptic cleft is the **postsynaptic membrane**, which is the surface membrane of the next neurone in the chain (the postsynaptic neurone). When a nerve impulse arrives at the presynaptic membrane of a synapse, it causes the release of a chemical **neurotransmitter**. Molecules of this neurotransmitter diffuse across the synaptic cleft and change the polarisation in the surface membrane of the postsynaptic neurone.

Figure 13.6 shows an **excitatory synapse**, i.e. one that causes an action potential in the postsynaptic cell. Notice that the end of the presynaptic neurone forms a swelling, called the **synaptic knob**, which contains small vesicles containing neurotransmitter. When an impulse arrives at the synaptic knob, it causes gated calcium ion channels in the presynaptic membrane to open. As a result, calcium ions ($Ca^{2+}$) flow into the synaptic

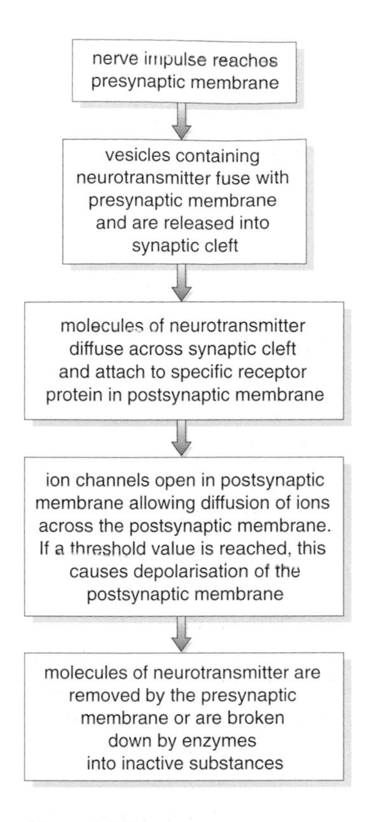


**Figure 13.7**
The flowchart summarises transmission of a nerve impulse across an excitatory synapse

knob along their electrochemical gradient, causing the vesicles to move to the presynaptic membrane and fuse with it. Molecules of the neurotransmitter are then released from the vesicles and diffuse across the synaptic cleft. They bind to specific receptor proteins in the post-synaptic membrane. These receptor proteins are attached to gated ion channels and change their shape when they bind with molecules of neurotransmitter, allowing ions to diffuse across the postsynaptic membrane. If the diffusion of ions reaches a **threshold value**, it causes an action potential in the postsynaptic membrane. Once started, the action potential is conducted along the postsynaptic neurone in the same way as shown in Figure 13.5. Figure 13.7 summarises transmission of an impulse across a synapse.

**Q** 5 **Use your knowledge of AS Biology to name the process by which neurotransmitter is released from the vesicles in the synaptic knob.**

The action of neurotransmitters is short-lived because molecules of neurotransmitters are quickly removed from the postsynaptic membrane. This happens because:

- molecules of neurotransmitter diffuse out of the synaptic cleft
- molecules of neurotransmitter are taken up by the membrane of the presynaptic membrane by endocytosis
- enzymes quickly break down molecules of neurotransmitter into inactive substances

There are many different neurotransmitters in the nervous system. You need to know only two.

- **Acetylcholine** is released by motor neurones onto muscle cells and by neurones in the parasympathetic division of the autonomic nervous system.
- **Noradrenaline** is released by neurones in the sympathetic division of the autonomic nervous system.

However, many substances are similar in shape to natural neurotransmitters and can fit into the specific protein receptors of post-synaptic membranes. **Agonistic substances** bring about the same effect as the neurotransmitter when they bind to specific protein receptors of post-synaptic membranes. Anatoxin is an example of an agonistic substance. It is produced by some algae and mimics the effects of acetylcholine. Since anatoxin is swallowed in contaminated water, its effects are strongest in the mouth, where anatoxin causes continuous salivation. **Antagonistic substances** also bind to specific protein receptor sites on postsynaptic membranes. Once this has happened, antagonistic substances prevent neurotransmitter molecules from binding with these receptor sites and so block the action of the neurotransmitters. High blood pressure can be treated using drugs called **β-blockers**. These drugs are antagonists of adrenaline-receptors on the surface membranes of heart muscle cells. Curare is another type of receptor antagonist, blocking the action of acetylcholine at the junction of nerves and

Table 13.3
Unidirectionality, summation and inhibition in synapses

muscles. This makes curare useful as a general muscle relaxant in patients undergoing major surgery. Table 13.3 summarises some aspects of synaptic transmission. You can find out more about agonists and antagonists in the assignment at the end of this chapter.

| Aspect of synaptic transmission | Explanation |
|---|---|
| Unidirectionality | Because neurotransmitters are released only by the presynaptic membrane, transmission at a synapse can only flow in one direction from the presynaptic neurone to the postsynaptic neurone. |
| Summation | Sometimes insufficient neurotransmitter is released from a single presynaptic neurone to reach the threshold needed to change the polarisation of the membrane of the postsynaptic membrane. In this case, simultaneous release of neurotransmitter from several presynaptic neurones is needed to change the polarisation of the postsynaptic membrane. Groups of rod cells do this when they synapse with individual relay neurones in the retina (see later in this chapter). |
| Inhibition | In some synapses, movement of ions causes the inside of the postsynaptic neurone to become more negative than its resting potential. This is called **hyperpolarisation** and inhibits stimulation of the postsynaptic neurone. Synapses like this are called **inhibitory synapses**. The AQA Specification A does not require you to know the mechanism of transmission by inhibitory synapses, but we will come across them again when we learn about the action of rod cells. |

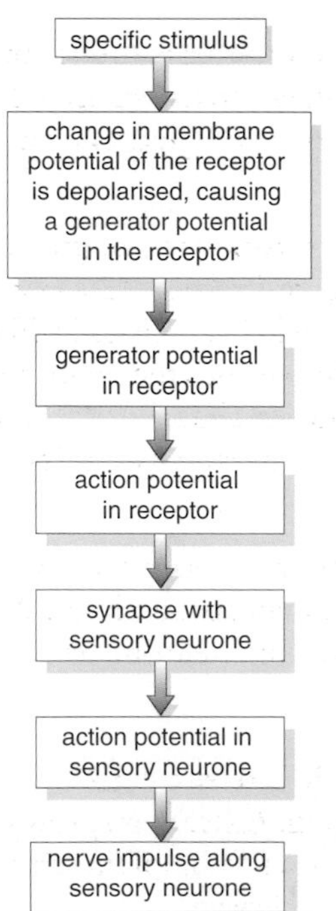


Figure 13.8
A flowchart showing the stages by which a receptor converts a stimulus into a nerve impulse

## Receptors

A **stimulus** is a change in the environment that is capable of causing a response by the nervous system. Very few neurones are directly sensitive to stimuli. This where receptors come in. A **receptor** is a transducer. This means that it converts a stimulus into a nerve impulse. As Figure 13.8 shows, a stimulus causes a change in the membrane potential of the receptor, resulting in a **generator potential** in that receptor. In turn, this generator potential results in an action potential along a sensory neurone.

Receptors are quite specific in the stimuli that will cause generator potentials. We need to be familiar with only two receptors: Pacinian corpuscles and the receptor cells in the retina of the eye. Each responds to only one type of stimulus: all other stimuli fail to cause a generator potential in these receptors.

### Pacinian corpuscles

Pacinian corpuscles are found beneath the skin, around joints and tendons, in the external genitalia of both sexes and in some internal organs. They are sensitive to changes in mechanical pressure. For example, Pacinian corpuscles in the skin of your feet enable you to detect pressure changes when you put on a pair of trainers. Figure 13.9 shows a light micrograph of a single Pacinian corpuscle in longitudinal section. It resembles a sliced onion, containing layers of connective tissue around the unmyelinated end of a myelinated neurone.

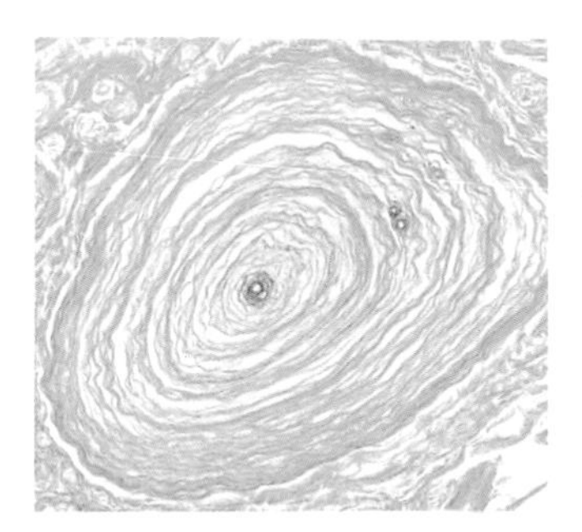

Figure 13.9
A light micrograph of a single Pacinian corpuscle in longitudinal section. The central dendrite is surrounded by layers of connective tissue, making it look like a sliced onion

We have seen the importance of ion channels in causing membrane potentials. The neurone inside a Pacinian corpuscle has **stretch-mediated $Na^+$ channels** in its surface membrane. These sodium channels are so called because they change their shape when stimulated by changes in pressure, so that their permeability to $Na^+$ ions changes.

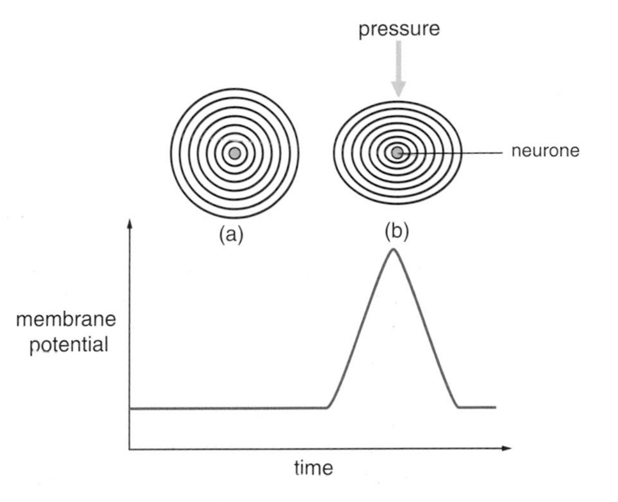


Figure 13.10
(a) In an unstimulated Pacinian corpuscle, the layers of connective tissue and the neurone that they enclose have a rounded cross-section. (b) When pressure is applied, the corpuscle is pressed flat. This flattening opens stretch-mediated $Na^+$ ion channels in the surface membrane of the neurone. As a result, $Na^+$ ions diffuse into the neurone and cause the generator potential that you can see in the graph

Figure 13.10 shows a cross-section through the layers of connective tissue and the neurone within a Pacinian corpuscle. It also represents the membrane potential of the neurone.

1 In its resting state (Figure 13.10a), a Pacinian corpuscle is round. In this condition, the stretch-mediated $Na^+$ channels are too small to let $Na^+$ ions pass through them and into the neurone. The neurone has a resting potential, represented on the graph.

2 When the layers of connective tissue and the neurone are compressed, the Pacinian corpuscle becomes flattened (Figure 13.10b). As a result, the surface membrane of the neurone is stretched. This causes its $Na^+$ ion channels to open and allow $Na^+$ ions to diffuse into the neurone. As a result, the membrane potential of the neurone changes. You can see this in the graph in Figure 13.10b, where a generator potential has been produced.

3 The generator potential causes an action potential. Once an action potential has been produced, a stimulated Pacinian corpuscle conducts an impulse along its length and stimulates a sensory neurone by releasing a neurotransmitter across a synapse.

**Q 6 When the pressure on a Pacinian corpuscle is released, the neurone does not immediately become round again. Instead, it springs into an elongated shape at right angles to Figure 13.10b. Explain why this enables you to feel that the pressure has been released.**

## The eye

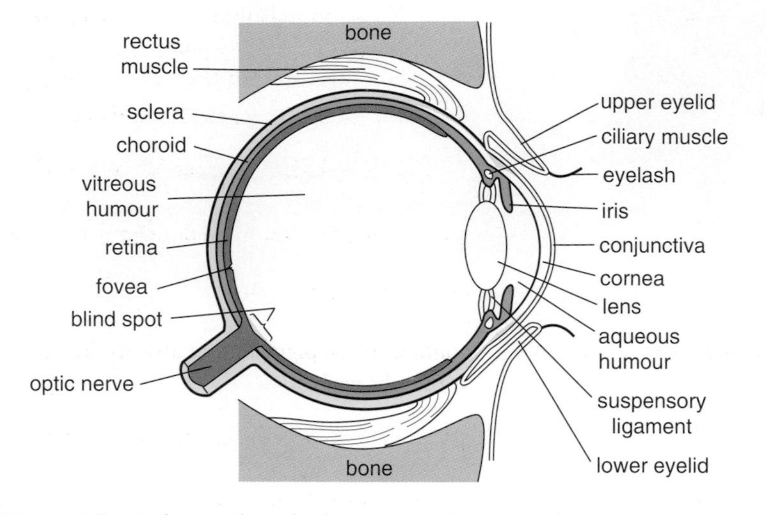


**Figure 13.11**
Structure of the human eye

Figure 13.11 shows that the human eye is a complex sensory organ, containing a variety of different tissues. Only one of these tissues, the **retina**, contains the receptors in which we are interested. Table 13.4 summarises the function of the different parts of the eye that can be seen in Figure 13.11.

Light is composed of particles, called photons. Each photon travels through a vacuum or transparent media as a light ray, which we represent as a straight line. When light rays pass from one medium to another medium of a different density, the angle at which they travel is changed. This is called **refraction**. For example, light is refracted when it passes from air outside a building into the glass of a window and again when it passes from the glass in a window into the air inside a room. When rays of light strike a non-transparent object, they are reflected back at a predictable angle.

**Q** 7 **Name the surfaces within an eyeball at which light will be refracted.**

There are millions of rays of light striking this page. They are reflected in countless directions and only some enter your eyes as you read. Because these rays of light are from an object that is close to your eyes, they are diverging (spreading out). If you look through a window at an object outside the room in which you are sitting, rays of light from distant objects enter your eyes. Because these rays of light are from a distant object, they are almost parallel. Figure 13.12 shows how we focus rays from near and distant objects by changing the shape of our lenses, in a process called **accommodation**. Contraction and relaxation of ciliary muscles changes the tension on the lens. When the lens is pulled thin, it bends light less. Our lenses are pulled thin when we focus the parallel rays of light from distant objects (**distant accommodation**). When the lens is allowed to revert to its fat shape, it bends light more. Our lenses are allowed to revert to their fat shape when we focus the diverging rays from near objects (**near accommodation**).

| Part of eye | Function |
|---|---|
| conjunctiva | • protects the cornea against damage by friction (tears from the tear glands help this process by lubricating the surface of the conjunctiva) |
| sclera | • protects eyeball against mechanical damage<br>• provides attachment surfaces for eye muscles |
| cornea | • refracts (bends) light more than any other part of the eyeball<br>• allows passage of light |
| choroid | • contains black pigment that absorbs light, thus preventing reflection of light within the eyeball<br>• contains blood vessels that supply other structures within the eyeball |
| ciliary body | • has suspensory ligaments that hold the lens in place<br>• secretes aqueous humour<br>• contains ciliary muscles that enable the lens to change shape, during accommodation (focusing on near and distant objects) |
| iris | • changes the size of the pupil, so controlling the amount of light passing to the inside of the eyeball<br>• contains pigment that absorbs light, reducing the amount of light passing to the inside of the eyeball |
| lens | • refracts (bends) light<br>• by changing shape, focuses light from near objects (near accommodation) and distant objects (distant accommodation) on to the retina |
| retina | • contains the light receptors – rod cells and cone cells<br>• contains relay neurones and sensory neurones that pass impulses along the optic nerve to the part of the brain that controls vision |
| fovea (yellow spot) | • a part of the retina that is directly opposite the pupil and contains only cone cells |
| blind spot | • the part of the retina at which the sensory neurones that form the optic nerve leave the eyeball; it contains no light-sensitive receptors |
| aqueous humour | • helps to maintain the shape of the anterior chamber of the eyeball |
| vitreous humour | • supports the lens<br>• helps to maintain the shape of the posterior chamber of the eyeball |

**Table 13.4**
The functions of the major parts of the human eyeball shown in Figure 13.11

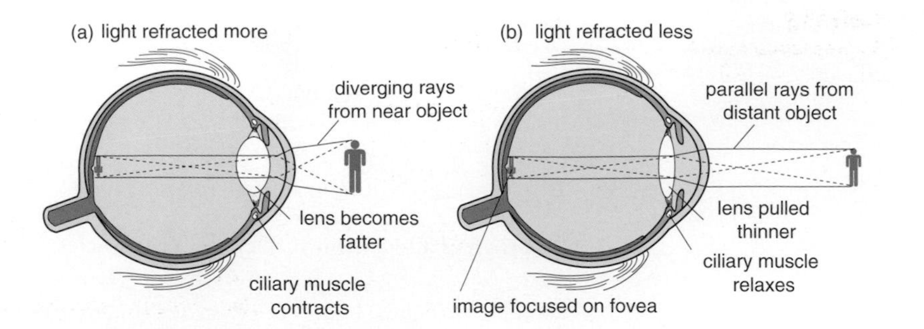


**Figure 13.12**
(a) During near accommodation, the ciliary muscles contract. This reduces the tension in the suspensory ligaments and the lens reverts to its fat shape. This bends light more, focusing the diverging light rays from a near object.
(b) During distant accommodation, the ciliary muscles relax. This increases the tension in the suspensory ligaments, pulling the lens much thinner. This bends light less, focusing the parallel rays from a distant object

If accommodation is successful, all the rays of light entering the eye from one part of an object will be focused at a single point of the retina. Unsuccessful accommodation results in blurred images on the retina. Some people produce clear images from distant objects but blurred images from near objects – they are long-sighted. Other people produce blurred images from distant objects but clear images from near objects – they are shortsighted. Both types of defect can be corrected with spectacles or contact lenses.

The human retina contains millions of light-sensitive receptor cells and the neurones with which they synapse. Figure 13.13 represents part of the human retina. Pigmented cells of the choroid are at the top of Figure 13.13. Although not strictly part of the retina, they have been drawn to remind you of their importance in absorbing light. The light receptors are shown in the layer of the retina below the pigmented cells. In this diagram, two types of receptor have been drawn. These are **rod cells** and **cone cells**. A layer of **bipolar relay neurones** lies beneath the layer of rod cells and cone cells and, finally, there is the lowest layer of sensory neurones (called **ganglion cells**). The axons of these ganglion cells make up the **optic nerve**, which passes impulses from the retina to the brain.

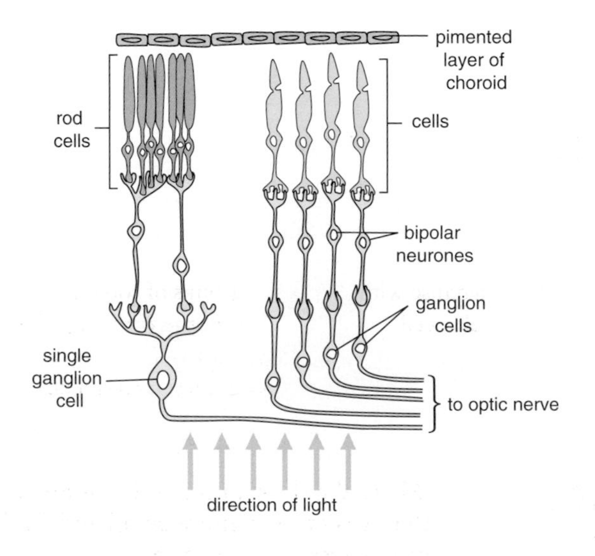


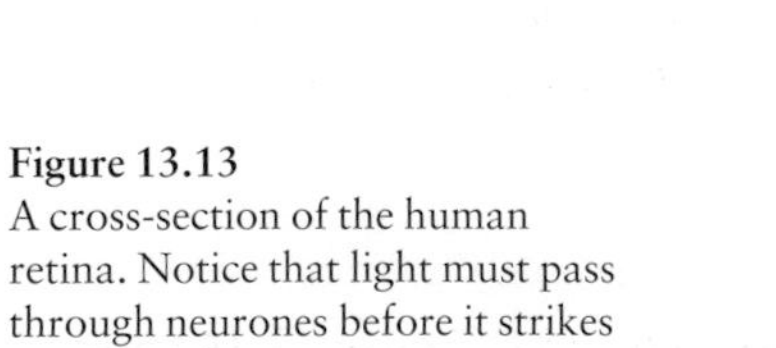

**Figure 13.13**
A cross-section of the human retina. Notice that light must pass through neurones before it strikes the retina

Table 13.5
A comparison of rod cells and cone cells

| Feature | Rod cells | Cone cells |
| --- | --- | --- |
| approximate frequency in a human retina | $120 \times 10^6$ | $6 \times 10^6$ |
| distribution | • evenly throughout the retina<br>• absent from the fovea<br>• the only type of light receptor at the periphery of the retina | • present mainly in the fovea (Figure 13.11) at a density of about $5 \times 10^4$ $mm^{-2}$ |
| shape of outer segment | • rod-shaped | • cone-shaped |
| sensitivity to light | • very sensitive to light, therefore operate even in dim intensities<br>• insensitive to colour (monochromatic vision) | • sensitive only to bright light, therefore operate only in bright light intensities<br>• sensitive to red light or green light or blue light |
| visual acuity | • produce poorly resolved images | • produce well-resolved images |
| light-sensitive pigments | • single pigment, called rhodopsin, in every rod cell | • one of three types of iodopsin pigment in any one cone cell<br>• each type of iodopsin is sensitive to red light or to green light or to blue light<br>• stimulation of different combinations of the three types of cone cell produces a perception of colour (trichromatic vision) |
| synapse with relay cell | • groups of rod cells synapse with one relay relay cell (retinal convergence) | • each cone cell synapses individually with a single relay cell |

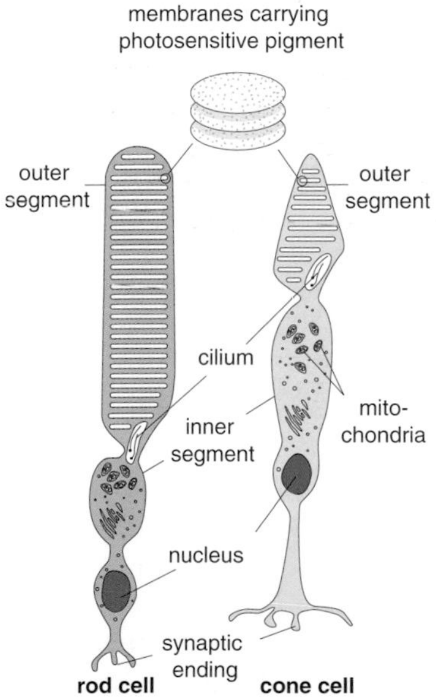

Figure 13.14
(a) Rod cells and (b) cone cells have an outer segment, with light-sensitive pigment held in stacks of membrane, an inner segment, containing the cell's nucleus together with organelles such as mitochondria and ribosomes, and a synaptic region

We need to look at the rod cells and cone cells first. Figure 13.14 shows their structure in more detail and Table 13.5 summarises some important differences between them.

Remember that receptors, such as rod cells and cone cells, convert a stimulus to a nerve impulse (transduction). Although rod cells and cone cells do this in a similar way, we shall only look at how transduction occurs in rod cells. In total darkness, the membrane of a rod cell is polarised, i.e. there is a potential difference between its cytoplasm and the surrounding fluid. As with neurones, this potential difference is caused by an imbalance in the concentration of $Na^+$ ions and $K^+$ ions between the cytoplasm and the surrounding fluid. In the dark, each rhodopsin molecule in a rod cell is made of a protein, called **opsin**, combined with ***cis*-retinal**. When *cis*-retinal absorbs light, it changes to a different isomer, called ***trans*-retinal**. In this form, the retinal cannot bind to opsin, so that the molecule of rhodopsin breaks down into *trans*-retinal and opsin. As a result, the surface membrane of the rod cell changes its permeability to $Na^+$ ions and $K^+$ ions, increasing the negativity of the cell's cytoplasm. It is this **hyperpolarisation** that causes a generator potential to be produced in the rod cell. As a result, the rod cell releases less neurotransmitter onto the relay cell with which it synapses.

$$\text{Rhodopsin} \xrightarrow{\text{light}} \text{opsin} + \textit{trans}\text{-retinal}$$

**Q** 8 **Does the neurotransmitter released by rods have an excitatory or inhibitory effect on relay cells?**

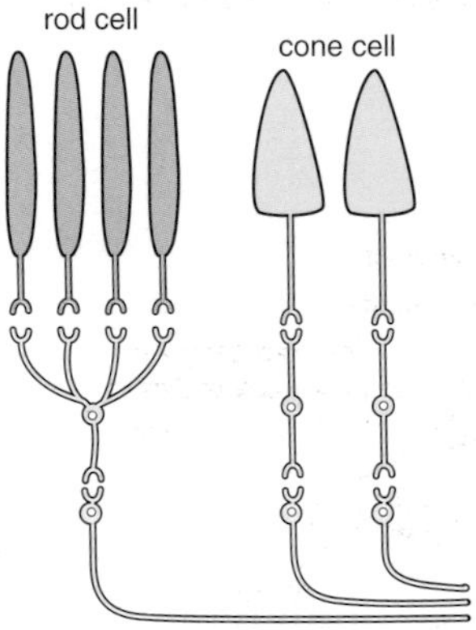

If reduction in the release of neurotransmitter from a rod cell reaches a threshold, an impulse passes through the relay cell, across a synapse and down the ganglion cell. The likelihood of reaching a threshold in the relay cell is increased because several rod cells synapse with each individual relay cell. This is called **retinal convergence**. Although it increases sensitivity to dim light, retinal convergence decreases **visual acuity**. Look at Figure 13.15, which is a simple representation of the synaptic connections in the retina. Two cone cells are shown. Each of these cone cells synapses with a different relay cell that, in turn, synapses with a different ganglion cell. Stimulation of these two cone cells would result in two impulses passing along the optic nerve to the brain. Four rod cells are shown in Figure 13.15. All four rod cells synapse with the same relay cell, which synapses with one ganglion cell. Stimulation of any combination of these four rod cells would result in a single impulse passing along the optic nerve to the brain.



**Figure 13.15**
Simplified diagram to show retinal convergence in the retina. Cone cells synapse individually with relay and sensory neurones, increasing visual acuity (the ability of the brain to resolve images). Several rod cells synapse to the same relay cell. As a result, they collectively cause a generator potential in the relay cell in dim light conditions (summation). However, this pattern of synapses results in poor visual acuity

Rhodopsin has to be resynthesised. Rod cells do this by combining retinal and opsin, using ATP produced by the mitochondria in their inner segments. This reaction occurs slowly compared to the breakdown of rhodopsin by light. In bright light, your eyes are **light-adapted**, meaning that most of your rhodopsin has been broken down into opsin and *trans*-retinal. If you go from bright light into dimly lit conditions, your vision will be poor until you have resynthesised enough rhodopsin for your rods to start working again. Your eyes have then become **dark-adapted**.

$$\textit{trans}\text{-retinal} + \text{opsin} \xrightarrow[\text{ATP} \rightarrow \text{ADP} + P_i]{} \text{rhodopsin}$$

**Q** 9 **Retinal is a derivative of vitamin A. Suggest the effect that a deficiency of vitamin A in the diet would have on vision.**

## The spinal cord and spinal reflexes

When learning about the retina, we saw how stimuli result in nerve impulses in sensory neurones. We can now learn how impulses travelling along sensory neurones result in a change in behaviour. The simplest behaviour patterns are called reflexes. A **simple reflex** is inborn (i.e. is not learned) and always results in the same, fixed response to a particular stimulus. Examples include dilating the pupil in dim light, salivating at the taste of food and withdrawing part of the body from a painful stimulus. The nervous pathway of a simple reflex is called a **reflex arc.** Sometimes a reflex arc involves parts of the brain, e.g. those that control the muscles in the iris and tension in the suspensory ligaments of the eye. Reflex arcs that involve the spinal cord, but not the brain, are called **spinal reflexes.**

The human **spinal cord** is a hollow tube of nervous tissue that runs from the brain, through a large space in the backbone, to the base of the spine. Thirty-one pairs of spinal nerves enter and leave the spinal cord along its length. Each spinal nerve contains large numbers of sensory and motor neurones. A region within the spinal cord, called the **grey matter,** contains large numbers of relay neurones. Figure 13.16 represents part of the spinal cord and one spinal nerve. It also shows one set of neurones involved in a reflex arc.

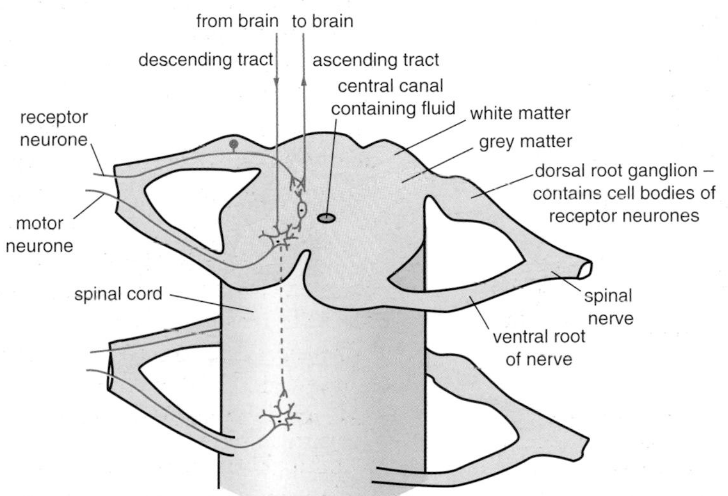


**Figure 13.16**
Cross-section through part of a human spinal cord. The grey matter contains unmyelinated relay cells. The white matter contains myelinated sensory and motor neurones. The position of three neurones involved in a spinal reflex is shown. Motor neurones always leave the spinal cord via the ventral root. Sensory neurones always enter the spinal cord via the dorsal root and have their cell bodies in the dorsal root ganglion

Imagine you have touched the hot plate of an iron whilst ironing your jeans. A simple reflex would occur in response to this stimulus: you would quickly pull your hand away from the iron. In this example, temperature and pain receptors in your skin detect the stimulus. As a result, they produce generator potentials and pass nerve impulses to sensory neurones just under your skin. These sensory neurones carry their impulses right along their length and into the spinal cord, via a spinal nerve. Sensory neurones always enter the spinal cord via the **dorsal root** of the nerve (you would normally call your dorsal surface the back of your body). The swelling in the dorsal root that contains the cell bodies of all these sensory neurones is called the **dorsal root ganglion.**

In the spinal cord, sensory neurones synapse with relay neurones that, in turn, synapse with motor neurones. All these synapses occur in the **grey matter** of the spinal cord.

The motor neurones that carry impulses to muscle cells in your biceps muscle leave the spinal cord via the **ventral root** of the spinal nerve (you would normally call your ventral surface the front of your body). Eventually, the motor cells release neurotransmitter onto muscle cells in your biceps, causing them to contract. As a result, you bend your arm and pull your hand away from the hot iron. This action is rapid and automatic – you did not have to think about it. In reality, large numbers of receptors, sensory neurones, relay neurones, motor neurones and muscle cells would be involved. For the sake of simplicity, only one of each is shown in Figure 13.16.

The withdrawal reflex above is obviously advantageous. By withdrawing your hand from a damaging stimulus, you protect your body. Luckily for us, the reflex arcs are there at birth – we do not have to learn to withdraw from harmful stimuli the hard way. However, we can learn to modify these reflexes. For example, in addition to withdrawing your hand from a hot iron, you might shout something like 'Ouch!'. This is a learned response and is controlled by nervous pathways that you developed during the learning process. In this case, additional relay cells carry impulses from the reflex arc up ascending tracts in the grey matter of the spinal cord to your brain. In your brain, new pathways of relay cells are formed that eventually stimulate motor cells to the muscles that control speech. Thus, simple reflexes can sometimes be adapted through learning.

**Q** 10 **Imagine you have spent several hours preparing a delicious meal. When you take the meal from the oven, the container burns your hand. Suggest the nervous pathway that prevents you from dropping the meal on the kitchen floor.**

## The autonomic nervous system

The **autonomic nervous system** is a part of the nervous system that controls internal glands and muscles, which are normally beyond your conscious control (Table 13.6). The autonomic nervous system has two divisions.

- The **parasympathetic division** of the autonomic nervous system generally has an inhibitory effect and helps relaxation. The motor neurones of the parasympathetic division release the neurotransmitter acetylcholine onto the glands or organs that they stimulate.
- The **sympathetic division** of the autonomic nervous system generally has an excitatory effect and helps the body to react to stress. The motor neurones of the sympathetic division release the neurotransmitter noradrenaline onto the glands or organs that they stimulate.

| Target organ or tissue | Effect of parasympathetic stimulation | Effect of sympathetic stimulation |
|---|---|---|
| Iris of eye | constricts pupil | dilates pupil |
| Salivary gland | stimulates saliva | inhibits saliva |
| Intercostal muscles | decreases breathing rate | increases breathing rate |
| Bronchi and bronchioles | constricts | dilates |
| Blood vessels | dilates – decreasing blood pressure | constricts – increasing blood pressure |
| Heart | decreases heart rate and stroke volume | increases heart rate and stroke volume |
| Gut | stimulates peristalsis | inhibits peristalsis |
| Sweat glands | no effect | increases sweat production |

Table 13.6
The effects of stimulation by the parasympathetic and sympathetic divisions of the autonomic nervous system are usually antagonistic

## Extension box 2

### Control of the heart rate

You learned about the structure and action of the heart in your AS Biology course. You learned that heart muscle is **myogenic**, i.e. it contracts of its own accord. You also learned how the activity of the sinoatrial node, atrioventricular node and Purkyne tissue coordinate the contraction of cardiac muscle to produce a heart beat. The autonomic nervous system is able to change the rate of contraction of the heart.

A part of the brain, called the **medulla**, contains two regions that control the rate at which the heart beats. **The cardioacceleratory centre (CAC)** speeds up the heart rate. Sympathetic neurones run from the CAC down a descending tract of the spinal cord and along a spinal nerve to the sinoatrial node, atrioventricular node and parts of the heart muscle itself. When the CAC is stimulated, nerve impulses travel along these sympathetic neurones and they release noradrenaline onto their target cells. This causes the heart rate to increase.

The medulla contains a second group of neurones, called the **cardioinhibitory centre** (CIC). Parasympathetic neurones from the CIC leave the brain in the vagus nerve and run to the sinoatrial node and atrioventricular node. When the CIC is stimulated, nerve impulses travel along these parasympathetic neurones and they release acetylcholine onto their target cells. This causes the heart rate to decrease.

In this way, opposing stimulatory sympathetic neurones and inhibitory parasympathetic neurones control the heart rate. They do so through simple reflexes in response to blood pressure. The reflexes are started by pressure receptors in the wall of the aorta and the wall of the carotid artery (passing from the aorta to the head), which send nerve impulses to the CAC and CIC in the medulla. Figure 13.17 shows these autonomic pathways involved in control of the heart beat.

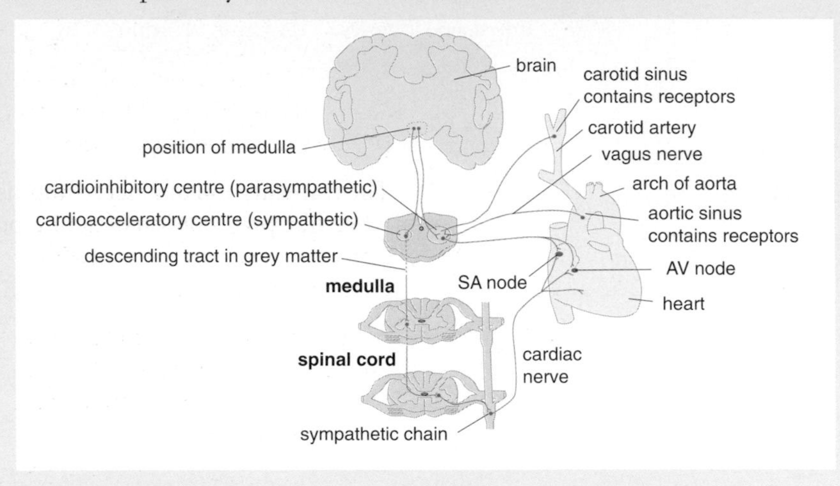


**Figure 13.17**
The autonomic nervous system controls the rate at which the heart rate. Pressure receptors in the walls of the aorta and carotid artery stimulate the CAC and CIC in the medulla of the brain. Sympathetic neurones from the CAC stimulate an increase in heart rate; parasympathetic neurones from the CIC stimulate a decrease in the heart rate

## Simple behaviour in animals

The behaviour of some animals consists almost entirely of simple reflexes. The adaptive advantage of this is often easy to see but the behaviour cannot be adapted. Two types of simple behaviour, kinesis and taxis, help to maintain animals in a favourable environment.

**Figure 13.18**
These woodlice cannot regulate their water loss. They show a non-directional response (kinesis) to air humidity

A **kinesis** (plural, kineses) is a non-directional response to a stimulus. The woodlice shown in Figure 13.18 cannot regulate their water loss. In dry conditions, woodlice increase both their rate of movement and their rate of randomly changing direction. As a result of this kinesis, woodlice are likely to move away from dry conditions and come to rest in damp conditions, where they are less likely to die from water loss.

**Figure 13.19**
A male silkworm moth showing its large antennae. 'Hairs' on these antennae contain millions of cells that respond to a substance released by female silkworm moths. The male silkworm moth shows a chemotactic response to females when the concentration of this substance is as low as 14 000 molecules/cm$^{-3}$ of air

A **taxis** (plural, taxes) is a directional response to a stimulus. Many photosynthetic protoctists swim towards light, i.e. they are positively phototactic. This behaviour has obvious adaptive advantages since it maintains these protoctists in an environment that is suitable for photosynthesis. The male silkworm moth shown in Figure 13.19 moves directly towards a particular chemical, i.e. he is positively chemotactic. His large, feathery antennae contain millions of cells that detect bombykol, a substance released by female silkworm moths. Even at concentrations as low as 14 000 molecules per cm$^{-3}$ of air, a male can detect this substance and flies directly towards its source – the tip of the female's abdomen.

## Summary

- The nervous system is made of three types of neurone: sensory neurones, relay neurones and motor neurones. Sensory and motor neurones can be myelinated, in which case they transmit impulses faster than unmyelinated neurones.
- A neurone has a resting potential, which is a voltage across its surface membrane. This results from an imbalance in ions, especially $Na^+$ and $K^+$ ions, across its membrane. When an impulse is passed along an excitatory neurone, this resting potential is reversed, producing an action potential.
- The transmission of a nerve impulse along a neurone depends on the wave-like propagation of action potentials along its length. In myelinated neurones, the action potentials 'jump' from one node of Ranvier to the next.
- Because the surface membrane of a neurone has a refractory period, impulses can only pass in one direction along a neurone. The refractory period also limits the frequency with which neurones can pass separate impulses.
- So long as a threshold level of stimulation is reached, a neurone transmits an impulse along its length at a constant and maximum strength. This is the 'all or nothing' principle.
- Nerve impulses pass from neurone to neurone across small gaps, called synapses. These are bridged by neurotransmitters that are secreted by small synaptic knobs at the end of neurones. Acetylcholine and noradrenaline are examples of neurotransmitters.
- Receptors convert specific stimuli into nerve impulses. They do so because a stimulus changes their resting potential, causing a generator potential in the receptor cells. Receptors pass their impulses across synapses to relay neurones or to sensory neurones.
- Pacinian corpuscles are sensitive to changes in pressure. Rod cells and cones cells are receptors in the retina of the eye that are sensitive to light.
- Rod cells produce a generator potential when light bleaches a pigment, called rhodopsin, which is present in their intracellular membranes. Since several rod cells synapse with an individual relay cell, they collectively cause depolarisation of the relay cell. This process, called summation, increases the sensitivity of the retina to light but reduces the visual acuity produced by rod cells.
- The spinal cord runs from the brain to the base of the spine. In a simple spinal reflex arc, sensory neurones synapse with relay neurones and relay neurones synapse with motor neurones within the grey matter of the spinal cord. These simple spinal reflexes protect the body from damage but can be modified as a result of learning.

- The autonomic nervous system controls internal glands and muscles over which we normally have little conscious control. The parasympathetic division of the autonomic nervous system secretes acetylcholine and helps relaxation whereas the sympathetic division of the autonomic nervous system secretes noradrenaline and helps in preparation for emergencies.
- Changes in the heart rate are controlled by the balancing action of the parasympathetic and sympathetic divisions of the autonomic nervous system.
- Simple behaviour in animals consists largely of simple reflexes. Kineses are non-directional responses to stimuli; taxes are directional responses to stimuli.

# Assignment

## Snakes, puffer fish and paralysis

You may not have a lot of sympathy for poisonous snakes – not many people have! But think for a minute about the problems that such a snake faces in catching and swallowing a large rat. Rats move fast but snakes are relatively slow. So it is a question of first stop your rat. In addition, a snake cannot chew its prey or bite pieces off. This means that once caught, the rat must be swallowed whole. Clearly a rat must be subdued before it is swallowed or the snake could come to considerable harm. Observations such as these suggested that snakes and other poisonous animals produce substances in their venom which in some way affect the nerves and muscles of their prey, often bringing about paralysis.

In this assignment we will look in more detail at the way in which substances produced by poisonous animals affect our nerves and muscles. You will need to use material from various parts of the specification as well as from this chapter to answer the questions. Before you start, it would be worth looking up the following key topics either in your notes or in your textbook:

- proteins and protein structure
- plasma membranes
- diffusion and active transport.

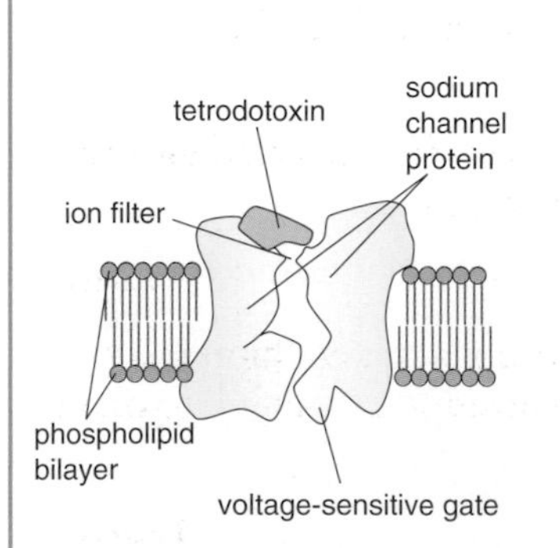


**Figure 13.20**
A voltage-sensitive sodium gate is one of a number of different sorts of protein in the plasma membrane surrounding an axon

Look at Figure 13.20, it shows a voltage-sensitive sodium gate in the axon membrane.

1 (a) Describe the part played by this voltage-sensitive sodium gate in producing an action potential.

*(2 marks)*

(b) The plasma membranes of nerve cells contain many other proteins like the one shown in the diagram. Some are sensitive to sodium ions, some to potassium ions and some to calcium ions. Use the diagram to suggest why each of these proteins is specific to a particular ion.

*(2 marks)*

2 (a) The puffer fish is regarded as a great delicacy in Japan. The restaurants that serve it, however, have to be specially licensed. This is because some of the organs in the body contain an extremely poisonous substance, tetrodotoxin. Unless the fish is prepared properly, you are very likely to be poisoned. Tetrodotoxin locks into sites at the entrance to the ion channel in the sodium gate. Use this information to suggest why one of the symptoms of being poisoned by tetrodotoxin is paralysis.

*(2 marks)*

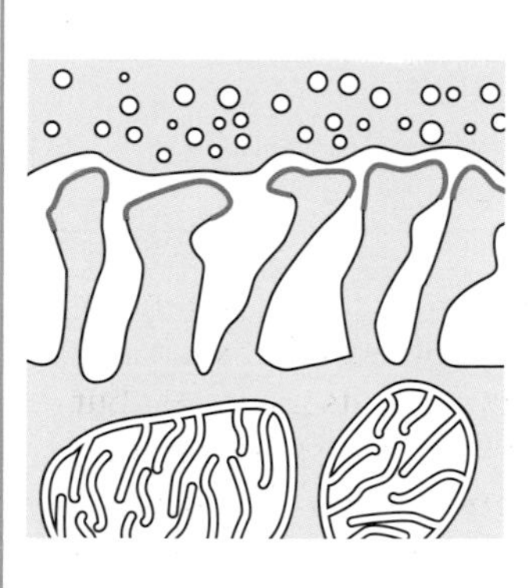

**Figure 13.21**
A neuromuscular junction is very similar to a synapse. This drawing has been taken from an electronmicrograph. It has been injected with bungaratoxin and then stained with bungaratoxin antibodies. In this drawing the position of the bungaratoxin antibodies are shown in red

We will have a look now at a neuromuscular junction. This is where a neurone meets a muscle. Neuromuscular junctions are very similar to synapses between two neurones, the only real difference is that transmission across a neuromuscular junction will lead to a muscle contracting rather than another nerve impulse. Figure 13.21 shows a neuromuscular junction.

3 Look at Figure 13.21. What is the evidence from this drawing that the feature labelled **A** is the presynaptic membrane?

*(1 mark)*

The krait is a small but extremely poisonous snake found in South-east Asia. In 1963 a small polypeptide consisting of approximately 70 amino acids was isolated from krait venom. This substance was called bungaratoxin, and was found to cause paralysis when very small amounts were injected into mice. When the tissues of these injected mice were treated with fluorescent antibodies and examined with a microscope, it was found that the toxin was found only at the neuromuscular junction.

4 Explain why bungaratoxin only has an effect if it is injected into an animal. It has no effect if it is simply swallowed.

*(2 marks)*

5 (a) Use your knowledge of antibodies to explain how fluorescent antibodies can be used to locate the position of bungaratoxin in the neuromuscular junction.

*(3 marks)*

(b) Use the information given in this question so far to suggest why animals bitten by a krait often die because the muscles associated with breathing become paralysed.

*(4 marks)*

6 Table 13.7 shows a number of other substances and describes briefly the way in which they affect the nervous system. For each substance, describe how the effect could lead to the consequence described in the last column of the table.

**Table 13.7**
The effects of a variety of poisons on the nervous system

| Substance | Effect | Consequence |
|---|---|---|
| Black widow spider venom | Causes large amounts of acetylcholine to be released | After a short period of time synaptic transmission stops |
| Organophosphate insecticides | Inhibit the enzyme acetylcholinesterase | One of the symptoms of accidental poisoning is the production of large quantities of tears and saliva |
| W-conotoxin produced by cone shells | Prevents $Ca^{2+}$ ions from crossing the pre-synaptic membrane | Paralysis |

*(4 marks)*

# Examination questions

1 (a) The graph shows the changes in the permeability of an axon membrane to sodium ions and to potassium ions during an action potential.

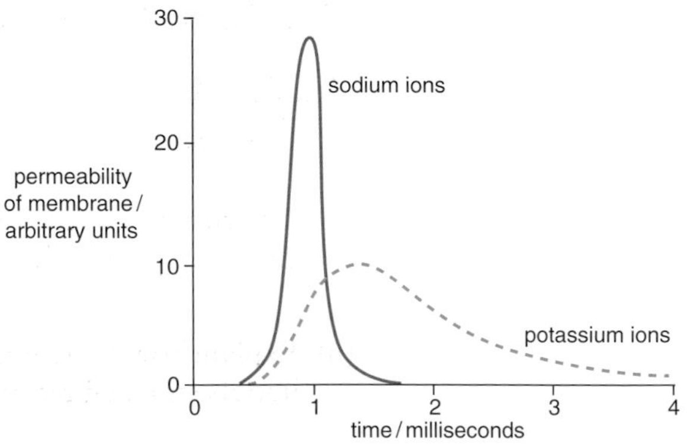


Use the information in the graph to explain how:

(i) at the start of an action potential, the potential difference across the membrane rapidly changes from negative to positive;

*(2 marks)*

(ii) the resting potential is restored.

*(1 mark)*

(b) Suggest why, during a period of intense nervous activity, the metabolic rate of a nerve cell increases.

*(2 marks)*

2 The earthworm, *Lumbricus terrestris*, is a nocturnal animal. It feeds at night on the soil surface and spends the day underground in a burrow.

(a) In the laboratory, *L. terrestris* can be shown to move directly from the dark towards light of low intensity.

(i) What sort of behaviour is being shown in this case?

*(1 mark)*

(ii) Explain the importance of this response in the life of *L. terrestris*.

*(1 mark)*

(b) In very dry conditions *L. terrestris* burrows deeper into the soil and curls into a ball. How does curling into a ball increase the animal's chance of survival?

*(2 marks)*

(c) If the worm is touched, it responds by contracting all its segments rapidly. How can this movement be related to the fact that the nerve cord of a worm contains a number of axons which have a very large diameter?

*(1 mark)*

3 The diagram shows rod cells and neurones from the retina of an eye.

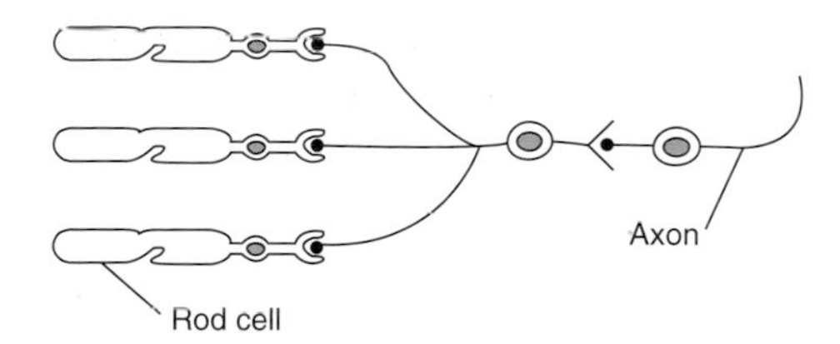


(a) Describe how light falling on a rod cell can give rise to a generator potential.

*(5 marks)*

(b) Explain how an action potential is produced and a nerve impulse transmitted along an axon.

*(6 marks)*

(c) With reference to the diagram, explain the following properties of a synapse:

(i) summation;

*(3 marks)*

(ii) transmission across a synapse will only occur in one direction.

*(3 marks)*

# Index